Environmental History

Volume 9

Series editor

Mauro Agnoletti, Florence, Italy

More information about this series at http://www.springer.com/series/10168

Mauro Varotto · Luca Bonardi
Paolo Tarolli
Editors

World Terraced Landscapes: History, Environment, Quality of Life

Springer

Editors
Mauro Varotto
Department of Historical and Geographic Sciences and the Ancient World
University of Padova
Padua, Italy

Paolo Tarolli
Department of Land, Environment, Agriculture and Forestry
University of Padova
Legnaro (PD), Italy

Luca Bonardi
Department of Philosophy "Piero Martinetti"
University of Milan
Milan, Italy

ISSN 2211-9019 ISSN 2211-9027 (electronic)
Environmental History
ISBN 978-3-319-96814-8 ISBN 978-3-319-96815-5 (eBook)
https://doi.org/10.1007/978-3-319-96815-5

Library of Congress Control Number: 2018957637

This Springer imprint is published by the registered company Springer Nature Switzerland AG
The registered company address is: Gewerbestrasse 11, 6330 Cham, Switzerland

Series Editor's Preface

Terraces are one of the most important examples of the world's agricultural heritage accumulated over some millenniums of human history. They represent remarkable landscape systems characterizing many regions of the world, especially in fragile and precious ecosystems, such as hills and mountains. Their diffusion dates back to the early development of agriculture; in Europe, they have been found in archeological sites of the copper age, but their presence is reported also in Central America, Africa, and Asia, making them a common heritage of many different countries and ethnic groups. As cultural landscapes, their development perfectly matches the definition given by the American geographer Carl Sauer in 1926: "The cultural landscape is fashioned from a natural landscape by a culture group. Culture is the agent, the natural area the medium, the cultural landscapes the result." Nevertheless, their features and distribution may also well fit into the description of agricultural landscapes given by the Italian agronomist Emilio Sereni in 1961: "The rural landscape is the form that man, in the course and for the ends of his productive agricultural activity, consciously and systematically imposes to natural landscape." This definition is important because it reminds us that the man, in order to survive, always had to modify the natural environment. Many generations of farmers through the centuries devoted a huge amount of hard work to turn steep slopes, often impossible to cultivate in other ways, into flat grounds suitable for seeding and planting. Modern agriculture, pushed by globalization, has contributed to marginalize these systems, favoring their abandonment or their transformation in uphill cultivations. Despite these trends, many of these land management forms survived and currently a new interest for these agricultural practices can be seen in agriculture and in science. This can be explained by the current social, economic, and environmental challenges that the world is facing, in front of changes occurring in almost every sphere of life.

According to the definition given by Eric Hobsbawm 1994, we are living in the "age of extremes," or the "short 20th century." The last hundred years can be clearly seen as a transition period in the economic, social, cultural, technological, and political fields, together with environmental change and its consequences. According to many scientists, the impact of these changes is due to the effect of unsustainable practices carried out by human activities. Examples of these practices are the incorrect, imbalanced, and unsustainable use of natural resources, as well as untenable development models, which do not consider long-term impacts or "side effects" of the activities conducted. Concerning the rural territory, the development model promoted in the last decades has not only shown to be ineffective to solve the economic problems of many rural areas, but also contributed to the loss of cultural values associated with rural communities and environmental degradation. This has brought to the loss of valuable land use systems and landscapes shaped by several generations of farmers, to the abandonment of millions of hectares of farmed land, and to urbanization processes, creating social degradation and increasing urban sprawl.

As one of the human activities that have a direct relationship with nature and environment, agriculture is often considered as the driver of the negative trend that is being followed, representing the greatest immediate threat to species, soil, and ecosystems. In fact, unsustainable farming practices result in land conversion leading to soil erosion and degradation, habitat loss, genetic erosion, inefficient use of water, pollution, impacting biodiversity and human life. Nevertheless, when agriculture is practiced in a sustainable way, it can preserve the landscape and biocultural diversity, protect watersheds, and improve soil health and water quality. The use of sustainable ecological practices, such as terracing, is a key feature distinguishing resilient agricultures developed over centuries, based on long experience and proven traditions. This kind of farming may be considered as less productive from modern-intensive systems, but it has ensured sustainable yield over time, thanks to time-tested technologies and traditional know-hows, using reduced external energy inputs. Terraces, with their incredible variety of forms, material, and design, can be surely considered one of the best examples of sustainable agricultural practices and of the co-adaptation of a community to difficult and changing environments. They provide many services to the ecosystem, a huge agrobiodiversity, ancestral knowledge transmitted through generations, and strong cultural and social values.

The identification and conservation of terraced systems can be described at international, regional, and national levels. Considering the United Nations, the cultural landscapes nominated in the UNESCO World Heritage List, developed in 1992, include some terraced landscapes listed as examples of outstanding universal values. Starting in 2002, also the FAO developed a program named Globally Important Agricultural Heritage Systems (GIAHS), expressly designed to protect traditional agricultural landscapes. Today, the FAO Agricultural Heritage program includes 50 landscapes across the world in Asia, Africa, Europe, Central and South America, and at least 17 of them have terraces. While UNESCO does not specifically address "agriculture" in the definitions of cultural landscapes, the FAO criteria are based on agriculture and consider culture, traditional knowledge, food production, agrobiodiversity, and landscape the main criteria for the designation of the sites. The GIAHS became an official program of the FAO only in 2015, after almost 13 years spent as a research project. The rapid growth of nominated sites occurred in the last few years confirms the growing interest for this topic form the agricultural sector.

At the regional level, it is worth mentioning the European Landscape Convention which offers an opportunity to include these systems in the national policies for landscape planning and conservation. Some European Countries, such as Slovenia, developed specific research programs focusing on terraced landscapes as cultural values, as also the Register of Slovenian Cultural Heritage, which is a central repository of data on heritage maintained by the Slovenian Ministry of Culture, where we can find examples of terraced landscapes. In Italy, specific legislation for landscape protection developed since 1922, but the National Code for Cultural Heritage developed in 2004 does not provide any specific protection for agricultural landscapes, while the return of the vegetation on abandoned farmed land is protected. In 2010, the Ministry of Agriculture developed the National Register of Historical Rural Landscapes and Traditional Agricultural Practices, with the aim of including them into rural development strategies. At the moment, 6 landscapes out of 15 listed in the register have terraces included. It is worth mentioning that also China, in 2012, as also Korea and Japan, developed a list of protected agricultural heritage sites, meant also to provide a tentative list for applying to the FAO GIAHS program. In the current situation, traditional planning and conservation tools only based on restrictions are not the best policy for maintaining terraced landscapes, while the inclusion of these systems in rural development strategies would be the best solution in combination with protection. High-quality food production associated with high-quality landscapes may also offer new opportunities for territories not suited for industrial production, especially when coupled with sustainable tourism. This might also favor the survival of rural communities who are the ones retaining the knowledge to maintain these systems.

This book collects a wide range of reviewed research material, showing a growing interest of the scientific community for this topic, providing also new research material. Thanks to the activity of the International Terraced Landscapes Alliance (ITLA), as well as to the passionate work of spontaneous groups of practitioners materialized in the last few years in many rural territories, the knowledge associated with terraces can survive. The hope is that a transdisciplinary scientific approach, together with local communities and the support of public policies, can effectively help to maintain an active role of these systems in the society and the economy. In this respect, the environmental history series of Springer really welcome this book, hoping that it will meet the same success of other works already published.

Florence, Italy

Mauro Agnoletti
Editor-in-Chief

Contents

Contributors

Ada Acovitsióti-Hameau Société Scientifique Internationale Pour L'étude Pluridisciplinaire de La Pierre Sèche, Brignoles, France

Kaodi Alum Department of Historical, Geographic Sciences and the Ancient World, University of Padova, Padua, Italy

Sabina Asins-Velis Centro de Investigaciones Sobre Desertificación-CIDE, CSIC, Universitat de València, Generalitat Valenciana, València, Spain

Lučka Ažman Momirski Faculty of Architecture, University of Ljubljana, Ljubljana, Slovenia

Jean-François Blanc Member of the Scientific and Executive Committee of International Terraced Landscape Alliance, Lumbin, France

Luca Bonardi Department of Philosophy "Piero Martinetti", University of Milan, Milan, Italy

Gerardo Brancucci Department of Architecture and Design, Polytechnic School, University of Genova, Genoa, Italy

Stefano Brugnaro Department of Civil, Environmental and Architectural Engineering, University of Padova, Padua, Italy

Lianet Camara University of Verona, Verona, Italy

Benedetta Castiglioni Department of Historical and Geographic Sciences and the Ancient World, University of Padova, Padua, Italy

Salvatore Ceccarelli Rete Semi Rurali, Scandicci, Italy

Lucio Cecchini Department of Agricultural, Food and Environmental Sciences, University of Perugia, Perugia, Italy

Margherita Cisani Department of Historical and Geographic Sciences and the Ancient World, University of Padova, Padua, Italy

Federica Corrado Polytechic of Turin, Turin, Italy

Alberto Cosner Archaeologist Freelance Professional, Trento, Italy

Angelica Dal Pozzo Department of Historical, Geographic Sciences and the Ancient World, University of Padova, Padua, Italy

Massimo De Marchi Department of Civil, Environmental and Architectural Engineering, University of Padova, Padua, Italy

Mourik Bueno de Mesquita Centro Bartolomé de Las Casas, Cusco, Peru

Fortuna De Rosa Interdepartmental Research Centre in Urban Planning "A. Calza Bini", University of Naples Federico II, Naples, Italy

Erwin Durbiano Polytechic of Turin, Turin, Italy

Francesco Ferrarese Department of Historical, Geographic Sciences and the Ancient World, University of Padova, Padua, Italy

Enrico Fontanari IUAV University Venice, Venice, Italy

Luigi Fusco Girard Interdepartmental Research Centre in Urban Planning "A. Calza Bini", University of Naples Federico II, Naples, Italy

Xuan Gao School of Tourism and Geography Science, Yunnan Normal University, Kunming, P.R. China

Giulia Giacchè Mixed Research Unit Spaces and Societies (ESO), University of Rennes, Rennes, France

Antonia Gravagnuolo Institute for Research on Innovation and Services for Development (IRISS), National Research Council, Naples, Italy

Antonio Ignacio Hernández Cordero Grupo de Investigación Geografía Física y Medio Ambiente, Instituto Oceanográfico y Cambio Global, IOCAG, Universidad de Las Palmas de Gran Canaria, ULPGC, Las Palmas, Spain

Yuanmei Jiao School of Tourism and Geography Science, Yunnan Normal University, Kunming, P.R. China

Luohui Liang Institute for Sustainability and Peace, The United Nations University, Tokyo, Japan

Raelene Marshall Dry Stone Walls Association of Australia, Melbourne, Australia

Anna Marson IUAV University Venice, Venice, Italy

Toshiya Okuro Graduate School of Agricultural and Life Sciences, The University of Tokyo, Tokyo, Japan

Juan Manuel Palerm Salazar Department of Graphic Expression and Architectural Projects, Universidad de Las Palmas de Gran Canaria, Las Palmas, Spain

Salvatore Eugenio Pappalardo Department of Civil, Environmental and Architectural Engineering, University of Padova, Padua, Italy

Domenico Patassini IUAV University Venice, Venice, Italy

José Manuel Pérez Sánchez Faculty of Anthropology, Universidad Autónoma del Estado de México, Toluca, Mexico

Davide Rizzo UP 2018.C102 INTERACT, UniLaSalle, Beauvais, France

Lidia Esther Romero Martín Grupo de Investigación Geografía Física y Medio Ambiente, Instituto Oceanográfico y Cambio Global, IOCAG, Universidad de Las Palmas de Gran Canaria, Las Palmas, Spain

Aarón Santana Cordero Department for Management of Science and Technology Development, Ton Duc Thang University, Ho Chi Minh City, Vietnam; Faculty of Environment and Labour Safety, Ton Duc Thang University, Ho Chi Minh City, Vietnam

Kazuhiko Takeuchi Graduate School of Agricultural and Life Sciences, The University of Tokyo, Tokyo, Japan

Paolo Tarolli Department of Land, Environment, Agriculture and Forestry, University of Padova, Legnaro (PD), Italy

Theano S. Terkenli University of the Aegean, Mytilene, Greece

Biancamaria Torquati Department of Agricultural, Food and Environmental Sciences, University of Perugia, Perugia, Italy

Carla Vargas Negrín Palerm & Tabares de Navas, Santa Cruz, Spain

Mauro Varotto Department of Historical and Geographic Sciences and the Ancient World, University of Padova, Padua, Italy

Sonia Venanzi Department of Agricultural, Food and Environmental Sciences, University of Perugia, Perugia, Italy

Chapter 1
Introduction

Mauro Varotto, Luca Bonardi and Paolo Tarolli

Abstract This volume collects the best scientific contribution presented in the 3rd World Conference on Terraced Landscapes held in Italy from October 6–15, 2016, offering a deep and multifaceted insight into the remarkable heritage of terraced landscapes in Italy, in Europe, and in the world (America, Asia, Australia). It consists of two parts: a geographical overview on some of the most important terraced systems in the world (first part) and a multidisciplinary approach that aims to promote a multifunctional vision of terraces, underlining how these landscapes meet different needs: cultural and historical values, environmental and hydrogeological functions, quality and variety of food, community empowerment, and sustainable development (second part). The volume offers a great overview on strengths, weaknesses, functions, and strategies for terraced landscapes all over the world, summarizing in a final manifest the guidelines to provide a future for these landscapes as natural and cultural heritage.

Since ancient times, terraces have been among the most evident human modifications to Earth's landscapes. They cover large areas from East Asia, to Europe, Africa, and the center-south of the Americas. Agricultural terraced landscapes are important anthropological signatures, and, in certain areas (e.g., China, Italy, Peru), they are considered an important cultural heritage. The main purpose of this agri-

M. Varotto (✉)
Department of Historical and Geographic Sciences and the Ancient World,
University of Padova, Padua, Italy
e-mail: mauro.varotto@unipd.it

L. Bonardi
Department of Philosophy "Piero Martinetti",
University of Milan, Milan, Italy
e-mail: luca.bonardi@unimi.it

P. Tarolli
Department of Land, Environment, Agriculture and Forestry,
University of Padova, Legnaro (PD), Italy
e-mail: paolo.tarolli@unipd.it

M. Varotto et al. (eds.), *World Terraced Landscapes: History, Environment, Quality of Life*, Environmental History 9,
https://doi.org/10.1007/978-3-319-96815-5_1

cultural practice was to cultivate land in high, steep-slope areas, retaining more water and soil, reducing soil erosion, promoting irrigation, and allowing machinery and plows to work in better conditions.

In Europe, the cultivation of the vine and the olive tree was a powerful push toward terracing, while Asia, Africa, and the Americas widely used terraces for cereal crops (corn, rice, sorghum, millet, etc., depending on the area). The resulting systems assume diverse shapes and sizes—from micro-adjustments to deep morphological changes of the slopes, from terracing small, isolated areas of a few thousand square meters to interventions brought, homogeneously and sometimes planned, to slopes of thousands of hectares. For centuries—especially under the impetus of population growth in the Modern Age—the areas occupied by terraces worldwide have been growing. However, during the twentieth century, this trend changed. Several terraced landscapes were abandoned, particularly the narrowest terraces that were impossible to work with machinery and those that could only be cultivated with cereals or left as meadows. This occurred in several areas of Europe, both mountainous and coastal, but other world regions were similarly affected. People started migrating toward cities, where the socioeconomic opportunities were higher than on farmlands. The abandoned terraced landscapes coincided with progressive land degradation. Several regions (e.g., Liguria in Italy) were characterized by increased surface erosion or small landslides. Almost everywhere, these processes happened under substantial political and cultural indifference, without even an atmosphere of open support for the societies affected. Only since the end of the twentieth century has this posture given way to the maturation of a different sensitivity, attentive, above all, to the cultural and environmental consequences of abandonment processes. This new perspective has progressively involved the most diverse fields—from the political-administrative to the academic, from the world of environmental and social voluntary workers to that of various agricultural organizations. This involvement has, admittedly, been more or less profound, depending on the region. However, recognizing the historical, geographical, environmental, ecological, and economic meanings terraced landscapes carry, and the frequent inertial maintenance of the anti-erosive component, has led to developing social and regulatory frameworks potentially oriented to terrace protection, recovery, and enhancement. In a growing number of cases, this has translated into positive refurbishment interventions, thanks also to implementing specific action programs.

At the same time, on a strictly productive level, terrace products are being reintegrated into commercial circuits, although the disadvantages inherent in an activity based on burdensome and unprofitable manual labor continue to prevail. Still, this trend is supported by growing demand for niche products strongly characterized on a qualitative basis. Experimenting with new cultural destinations and researching innovative materials, tools, and technical means frequently accompany this reintegration. However, the places of rebirth are flanked by many terraced areas that reached the twenty-first century in a profitable state, due to their marked specialization.

To understand the multiple issues these developments generate, in 2010 and 2014 the first and second International Conferences on Territorial Landscapes took place, respectively, in Mengzi in the People's Republic of China (Peters and

Junchao 2012), and in Cusco, Peru (Tillmann and Bueno de Mesquita 2015). This was followed by the third meeting in Italy, held in Venice, Padua, and ten other Italian terraced locations on October 6–15, 2016 (www.terracedlandscapes2016.it). This volume contains a scientific selection of over 150 contributions presented in different forms in these venues.

Given the impossibility of returning an analytical picture of such a widespread phenomenon, the first part of the book gives an overview of some areas significantly affected by the presence of terraces and, at the same time, representative of terrace diversity. This overview is also intended to relate the different scientific approaches to terracing. The prevalence of contributions related to the European area merely reflects the precocity and the number of studies in this region. It certainly does not show that terracing is comparatively less important to the agricultural landscape of the other continents.

The book's second portion has an interpretative nature—highlighting the problems linked to the evolution outlined above and inviting readers to a complex interpretation of terraced landscapes and their functions. Space is made here to reflect on criteria for mapping often unrecognizable realities, such as diagnosing the hydrogeological risk of terrace abandonment and the natures of losing both places of cultural heritage and the knowledge that accompanies them. Similarly, the need for and methods of integrating terraces into urban planning instruments and territorial development policies are analyzed.

Conceived as devices for erosion control, for water management, and for leveling cultivation surfaces exclusively for agricultural production, the terraces are now invested with additional meanings far from their original motives.

Envisioning opportunities offered by a new, multifunctional view of terracing, in addition to relaunching agricultural productivity, the book's contributions focus on safeguarding biodiversity, including agriculture, attracting tourists, promoting landscapes and their products, and considering the environmental values that accompany contemporary discourse. As suggested by the Honghe Declaration (2010), "terraces illustrate the close integration of humans and nature, and the need to safeguard biodiversity and cultural diversity."[1] From an environmental point of view, the terracing, realized in response to the problems of soil erosion, can be considered an *ante litteram* and a voluntary application of the sustainability concept, which, today, can take on the wider meanings of ecological, economic, and social sustainability.

The book's general message is that the terraced landscapes need to be protected and well-managed (also using recent advances in remote sensing monitoring). These actions can help mitigate environmental issues, such as soil erosion and landslides, and can also improve socioeconomic benefits, offering new job opportunities for younger generations.

---

[1] http://www.paesaggiterrazzati.it/wp-content/uploads/2016/12/Honghe-Declaration_English_2010l.pdf.

The multifunctionality of terraces can only be matched by a plurality of scientific skills, drawn on to understand problems and search for the most appropriate solutions. The marked diversity characterizing the disciplinary origins of this volume's authors is the fruit of a precise choice and of a bond deriving from the multifaceted nature of the object investigated. It also reflects the composite character of the participants in the third meeting of the International Terraced Landscapes Alliance in 2016. The meeting was an occasion of debate among over 250 people with different experiences, skills, and backgrounds: farmers, administrators, environmental and cultural associations, universities and the research world, drystone professionals and artisans, non-governmental organizations, and simple enthusiasts or terrace lovers. For this reason, the present volume can also be read as a broad and indispensable theoretical premise to the Proceedings (Alberti et al. 2018) and to the Manifesto "Choosing the Future for Terraced Landscapes" shared by the participants in the Congress and included at the end of this book. They are referred to for a more detailed view of the terracing problem on local and regional scales, as well as for an analysis of the prospects and directions for future action.

References

Alberti F, Dal Pozzo A, Murtas D, Salas MA, Tillmann T (eds) (2018) Terraced landscapes: choosing the future, Regione del Veneto

Peters HA, Junchao S (eds) (2012) First terraced landscape conference (Honghe, China). Paper collection. Kunming-Yunnan People Publishing

Tillmann T, Bueno de Mesquita M (eds) (2015) II Congreso Internacional de Terrazas. Encuentro de culturas y saberes de terrazas del mundo. JICA-CBC

Part I
Terraced Landscapes in the World: A General Overview

Chapter 2
Terraced Vineyards in Europe: The Historical Persistence of Highly Specialised Regions

Luca Bonardi

Abstract This contribution analyses the spread, origin and evolution of the most important European terraced viticulture complexes. Found in Mediterranean, Atlantic and continental areas, the terraces dedicated to viticulture owe their localisation to recurring geographical–environmental and historical movements in the different regions. Proximity to market outlets and to waterways were the key elements in their origins and distribution. The reasons behind the construction of the great viticulture terraced areas then guided the evolution of their twentieth-century history. In terms of the elements involved, the most influential inherited factors appear to be strong land fragmentation and the early entry into favourable commercial circuits.

2.1 Methodological Introduction

Through their historical origin and the reasons behind their growth, the terraces dedicated to viticulture form, within the more general framework of agrarian terracing, a landscape category characterised by specific models of development.

Few of these specific elements emerge from the studies on terraced landscapes carried out up until now, mostly directed at analysing, according to each case, the environmental, historical, technical and economic aspects from a regional or local perspective (e.g. Terranova 1989; Blanc 2001; Scaramellini and Varotto 2008; Queijeiro et al. 2010; Barbera et al. 2009; Formica 2010; Schultz 2010; Constans 2010). Even at this level, however, research dedicated specifically to viticulture is rare, both where viticulture has been the principal mover behind the building of the terraces (as is the case for much of Alpine and Central European terracing) (Petit et al. 2012; Scaramellini 2014; Lorusso forthcoming), and where it shares this role with other cultivation. The author has previously carried out a number of deeper

L. Bonardi (✉)
Department of Philosophy "Piero Martinetti", University of Milan, Milan, Italy
e-mail: luca.bonardi@unimi.it

M. Varotto et al. (eds.), *World Terraced Landscapes: History, Environment, Quality of Life*, Environmental History 9,
https://doi.org/10.1007/978-3-319-96815-5_2

studies, from a landscape angle, into the comparative relationship between terracing and viticulture (Bonardi 2010, 2014).

This chapter, in contrast, uses a historical and environmental perspective to bring together the geographical–territorial and historical factors that have influenced the localisation and evolution of terraced viticulture. In the final section, we seek to understand how such factors are basic to the original pathways that have differentiated the recent history of terraced viticulture from that of other agricultural production.

As far as localisation aspects are concerned, the study is based upon field research carried out in the course of the last fifteen years. This research was preceded by a selective study of the available photo-cartographic material.

Alongside the analysis of the historical literature and scientifically based material on the different regions, analysis of the European wine trade has proved to be fundamental in verifying the hypothesis of "convergent evolution" of viticulture areas strongly directed towards commerce.

Only those terraces built with stone (predominantly dry-stone, unmortared) have been considered here, while those of earth embankments have not been included.

The study also considers the vast single systems, stretching in a cohesive form over tens, hundreds or thousands of hectares, in which viticulture plays the central role to this day. By way of contrast, the study does not analyse those irregular scattered vineyards of limited size, existing just about everywhere, as much in areas dominated by non-terraced viticulture (e.g. the vineyards of Alsace or the Alto Adige), as in those which are terraced but oriented predominantly towards other production, such as the western Ligurian terraces dedicated above all to olive growing.

Likewise, the study only touches lightly on those contexts in which the terracing, while of a relevant complex, appears scattered on the hillsides like "leopard spots" (e.g. Chianti, Valpolicella, Ischia, Etna, Majorca, Canton Ticino), and those where the only evidence today is of abandoned viticulture, as in various Mediterranean islands (Aeolian Islands, Cyclades), and in some Italian Alpine valleys such as Val d'Ossola (Moschini 2017) and Val di Susa (Bartaletti 2006).

2.2 Geographical Background

Terracing is found, although with varying degrees of intensity, across all continents, intended to support very different crops, both foodstuffs and others. In Asia, Africa and South America, the building of terraces is to a great extent linked to seed crops, in particular to cereal production. Intensive terracing at high altitudes is found in the Bolivian and Peruvian Andes, in East Africa (particularly in Ethiopia), in various regions of the Middle East and the Arabian peninsula, in the north of the Indian subcontinent and northern Indochina, in southern China, in Indonesia and in the Philippines.

In common with these contexts, important regional density of terracing is found in Europe south of latitude 50°N. Here the cultivation of vines, together with that of the olive in Mediterranean areas, has led to the existence of whole areas given over to closely built terracing. The most important examples are found in the Iberian peninsula, in the outlying Atlantic islands of the Azores, Madeira and, above all, the Canaries, in France, Switzerland, Germany, Austria and Italy. Smaller complexes are found in Slovenia, Croatia, Romania, Greece and Malta. Outside the European area, occasional or less important examples are found in Cyprus, Lebanon, Israel and Georgia. Viticulture terracing is also found in Australia, New Zealand and Brazil, linked to European emigration.

In the course of the twentieth century, all European regions experienced, with differing levels of intensity, contraction of the areas given over to terraced viticulture. Owing to this process, in a number of areas where terracing actually represented the main form of land-use under vines, productive terracing has today been marginalised (e.g. along the Luxembourg stretch of the Moselle). The principal areas of European terraced viticulture are brought together in Table 2.1. For the interpretative purposes that follow, account has also been taken in each of those areas of the geographical–environmental context in which they are found. Their localisation is given in Fig. 2.1.

As appears from Table 2.1, the principal terraced viticulture areas are found in four environments: fluvial, lacustrian and maritime, reduced to one from an historical–economic perspective, and Alpine.

What are the historical reasons that link these specific contexts to the practice of viticulture? And in particular, what compels these contexts towards terraced viticulture?

As we shall see, such relationships, apart from causing the characteristic typologies of the landscape, combine to explain the historical events of much of European viticulture practised by means of terracing.

2.3 The Compelling History of Viticulture Terracing

The environmental preconditions that dictate the building of terracing in the interests of the slopes' sustainable agricultural productivity, above all in the face of erosion, has been amply discussed from both the historical perspective (Frapa 1984, 1997; Provansal 1990) and the hydrological (Terranova et al. 2006; Consell de Mallorca 2007; Agnoletti et al. 2012; Tarolli et al. 2014; Cevasco et al. 2014; Camera et al. 2015). The passage from a potential necessity on every agriculturally exploited slope to the concrete existence of terracing is, however, the result of specific historical possibilities.

To understand the processes that have led to the construction in Europe of highly specialised terraced sites, both those predominantly intended for the cultivation of vines and olives and those designed on the Brenta Canal for the cultivation of

Table 2.1 Principal terraced viticulture landscapes (cultivated surface > 50 ha)

Wine-growing region	Country (region)	Wine-growing terraced area (ha) (source)	Geographical background[b]
(1) Alto Douro	Portugal (Nord)	3502 (Lourenço-Gomes et al. 2015)	Fluvial
(2) Ribeira Sacra	Spain (Galicia)	1100 (Queijeiro et al. 2010)	Fluvial
(3) Banyuls-Collioure	France (Languedoc-Roussilion–Midi Pyrénées)	700[a]	Maritime
(4) Vallée du Rhône	France (Auvergne–Rhone Alpes)	140[a]	Fluvial
(5) Lavaux	Switzerland (Vaud)	736 (OCVP 2015)	Lacustrian
(6) Valais	Switzerland (Valais)	1511 (Parvex and Turiel 2001)	Alpine
(7) Mosel	Germany (Rheinland-Pfalz)	75[a]	Fluvial
(8) Wachau	Austria (Nieder-Österreich)	360 (Kieninger et al. 2016)	Fluvial
(9) Bassa Valle d'Aosta	Italy (Valle d'Aosta)	80 (Moreno 2012)	Alpine
(10) Valtellina	Italy (Lombardia)	926 (ASR 2015)	Alpine
(11) Val di Cembra	Italy (Trentino-Alto Adige)	600[a]	Alpine
(12) Cinque Terre	Italy (Liguria)	100 (Besio 2002)	Maritime
(13) Costiera amalfitana	Italy (Campania)	120[a]	Maritime
(14) Costa Viola	Italy (Calabria)	130 (Nicolosi and Cambareri 2007)	Maritime
(15) Pantelleria	Italy (Sicilia)	500 (www.cervim.org)	Maritime

[a]Surface area calculated by the author on ortho-cartographic basis (regional and national geoportals)
[b]By fluvial is meant the presence of a river that is at least partly navigable

tobacco (Chemin and Varotto 2008), or on the Amalfi Coast for citrus fruits (Ferrigni 2011), we must first consider the enormous labour needed to build them.

Estimates of the time required to build terracing have been made by Quaini (1973) for Liguria, and by Blanchemanche (1990) and Beauchamp (1992) for the south of France.

Added to the initial labour of construction is the work imposed by the terracing's morphology, by the maintenance of the stone structure, by the capture, distribution and draining of water, and by the replacement of soil carried away by surface erosion. The costs represented by such activities are markedly higher than those incurred on level ground or by other methods of slope cultivation. There is an

Fig. 2.1 Localisation of the principal European terraced viticulture landscapes

interesting quantitative evaluation of these aspects in Gugerell (2009), who, however, points out that the incidence of such costs varies according to different technical and morphological factors, including the gradient of the slope and the initial condition of the terrain.

In each case, the building of entire terraces of vast dimensions, extending over hundreds or thousands of hectares, can only be economically justified for cultivation capable of guaranteeing, relatively quickly and in the long term, a sufficient return, or at least its strong probability. Apart from specific exceptions, the cultivation of the olive and, above all, of the vine, stretching for climatic reasons over a vast area, are those which best meet this requirement.

At least partially, viticulture everywhere is geared towards satisfying market demands, be they local and regional or, particularly for higher-cost production, national and international. Thanks to these attributes, for many areas the vine has for centuries represented the principal means of opening up economically and gaining access to monetary trade. For Valtellina, one of the most important terraced wine-growing areas of the continent, the role of viticulture in this sense was made clear by Romegialli (1834: 34), when he recalled how, "in this part of the country, and it is the largest part, where the vines are rooted, the inhabitants' labour is concentrated mainly on their cultivation, as wine is the only means of bringing money into the province".[1] Although examples of extreme terracing can be

[1]"In quella parte di paese, ed è la massima, dove allignano le vite, lo studio degli abitanti è volto principalmente alla loro coltivazione, per essere il vino l'unico mezzo da introdurre denaro nella provincia".

met almost everywhere, it is the economic value of the vine that justifies on a grand scale the "sparing of no effort, to carry soil many times up the steepest mountain or hill and support it by means of stone walls that, often collapsing, then need to be rebuilt"[2] (Romegialli 1834: 34).

As far as the terraced regions are concerned, the economic significance of wine is obviously interwoven with its qualitative importance, also guaranteed by the favourable soil temperature (the pedoclimate) along well-drained, sun-facing slopes.

Within this framework, the vine is therefore the driving force towards the improvement of the terrain gained through the work of terracing. Such changes were carried out through various models of ownership management, including systems of rent, recourse to paid labour and the direct work of small proprietors. The first of these above all has established roots in much of the terraced vineyards of Europe, in particular with long-term agrarian contracts, such as to justify the tenant farmer's investment in improvements. This is the case in the *pastinato* of the coasts of Trieste (Iona 1986) and Amalfi (Del Treppo and Leone 1977), of the analogous *livello* in Valtellina (Scaramellini 2014), of the *bail à complant* in France (Gauchet 2001) and of the *rabassa morta* in Catalonia (Carmona and Simpson 1998).

Small-proprietorship contracts, in which the importance of the work takes precedence over that of the terrain, often agriculturally unproductive when the contract was drawn up, and over that of capital investment (minimal or non-existent), mostly apply to areas that can be managed at a family level and are, therefore, of modest dimensions.

From Late Medieval Period, but in some places only much later, for example between the eighteenth and nineteenth centuries on the Calabrian Viola Coast (Di Fazio et al. 2005), it was by means of this route that the foundation was laid for the successive development of small and very small land holdings, one of the most characteristic, and problematic, elements in today's agricultural terracing.

Innumerable adjacent holdings brought under cultivation in this way have produced vast and strongly specialised cohesive systems, which represent the dominant model of European terraced viticulture. Monoculture has become the norm in these, although in many contexts, partially also dictated by self-subsistence, this was originally diluted by the vine being combined with other cultivation. This is clearly illustrated by the situation that could still be seen in Alto Douro at the end of the eighteenth century, in an extraordinarily expansive stage of the region's vineyards and their wines' established success at a continental level: "in some areas where it was possible to combine vineyards with grain cultivation […] the vineyard was

[2]"Non risparmia per esso fatiche e dispendj, fino qualche volta, a portare il terreno sul più ripido del monte o del colle, e sostenerlo a forza di muriccioli che, spesso crollati, convien poscia ricostruire".

formed only by vines planted *em pilheiros*, on the wall of the terrace, leaving the inbetween land called *geio*, free for cultivating cereals" (Pereira and Morais Barros 2016: 135).[3]

In other situations, on the properties of the aristocracy and gentry, very often city-based and sometimes already engaged in the wine trade, monoculture, only fully established between the nineteenth and twentieth centuries, was conducted by means of paid labour, above all during the stage of building the terraces. The existence of specific skilled trades specialising in their construction, or on specific elements of it (Blanchemanche 1990: 167), is documented in different parts of Europe, as in Provence and Majorca (Blanc 2001). Likewise, there is evidence of temporary migrations of a skilled workforce, such as, in the mid-nineteenth century, the Ligurian workers in Canton Ticino, specifically employed in the construction of viticulture terracing (Ceschi 1999).

The establishment of strongly specialised wine-growing areas came about to a great extent through a localised system that included the terraced areas. Alongside the role of the pedoclimatic characteristics, central to high-quality production but equally important for that of a lower quality (Gauvard 1996), two elements were very significant. On the one hand, the difficulties of communication and transport by land in pre-nineteenth-century Europe, and on the other the problems of preserving the wine, which deteriorates easily, not least in transport. Right up to the modern age, and sometimes even beyond, these are the most important governing factors in European viticulture geography. As was clearly shown by Dion (1959), this has often restricted viticulture to the peri-urban areas, when not directly urban (Bonardi and Cavallo 2014), so placed as to be close to important centres of consumption. One only need think of the great vineyards of the Paris region, today reduced to a few tens of hectares, which were dedicated largely to the capital's consumption (Dion 1959). Or, in Italy, of the vast viticulture areas that, still in the nineteenth century, could be found in Alto Milanese, in Rome's urban perimeter, and close to other great cities. The position of terraced vineyards such as those of

[3]On the other hand, such a description closely recalls that of the model of terraced vine cultivation observed by Montaigne in Tuscany two centuries earlier. Talking of the countryside of Lucca, he wrote, "On ne peut trop louer la beauté et l'utilité de la methode qu'ils ont de cultiver les montagnes jusqu'à la cime, en y faisant, en form d'escaliers, de grand degrés circulaires tour autour, et fortifiant le haut de ces degrés, tantôt avec des pierres, tantôt avec d'autres revêtemens, lorsque la terre n'est pas assez ferme par elle-même. Le terre-plain de cet escalier, selon qu'il se trouve ou plus large ou plus étroit, est rempli de grain. Et son extrémité vers le vallon, c'est-à-dire, la circonférence ou le tour, est entourée de vignes; enfin, partout où l'on ne peut trouver ni faire un terrein uni, comme vers la cime, tous est mis en vignes" (One cannot praise too much the beauty and utility of the method which they have of cultivating mountains up to the top, making thereof in the form of staircases great circular steps all about them and reinforcing the higher parts of these steps both with stones and with other revetments when the earth is not solid enough on its own. The level ground of this staircase, according as to whether it is wider or narrower, is full of grain and its extremity towards the valley, that is to say the circumference of the circuit, is encircled by vines. Finally, wherever one cannot find or make a continuous land for cultivation, as towards the crest, everything is set to vines) (de Montaigne 1889: 444).

Lavaux, of Wachau, of the Moselle and the Rhine, and of the Rhone's Haute Côte, are part of this type of framework.

Clearly, the peri-urban localisation strategy could only represent a partial solution to the overall problem of wine, leaving unanswered the problem, among others, of the demand from non-producing areas, in particular from central and northern Europe.

The location and very existence of part of European terraced viticulture represents a response to this last problem.

2.4 How Water Made Wine

The fluvial, lacustrian and maritime geographical frameworks (Table 2.1), which embrace about 70% of the areas being considered, define as a whole a macro-category centred upon a strong relationship between terraced viticulture and a water environment.

In certain continental contexts, such as those of the Wachau, the Moselle, the Rhine and other areas of German terraced viticulture, the location along the sides of great river valleys can be partly explained by the influence of the micro-climate created by the great rivers. Nevertheless, it is above all their role as important waterways that should be looked at to gain a comprehensive understanding of this relationship.

As can be easily realised, the waterways that influenced terracing are almost always navigable, even if occasionally with difficulty and the need for operations to improve the river's course.[4]

In a number of cases, water transport was, and at times still is, the only way of reaching the vineyards. Circumstances of this kind are found on the Calabrian Costa Viola, at Ischia, in the Cinque Terre, in Ribeira Sacra and in Alto Douro.

Above all, however, it was the possibility of transporting wine for long distances that determined the central position of navigable routes in the development of viticulture. Rivers, canals and lakes, apart from serving the neighbouring urban markets, were often the first artery of a more extensive network that used the sea as the principal means of international trade. The position of the great wine-growing regions of Europe, terraced and not, above all in France, Spain, Portugal and Germany, is evidence of the undoubted role played here by waterways.

Even more so, the importance of transport by water explains the directly coastal and micro-insular and insular position of many extensive terraced viticulture systems on the Mediterranean or Atlantic: Cinque Terre, the Amalfi coast, the

---

[4]Linked to the necessity for expanding viticulture are the works carried out at the end of the eighteenth century to alter the Portuguese course of the Douro, in the stretch through the Valeira gorges. These works enabled major viticulture exploitation of the slopes in the upper reaches of the valley (d'Abreu 2007). For the advantages given to viticulture around Vienne by the eighteenth-century engineering works on the course of the Loire, see Gadille (1978).

Calabrian Costa Viola, Pantelleria, the Azores, the Canaries, etc. This relationship, however, is not restricted to wine-growing, but is also part of other specialised growing, such as that of olives and citrus fruits (Ferrigni 2011).

Nor should it be forgotten that the sea carried considerable quantities of wine products destined for the direct provision of the military and commercial navies. It is significant that an amply terraced viticulture such as that around Etna had reached its greatest expansion (eight thousand hectares of cultivated ground in the whole of the province of Catania) between the eighteenth and nineteenth centuries, thanks to trade with the French, Austro-Hungarian and British navies (www.cervim.org).

Provisioning the British navy, for whom were intended large quantities of wine coming from the terraced regions, seems a key factor in the development of viticulture in the Mediterranean and Atlantic areas. For different terraced regions, the demand coming from Britain represented the principle driving force for bringing land under cultivation, but, as the markets closed, it also represented productive decline and abandonment. The history of terracing and the wines of Douro and Banyuls-Collioure fit amply into this scheme of things, together with those of other regions producing fortified wines such as Madeira, sherry and Marsala. The history of the area of Cap de le Nào, in the province of Alicante, shows the importance of the British market. As shown by Courtot (1990), the great development of terracing in this region, linked to the production of raisins to satisfy the demands of the British market, fell into crisis as soon as British importers found markets to supply them in Greece and Turkey that were more conveniently and easily reached after the development of steam navigation.

Analogous dynamics, part of a changeable geopolitical and commercial picture, are behind the rapid development and subsequent crises of the terraced viticulture of Tenerife, in the Canaries, between the seventeenth and eighteenth centuries (Unwin 2005).

The history of European viticulture is one that constantly interweaves the destinies of different regions of production. Tensions and conflicts between states, the politics of customs duties and tax systems interventions have determined the continuous market repositioning of those wine-producing regions strongly directed towards trade, including, almost always, the terraced ones.

The localised constraints outlined here were broken in the nineteenth century by the improvement of the network of road communications and, even more, by the establishment in the second half of the century of communication routes opened up through the railways (Fig. 2.2).

2.5 An Apparent Exception: The Wines of the Alps

The second key to localisation of terraced wine-growing is to be sought in the position of the Alps, straddling Europe between its heart and the Mediterranean, and in the specific geological, geomorphological and climatic characteristics of some of its areas. The existence of important terraced complexes in the Alps can be

Fig. 2.2 Terraces market oriented in Alto Duero. *Photo* L. Bonardi

explained by the proximity of outlet markets that, in a geographical sense, were almost as significant as those along the great river valleys. To understand how the Alps could play a role of this type, we must above all examine the distribution of the terraced vineyards in the region. With the sole exception of Canavese, in Piedmont, which has particular morphological characteristics, the vineyards are found along the south-facing slopes of the great internal valleys that run from east to west[5]: Val d'Aosta, Valtellina, Val di Cembra, Vallese. The same arrangement is also found today in contexts of modest production, such as the vineyards of Val di Susa and Val Venosta. In addition, the limited but mainly south-facing stretches formed at the confluence of valleys of a north–south orientation make up the most intensively terraced areas in those valleys. An example of this type is the viticulture terracing, today mostly abandoned, of Val Chiavenna (Aldighieri et al. 2006). Analogous situations are also met outside the Alps, as at the confluence of the lateral valleys of the Rhone, giving rise to a number of terraced areas around Condrieu (Gadille 1978).

The absence of terracing on the facing side, looking north, causes in many cases conspicuous landscape asymmetry but also, in the past, some economic integration

[5]Obviously, this does not exclude the development of viticulture in valleys of contrary orientation, as in the important case of the Adige valley.

between the slopes. One of the best known examples is the cultivation of a coppice of chestnuts on the Valtellina Orobie Alpine slope, which were used to provide stakes for supporting the vines grown in the Rhaetic Alpine terraced vineyards (Lorusso forthcoming).

The inland position and longitudinal orientation of these valleys determine a common climatic picture for many stretches that have terraced viticulture. These zones receive low rainfall, significantly less than that received by the neighbouring Alpine sectors that have, however, a different exposure to the flow of the prevailing seasonal currents. The most representative cases of this influence are those of the eastern sector of terracing of Valtellina, Vallese, and, to a more limited extent, Val d'Aosta.[6] The low rainfall and strong exposure to the sun guaranteed by the southern aspect represent the best possible conditions for viticulture, even if in situations not devoid of water stress. Therefore, it is not by chance that, with the sole exception of the slopes along the Brenta Canal, once given over to intensive tobacco cultivation, all the major Alpine terraced systems were exclusively dedicated to the production of wine.

These environmental conditions gave rise to systems of generous proportions, thanks to the relative proximity of important market outlets, represented both by neighbouring urban centres, as in the case of Vallese, and by regions beyond the Alps. The terraced viticulture of Valtellina (Scaramellini 2014), of Ossola (Moschini 2017) limited today to few productive fragments and of Val d'Aosta (Moreno 2012: 171–172) was historically dependent on their position alongside Switzerland and so of export to the Swiss cantons. This factor was particularly, and more importantly, seen in Valtellina, whose production, thanks to its centuries-long affiliation with the Grigioni, penetrated deep into the German, Alpine and continental world (Scaramellini 2014).

Not very different, along the side of the western Alps, the once flourishing viticulture of Val di Susa is explained by the protectionist policy imposed in late mediaeval times by the local controlling class, "against the wines of Lombardy and the Po to favour the sale of wine produced in the area both locally but above all in the transalpine regions (in the direction of Briançon and its area), reaching at the beginning of the modern age a spread of viticulture (with terracing as well as intensive planting in the valley bottoms) that had a certain anomaly in view of the environmental context"[7] (Varanini 2003: 657).

Alongside the morpho-climatic and geographical location aspects, it is probable that the development of terraced vines on the Alpine watershed had been at some point also supported by climatic conditions. In this sense, a relatively significant

---

[6]The most significant minimal rainfall of this valley, around 500 mm annually, is measured in the area of Aosta, therefore further to the west compared to the location of what is today the most important productive terraced system of the region.

[7]"contro i vini lombardi e 'padani' per favorire lo smercio in sede locale, ma soprattutto verso le regioni transalpine (in direzione di Briançon e della sua regione) del vino prodotto in zona, arrivando agli inizi dell'età moderna ad una diffusione della viticoltura (con terrazzamenti, oltre che con intensivi impianti in fondovalle) certo anomala rispetto al contesto ambientale".

Fig. 2.3 Transport of wine barrels over Bernina Pass at the turn of twentieth century

role may have been played by the worsening of climatic conditions that, above all in the second half of the sixteenth century, led to a deterioration, both qualitative and quantitative, in wine production in central Europe (Petit et al. 2012; Vinea Wachau 2014). This could have triggered a major recourse to wines coming from more "sheltered" southern Alpine lands and, therefore, an incentive to bring new ground under cultivation in these regions.

Together with these particular elements should not be forgotten the integration of the terraced lands, both Alpine and those beside fluvial and maritime transport routes, in the more general dynamics that have involved European viticulture in the course of the modern and present age. The principal factors underlying the expansive processes are those generally connected to demographic growth, and so of consumption, and those of a progressive growth, from mediaeval times, of the symbolic value of wine as an indication of social status (Pini 2003), particularly in central and northern Europe (Unwin 2005).

To the decrease in the acreage of terraced viticulture from the late nineteenth century, various factors, peculiar to the world of terraced agriculture in general, contributed (Fig. 2.3).

2.6 Convergent Legacies

The historical processes outlined above gave rise to a group of common legacies that controlled the history of terraced wine-growing areas in the course of the twentieth century. Those belong to the picture of general reduction in Europe of the use of terracing, indicated by slight chorological variations and varying intensity according to the region.

Such contraction occurred in connection with a general productivity crisis in farming on the slopes, coinciding both with mechanisation of agriculture on level ground and with emigration towards urban or industrial centres, or, for the Mediterranean coasts, to tourist centres. The contraction was brought about in real terms by the replacement of terracing with other forms of management through the possibilities offered by mechanisation, by expansion of residential areas and, above all, by abandonment of the land. This last phenomenon assumed almost everywhere astounding dimensions, with reductions that in Italy often exceeded 75% of land previously under cultivation. In Calabria's Costa Viola, the extent of cultivated terracing has dropped from 612 ha in 1929 to 130 ha today (Nicolosi and Cambareri 2007); in the Cinque Terre, an area among the most hit by abandonment, the passage has been from a maximum of 1200 ha of cultivated land in the course of the twentieth century to around 100 ha today (Besio 2002). This trend of contraction also continues in our times, if more slowly, in spite of various operations of support for marketing and of heritage conservation, among them the inclusion of various areas in UNESCO's list of World Heritage Sites. In Lavaux, the loss between 1993 and 2015 was 40 ha (OCVP 2015). Between 2006 and 2015, the wine-growing area of the Vallese, terraced and not (but the former was probably worst hit), decreased by an annual rate of between 4 and 8% (Canton du Valais 2016).

As has been said, elsewhere the dominant phenomenon has by contrast been the movement towards other ways of managing the slopes. In Alto Douro, terracing has been replaced by what is known as *patamares* and by vertical planting. The latter is also responsible for the major transformation of the wine-growing landscape in the Rhine and Moselle valleys.

To the reasons for the decline in terracing cited above must be added that of the significant diseases that attacked the vine in Europe from the mid-nineteenth century. *Oidium*, *peronospora* and above all *phylloxera*, while not permanently changing the productive assets on a national scale (Bonardi 2014; Jacquet and Bourgeon 2010), sometimes began or hastened the process of abandonment at a local or regional level.

As appears in the course of specific research studies (APARE 1983; Bonardi and Varotto 2016), one of the causes of these areas' economic weakness is the fragmentation of ownership. Among the most macroscopic examples can be cited the distribution values of ownership in Val d'Aosta, where 80% of the total cultivated land (equivalent to 98% of the businesses) is held in properties of less than one hectare, and those of Ribeira Sacra, with 93% of the cultivated land subdivided between farms of less than two hectares (www.cervim.org). The latter situation is very similar to that of Calabria's Costa Viola, where the average size of the holdings is less than two and a half hectares (Di Fazio et al. 2005).

This phenomenon has its origins, as has been seen, in the processes of the actual construction of these holdings and, specifically, from those aspects of management of the ownership, common to so many areas, that from the beginning of the slopes' agricultural exploitation has favoured strong parcellisation of the land ownership. Overcoming the problematic legacy through systems of consolidation represents a major challenge for these areas today.

Notwithstanding these findings, the reduction of terraced wine-growing areas, as with those olive-growing, has in general been less drastic in comparison to other crops, including cereal and fruit growing, once very widespread.

It is clear that the ubiquitous establishment of a market economy has in the first-place penalised subsistence farming and, more generally, mixed cropping systems. By contrast, highly specialised productive systems, as exactly is terraced viticulture, open to the international markets, found themselves in the twentieth century in a relatively advantageous position. The centuries of inclusion in commercial circuits, guaranteed by their position along favourable communication routes or alongside market outlets, placed these areas very early on in a competitive system on a continental, and in some cases world, scale. In such a context were developed both forms of protection and improvement of the quality of production and pronounced specialisation. Among the last are the sweet or aromatic wines that mark the areas of Alto Douro (*Porto*), Banyuls (*Vin doux naturel*), Pantelleria (*Moscato di Pantelleria* and *Passito di Pantelleria*) and the Cinque Terre (*Sciacchetrà*).

Specialisation and quality perfecting in the course of the twentieth century have built an advantage that benefits the best wine estates of these economies, even if built on terracing, compared to others. From this perspective, it does not seem a matter of chance that the area of the largest single terraced viticulture in Europe corresponds to the world's first viticulture designation, known as the "Pombaline Demarcations", instituted from 1756 to control the quality of the wines of Alto Douro[8] (Bianchi de Aguiar 2010). The first territorial delimitation of a sweet wine, on the other hand, dates back to 1909, that of the terraced area of Banyuls (Ferrer 1930).

From the same perspective of protection of the quality, and brand, of wine production is probably to be understood the system of vintage announcements applied for centuries in Valtellina following a rare model, for capillarity and historical persistency, in Italian viticulture. The explanatory layout of the process that leads from the geographical–environmental determinant to the survival of today's terraced viticulture is summarised in Fig. 2.4.

---

[8]The systems of demarcation instituted in the first half of the eighteenth century in Chianti and Tokaj were of different origins.

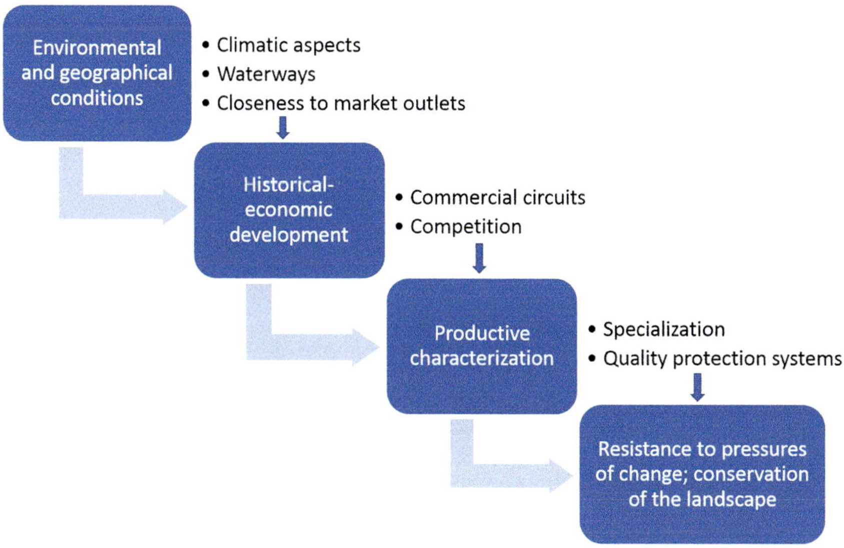

Fig. 2.4 Geographic variables and historical–economic development of terraced viticulture

2.7 Conclusions

This work has highlighted the historical and geographical reasons, convergent and interconnected, that governed the localisation and evolution of the principal terraced wine-growing areas. These are part of the wider picture of the relationships between the sites of European viticulture with the centres of wine consumption. The problems linked to the transport of the wine over long distances, connected with those of its preservation, were addressed historically by a topographical approach with the solution of placing the vineyards along waterways, from time to time represented by rivers, lakes, the Mediterranean Sea and the Atlantic Ocean. By guaranteeing the best means of transport for a product peculiarly liable to deterioration, such routes determined the economic fortune, often long-lasting, of the wine-growing regions developed along them. For Alpine localisation, an analogous solution was to some extent guaranteed by the existence of climatically favourable areas and the position alongside market outlets to Central Europe.

Much of the extent of European terraced vineyards can be explained by one or the other condition, which gave rise to the early opening up of the economies of these regions.

In fact, as Unwin (2005: 144) has already pointed out for the mediaeval period, "the idea of an independent Mediterranean peasantry, producing its own subsistence needs on its own land, is largely a myth". This concept is certainly applicable to the large terraced areas, even if in a number of cases only from the modern age, and weakens the traditional vision of terracing as a holding and representation of self-sufficiency. In all probability, the economic dimension of these regions, strongly specialised from an early date, and the search for qualitative solutions, later brought about the positive major resistance in the face of the crises that struck European terraced viticulture in the course of the last century.

References

Agnoletti M, Emanueli F, Maggiari G, Preti F (2012) Il disastro ambientale del 25 ottobre 2011 nelle Cinque terre. In: Agnoletti M, Carandini A, Santagata W (eds) Florens 2012—Studi e ricerche. Essays and researches, pp 25–39. Bandecchi e Vivaldi

Aldighieri B, Bonardi L, Comolli R, Conforto A, Mariani L, Mazzoleni G, Rizzotti T (2006) Vine-growing in Valchiavenna (SO): the Pianazzola project. Bollettino della Società Geologica Italiana 6:17–27 Special issue

APARE (1983) Des Agriculteurs en terrasses. Analyse, synthèse. Avignon: A.R.E.E.A.R. Languedoc Roussilon—Ministère de l'Agriculture

ASR Lombardia (Annuario Statistico Regionale) (2015) http://www.asr-lombardia.it/ASR/lombardia-e-province/agricoltura/produzione-agricola-zootecnia-e-risultati-economici/tavole/9853/2015/. Accessed 20 Feb 2017

Barbera G, Cullotta S, Rossi-Doria I, Ruhl J, Rossi-Doria B (2009) I paesaggi a terrazze in Sicilia. Metodologie per l'analisi, la tutela e la valorizzazione. ARPA, Sicilia

Bartaletti F (2006) Pendii terrazzati nelle Alpi Cozie: i casi di Chiomonte e Bardonecchia. Geotema: Paesaggi terrazzati 29:25–34

Besio M (2002) The conservation of built landscape and the maintenance of a collective architecture. In: World heritage expert meeting on vineyard cultural landscapes, Tokaj, Hungary, 11–14 July 2001, pp 41–42. Secretariat of the Hungarian World Heritage Committee in collaboration with the UNESCO World Heritage Centre

Bianchi de Aguiar F (2010) La spécificité des paysages du Douro. In: Pérard J, Perrot M (eds) Paysages et patrimoines viticoles. Rencontres du Clos-Vougeot 2009. Centre Georges Chevrier, pp 91–97

Blanc J-F (2001) Terrasses d'Ardèche. Paysages et Patrimoine. chez l'auteur

Blanchemanche Ph (1990) Bâtisseurs de paysages. Editions de la Maison de Sciences de l'Homme

Bonardi L (2010) Les paysages viticole en terrasses: des espaces de convergence. In: Pérard J, Perrot M (eds) Paysages et patrimoines viticoles. Rencontres du Clos-Vougeot 2009. Centre Georges Chevrier, pp 129–140

Bonardi L (2014) Paesaggi e peculiarità dei terrazzamenti viticoli. In: Bonardi L, Caligari A, Foppoli D, Gadola L, Grossi D, Stangoni T, Vanoi G (eds) Paesaggi valtellinesi: trasformazione del territorio, cultura e identità locale. Mimesis, pp 71–82

Bonardi L, Cavallo F (2014) Résistence et renaissance de la viticulture dans la lagune de Venise. In: Pérard J, Perrot M (eds) Le vin en heritage, anciens vignobles, nouveaux vignobles. Rencontres du Clos-Vougeot 2014. Centre Georges Chevrier, pp 85–107

Bonardi L, Varotto M (2016) Paesaggi terrazzati d'Italia. Eredità storiche e nuove prospettive. Angeli

Camera C, Apuani T, Masetti M (2015) Modeling the stability of terraced slopes: an approach from Valtellina (Northern Italy). Environ Earth Sci. https://doi.org/10.1007/s12665-015-4089-0

Canton du Valais (2016) Année vitivinicole 2015—Rapport annuel. Département de l'économie, de l'énergie et du territoire, Office cantonal de la viticulture. https://www.vs.ch/documents/180911/1045634/Ann%C3%A9e+vitivinicole+2015.pdf/331daa33-a42e-4e26-a7d4-98bf4cc02296. Accessed 12 Dec 2017

Carmona J, Simpson J (1998) A vueltas con la cuestión agraria catalana: el contrato de "Rabassa morta" y los cambios en la viticultura, 1890–1929. In: Documento de Trabajo 98-07 Depto. de Hist. Económica e Instituciones Serie de Hist. Económica e Instituciones 01. Universidad Carlos III de Madrid
Ceschi R (1999) Nel labirinto delle valli. Uomini e terre di una regione alpina: la Svizzera italiana. Casagrande
Cevasco A, Pepe G, Brandolini P (2014) The influences of geological and land use settings on shallow landslides triggered by an intense rainfall event in a coastal terraced environment. Bull Eng Geol Env 73:859–875
Chemin A, Varotto M (2008) Le "masiere" del Canale di Brenta. In: Scaramelllini G, Varotto M (eds) Paesaggi terrazzati dell'arco alpino. Atlante. Marsilio, pp 97–101
Consell de Mallorca (2007) Actes de les jornades sobre terrasses i prevenció de riscs naturals. Actas de las jornadas sobre terrazas y prevención de riesgos naturales, Mallorca, 14, 15 i 16 de setembre de 2006. Departament de Medi Ambient, Consell de Mallorca, Palma de Mallorca
Constans M (2010) Le patrimoine paysager viticole de Banyuls, entre reconstruction et destruction. In: Pérard J, Perrot M (eds) Paysages et patrimoines viticoles. Rencontres du Clos-Vougeot 2009. Centre Georges Chevrier, pp 181–199
Courtot R (1990) La place des terrasses dans les systèmes agricoles méditerranéens. In: Provansal M (ed) Méditerranée 71. L'agriculture en terrasses sur les versants méditerranéens; histoire, conséquences sur l'évolution du milieu, p 42
d'Abreu C (2007) Navegação no rio Douro: o sonho (re)corrente de Castela. Douro 22:37–78
de Beauchamp Ph (1992) L'architecture rurale des Alpes-Maritimes. Édisud
de Montaigne M (1889) Journal du voyage de Michel de Montaigne en Italie, par la Suisse et l'Allemagne en 1580 et 1581. Lapi
Del Treppo M, Leone A (1977) Amalfi medievale. Giannini
Di Fazio S, Malaspina D, Modica G (2005) La gestione territoriale dei paesaggi agrari terrazzati tra conservazione e sviluppo. AIIA
Dion R (1959) Histoire de la vigne et du vin en France des origines au XIX siècle. Doullens
Ferrer G (1930) Le vignoble de Banyuls-sur-Mer. Revue géographique des Pyrénées et du Sud-Ouest 2:185–192
Ferrigni F (2011) Dal Ducato alla Costiera Amalfitana. In: Ferrigni F (ed) Le regole del vernacolo. Viaggio nel patrimonio edilizio minore della Costiera Amalfitana e dell'Irpinia. www.univeur.org. Accessed 15 May 2017
Formica C (2010) Paesaggi terrazzati tra storia e sostenibilità. Il litorale campano e l'isola d'Ischia. Studi e ricerche socio-territoriali 0:25–56
Frapa P (1984) Le terrasses de culture entre le passé et l'avenir. Forêt méditerranéenne 2:129–130
Frapa P (1997) Restanques, faïsses ou banquettes: l'homme, l'eau et la pente. Études vauclusiennes 58:59–66
Gadille R (1978) L'héritage d'une viticulture antique, vignes et vins de Côte-Rôtie et Condrieu. Revue de Géographie de Lyon 53:7–22. https://doi.org/10.3406/geoca.1978.1213
Gauchet S (2001) Des pratiques agricoles aux pratiques paysagères. Aménagement et Nature 141:55–66
Gauvard C (1996) La France au moyen âge: du Ve au XVe siècle. PUF
Gugerell K (2009) Wine-Growing in the Wachau—red and white resistance. In: Splechtna B (ed) Proceedings of the international symposium: preservation of biocultural diversity—a global issue, BOKU University, Vienna, 6–8 May 2008, pp 107–113. University of Natural Resource Management and Applied Life Sciences, Wien
Iona ML (1986) La città di Maria Teresa. Pragmatismo e pianificazione. UR: urbanismo revista 4:21–25
Jacquet O, Bourgeon J-M (2010) Crise du phylloxéra et mutations du paysage. In: Pérard J, Perrot M (eds) Paysages et patrimoines viticoles. Rencontres du Clos-Vougeot 2009. Centre Georges Chevrier, pp 151–162
Kieninger PR, Gugerell K, Penker M (2016) Governance-mix for resilient socio-ecological production landscapes in Austria—an example of the terraced riverine landscape Wachau. In:

UNU-IAS, IGES (eds) Mainstreaming concepts and approaches of socio-ecological production landscapes and seascapes into policy and decision-making (Satoyama initiative thematic review vol 2). United Nations University Institute for the Advanced Study of Sustainability, Tokyo, pp 36–49

Lorusso D (forthcoming). Fa vini potentissimi e assai. Geografia e storia della viticoltura in Valtellina. In: Panzera F (ed) Viti&Vini. La vite e il vino nella nostra cultura, con uno sguardo a Ticino, Vallese e Valtellina. Bollettino Storico della Svizzera Italiana, 14, Edizioni Salvioni

Lourenço-Gomes L, Pinto L, Rebelo J (2015) Wine and cultural heritage. The experience of the Alto Douro Wine Region. Wine Econ Policy 4:78–87

Moreno D (2012) Valle d'Aosta. In: Agnoletti M (ed) Environmental history, vol 1. Italian historical rural landscapes. Spriger. https://doi.org/10.1007/978-94-007-5354-9_6167-173

Moschini C (2017) La vite in Val d'Ossola. Una coltivazione dal grande passato. Silvana

Nicolosi A, Cambareri D (2007) Il paesaggio terrazzato della Costa Viola. In: Atti XXXVI Incontro di Studio Ce.S.E.T. Firenze University Press, pp 179–194

OCVP—Office Cantonal de la Viticulture et de la Promotion (2015) Registre Cantonal des Vignes 2015. Department de l'economie et du sport—Service de l'agriculture, Canton de Vaud. http://www.vd.ch/fileadmin/user_upload/themes/economie_emploi/viticulture/fichiers_pdf/15_12_16_RCV_2015.pdf. Accessed 30 Oct 2016

Parvex F, Turiel A (2001) Ameliorations des structures agricoles et paysageres dans les perimetres viticoles. Sauvegarde des murs en pierres seches et du vignoble en terrasses valaisan. Etude exploratoire, rapport final. Service de l'agriculture de l'etat du valais, Offices des améliorations foncières et de la viticulture. http://www.pierreseche.com/a_lire/rapport_final_SEREC.pdf. Accessed 01 Sept 2016

Pereira GM, Morais Barros A (2016) Port Wine and the Douro region in the early modern period. RIVAR 3:127–144

Petit C, Werner K, Höchtl F (2012) Historic terraced vineyards: impressive witnesses of vernacular architecture. Landsc Hist 33:5–28. https://doi.org/10.1080/01433768.2012.671029

Pini AI (2003) Il vino del ricco e il vino del povero. In: Archetti G (ed) La civiltà del vino. Fonti, temi e produzioni vitivinicole dal Medioevo al Novecento. Centro Culturale Artistico di Franciacorta e del Sebino, pp 585–598

Provansal M (ed) (1990) L'agriculture en terrasses sur les versants méditerranéens; histoire, conséquences sur l'évolution du milieu, Séminaire Aix-en-Provence, 3 février 1990. Méditerranée 71

Quaini M (1973) Per la storia del paesaggio agrario in Liguria. Atti della Società ligure di storia patria 86:201–360

Queijeiro J, Vilanova M, Rodriguez I, de la Montaña J (2010) Paysages viticoles et terroir dans l'OAC Ribeira Sacra (Galice, NO de l'Espagne). In: VIII international terroir congress, chap 6, pp 91–96. http://terroir2010.entecra.it/atti/VIIICongress_sess6.html. Accessed 01 Sept 2016

Romegialli G (1834) Storia della Valtellina e delle già Contee di Bormio e Chiavenna, vol 1. Cagnoletta

Scaramellini G (2014) Coltura della vite, produzione e commercio del vino in Valtellina (secoli IX-XVIII). Territoires du vin 6. http://revuesshs.u-bourgogne.fr/territoiresduvin/document.php?id=1765

Scaramellini G, Varotto M (eds) (2008) Paesaggi terrazzati dell'arco alpino. Atlante. Marsilio

Schultz HR (2010) Paysages et vignobles des vallées du Rhin et de la Moselle. In: Pérard J, Perrot M (eds) Paysages et patrimoines viticoles. Rencontres du Clos-Vougeot 2009. Centre Georges Chevrier, pp 141–150

Tarolli P, Preti F, Romano N (2014) Terraced landscapes: from an old best practice to a potential hazard for soil degradation due to land abandonment. Anthropocene 6:10–25. https://doi.org/10.1016/j.ancene.2014.03.002

Terranova R (1989) Il paesaggio costiero terrazzato delle Cinque Terre in Liguria. Studi e Ricerche di Geografia 12:1–58

Terranova R, Bernini M, Zanzucchi F (2006) Geologia, geomorfologia e vini del Parco Nazionale delle Cinque Terre (Liguria, Italia). Boll Della Soc Geol Ital 6:115–128

Unwin T (2005) Wine and the vine. An historical geography of viticulture and the wine trade. Routledge

Varanini GM (2003) Le strade del vino. Note sul commercio vinicolo nel tardo Medioevo (con particolare riferimento all'Italia settentrionale). In: Archetti G (ed) La civiltà del vino. Fonti, temi e produzioni vitivinicole dal Medioevo al Novecento. Centro Culturale Artistico di Franciacorta e del Sebino, pp 635–663

Vinea Wachau (2014) Wachau souterrain—geology and wine. http://www.vinea-wachau.at/en/vinea-wachau/geology-and-wine/. Accessed 15 Jan 2017

www.cervim.org. Accessed 10 June 2016

Chapter 3
Italian Terraced Landscapes: The Shapes and the Trends

Mauro Varotto, Francesco Ferrarese and Salvatore Eugenio Pappalardo

Abstract Since the 1980s, Italian terraced landscapes have become the object of scientific attention, with a significant increase in systematic studies only in the last twenty years. However, the state of knowledge is still fragmentary. The studies were initially concentrated in limited areas considered particularly significant from an environmental or historical point of view. Even today, a detailed, national map of terraced landscapes is still lacking. Starting from this "state of the art" formed by extremely differentiated knowledge levels, the project "Mapping Terraced Landscapes in Italy" (MAPTER) began, thanks to the collaboration of several research centers at the third Meeting of the International Terraced Landscapes Alliance (ITLA) (October 6–15, 2016). MAPTER collected and attempted to harmonize, for the first time, the available data on both local and regional scales, integrating them with further surveys for uncovered areas, to produce an initial estimate of national terraced systems. This contribution delineates, first, the project's outcomes, presenting the initial data concerning the extent and geographical distribution of Italian terraced systems. The second part of the contribution includes observations on the new processes of returning to abandoned terraced lands. These observations have emerged from a survey (the Livingstones Project) promoted by the Italian Alpine Club, and they are used here to identify practices of a virtuous "third way" for managing rural mountain areas, far from marginality, abandonment, and productive intensification.

M. Varotto (✉) · F. Ferrarese · S. E. Pappalardo
Department of Historical and Geographic Sciences and the Ancient World, University of Padova, Padua, Italy
e-mail: mauro.varotto@unipd.it

F. Ferrarese
e-mail: francesco.ferrarese@unipd.it

S. E. Pappalardo
e-mail: salvatore.pappalardo@unipd.it

M. Varotto et al. (eds.), *World Terraced Landscapes: History, Environment, Quality of Life*, Environmental History 9,
https://doi.org/10.1007/978-3-319-96815-5_3

3.1 The State of the Art: A Growing Interest in the Last 20 Years

Terraced landscapes in Italy were, with few exceptions, long neglected by twentieth-century academic studies (Scaramellini 2010). However, this trend saw the first signs of reversal between 1980 and 1990, likely due to increasing terrace abandonment and consequent slope instability (Terranova 1989; see Tarolli et al. 2014, Bonardi and Varotto 2016).

Today, the growing awareness of terraced landscapes' value goes hand-in-hand with their degradation and disappearance, which has, in turn, given rise to scientific and institutional initiatives to counter this decay. Over the past few decades, scientific surveys focusing on terraced landscapes have increased due to various European cooperative projects, such as the pioneering "Terrasses" program in France (1982–1989), the Swiss recovery program "Proterra" (1996), and the SUDOE cooperation project "PATTER—Heritage of terraces in the western Mediterranean" (1999–2001). These projects were followed by the Mallorca Council's European project "TERRISC—Récupération des champs de terrasses prévention et des risques naturels" (2004–2006) and finally by the Interreg IIIB Alpine Space Project "ALPTER—Terraced landscapes of the Alps" (2005–2008) (www.alpter.net).

Society has also contributed to raising awareness about Europe's terraced landscapes through various movements, associations, and local and international initiatives. Among the first were the "Dry Stone Walling Association (DSWA)", born in Great Britain in 1968, the "Société scientifique internationale pour l'étude pluridisciplinaire de la Pierre Sèche (SPS)," founded in France in 1997, and the "Center for research and studies for the development of viticulture in the mountains (CERVIM)", founded in Aosta in 1987. The International Alliance of Terraced Landscapes was founded in 2010, following the first World Conference on Terraced Landscapes in Honghe (China), and an Italian branch was added in 2011. Both national and international organizations have been supplemented by local initiatives, such as those of museums and eco-museums (e.g., in Italy, the Ecomuseum of the Terraces of Cortemilia in 1995 or the Ethnographic Museum of Brenta Valley in 2004), and, more recently, dry-stone wall handicraft courses and schools, as in the Province of Trento (2016).

After the first inscription of a terraced landscape on UNESCO's World Heritage List (Ifugao in the Philippines, 1995: see Guimbatan and Baguilat 2006), several European vineyard terraces were also recognized. Among these were two Italian terraced landscapes, both inscribed in 1997: Cinque Terre (with a National Park later instituted in 1999) and the Amalfi Coast. By the end of 2013, 17 European agrarian landscapes were included on the list, with 12 of them related to vineyards and five specifically related to terraces. Another five candidates were also added to UNESCO's Tentative List (Bonardi 2011).

A number of contemporary European official documents stimulated giving right value to the rural landscapes of the Alps and beyond—the Alpine Convention

(1991) and its Mountain Farming Protocol, the European Landscape Convention (2000), the recent National Register of Historic Rural Landscapes, promoted by the Italian Ministry of Agricultural Food and Forestry Policies (2012), and the measures of the Common Agricultural Policy specifically oriented for restoring and enhancing cultural landscapes (Asins Velis and Romero González 2015).

Despite growing attention and initiatives, knowledge of the extent, distribution, and conservation of Italian terraced landscapes still appears fragmented and incomplete. Some studies are detailed but are also limited to areas considered most significant—those recognized by UNESCO (Cinque Terre and the Amalfi Coast) or other important terraced areas, such as Valtellina in Lombardy, Brenta Valley in Veneto, Cembra and Terragnolo Valleys in Trentino, or Costa Viola in Calabria (see for example Brancucci et al. 2000; Italia Nostra 2004; Di Fazio et al. 2005; Bonardi 2008; Brandolini et al. 2008; Sarzo 2009; Gravagnuolo 2016). Regional censuses and mapping have, so far, been carried out in only five regions: Liguria, Tuscany, Sicily, Trentino, and Veneto (see Varotto 2008; Barbera et al. 2010; Agnoletti et al. 2015; Alberti and Lodatti 2016). Such studies have adopted different mapping methodologies, from aerial and satellite photos to field surveys, or, in the best cases, micro-relief analysis obtained from LIDAR data (e.g., in Trentino, which probably represents the most detailed survey of terraced systems at the national level: see Osservatorio del Paesaggio Trentino 2017). These local and regional studies have not yet been matched at the national level. A detailed map of national terraced heritage is still lacking, despite the fact that recent hydrogeological diseases, caused by climate change, have repeatedly highlighted the need for monitoring and safeguarding these crop systems. Bonardi (2008, 2010, 2016) carried out an initial estimate of the extent and distribution of Italian terraces, distinguishing between great or "regional" terraced areas for commercial production (such as Valtellina), intermediate or "sovralocal" systems (such as Cembra or Brenta Valleys), and "local" terraced areas built for subsistence agriculture (Bonardi 2010). Even without exact quantification as to the extent of terraced landscapes in the Alps and in Italy (Bonardi suggests more than 200,000 ha: see Bonardi and Varotto 2016), several measurement and classification methods have been proposed and tested in sample areas based on the extent and intensity of terracing found there (Scaramellini 2005; Varotto and Ferrarese 2008; Ferrarese et al. in this book).

3.2 An Initial Survey on Extent, Distribution, and Land Use: The MAPTER Project

The MAPTER project was launched in 2016 with the aim of giving order and coherence to a cognitive landscape made up of highly differentiated knowledge levels, in terms of both information quality and geographical coverage. The project unified researchers and universities for an initial national survey of terraced areas, leading up to the third Meeting of the International Terraced Landscapes Alliance (ITLA) on October 6–15, 2016. MAPTER collected and harmonized georeferenced data available on local and regional scales, integrating them, where possible, with further surveys on the most significant uncovered areas, to achieve the first quantification of the extent and distribution of Italian terraces. (For further information on the methodology, see Ferrarese et al., in this volume). This is, however, still a provisional result; it requires new additions since it combines great precision (e.g., research based on LIDAR survey in Trentino: Osservatorio del Paesaggio Trentino 2017) with rigorous, but less precise, regional, morphometric investigations (e.g., surveys of aerial photos and on-site surveys, such as in Sicily and Veneto: see Barbera et al. 2012; Varotto 2008), and information gathered in areas like Calabria and Sardinia where an accurate and complete survey is still missing, and the data, therefore, is underestimated.

Combining the abovementioned research at regional, sub-regional, or local levels has allowed the first georeferentiation and quantification of terraced areas on a national scale: 169,153 ha, still recognizable on land or in aerial photos. This is a partial datum, since a vast portion of the terraced area is hardly visible or recognizable as a long-abandoned heritage, and there are not yet accurate surveys for some Italian regions (Lombardy, Calabria, Sardinia; see Table 3.1). It can, therefore, be assumed that the maximum extent reached by terracing in Italy, before the abandonment processes of the twentieth century's second half, could have been between 200,000 and 300,000 ha (as also suggested by Bonardi 2016). This would have been the result of desperate and heroic slope-colonization for agricultural purposes, mostly linked to the demographic growth of the Italian mountains between the mid-eighteenth and early twentieth centuries. The terraced landscape is, therefore, a built heritage that today is already more than half abandoned or largely unused, covered by part of the 5 million ha of woodland that grew between 1960 and 2010 (Fondazione Montagne Italia 2015).

Observing this initial mapping's distribution structure (Fig. 3.1 and Table 3.1), the largest terraced region, in extent, is Sicily, with over 63,000 ha surveyed, followed by Liguria with 42,000 and, more detached, Tuscany, Campania, and Lazio. This classification confirms the predominantly Apennine and, specifically, Tyrrhenian nature of the Italian terraces, already suggested by Bonardi 2016 and now more strikingly highlighted by the emergence of vast terraced areas, even in previously underestimated regions such as Lazio. This also helps to explain the prevailing orientation of terracing toward the southwest (average of the azimuthal directions of 188°), probably determined by climatic factors related to atmospheric conditions and less cold winter stress indexes.

However, considering the percentage of terraces on the regional surface, it is Liguria that boasts the highest ratio (almost 8%), followed at a long distance by Sicily (2.46%) and Tuscany (about 1%).[1] Though characterized by important sub-regional terraced systems in lower Valle d'Aosta, Valtellina (Lombardy), Lagarina Valley, Cembra, and Terragnolo (Trentino), or in the Venetian Prealps (Veneto), Alpine

[1]Nevertheless Agnoletti et al. (2015): 4572 estimate over 102,000 ha of terraced lands in Tuscany (about 4.5% of the regional surface), mostly occupied by olive trees.

Table 3.1 Ranking of Italian regions by size of the terraced areas surveyed. The letter "a" indicates the regions where the census was carried out only partially

Classification	Region	Terraced surface (ha)
1	Sicilia	63,554
2	Liguria	42,636
3	Toscana	22,730
4	Campania	11,357
5	Lazio	5671
6	Lombardia[a]	4974
7	Trentino-A. Adige[a]	4815
8	Veneto	2688
9	Valle d'Aosta	2364
10	Piemonte	2324
11	Calabria[a]	1753
12	Puglia	1731
13	Friuli-V. Giulia	1142
14	Abruzzo	1078
15	Marche	141
16	Molise[a]	76
17	Umbria[a]	49
18	Sardegna[a]	26
19	Emilia Romagna	10
20	Basilicata[a]	1
	Total	169,127

Source MAPTER project, 2016; elaboration: Francesco Ferrarese, Salvatore Eugenio Pappalardo, Mauro Varotto

terracing is not as extensive as Apennine terracing. (Lombardy, the first Alpine region in the ranking, is only in sixth place). Here, terraces reach the highest altitudes of the peninsula, with the municipality of Argentera (province of Cuneo) holding the record for the average altitude (over 1800 m above sea level). More than half of the Italian terraced areas, however, are located at lower altitudes, within 300 m asl, highlighting the importance and spread of the phenomenon—especially in coastal areas, in the insular regions, and in the foothill areas near valleys or plains also characterized by important urban phenomena (Figs. 3.1 and 3.2).

Without mapping the dry-stone retaining walls for most of the areas surveyed, it is not possible to quantify the overall wall length, which has been georeferenced in detail only for some areas (e.g., Cinque Terre, Terragnolo, Cembra, Canale di Brenta, Pont Saint Martin). In the absence of an overall survey, each region or terraced area has its own primacy. For example, Liguria boasts 40,000 km of dry-stone walls—enough to wrap around the earth—and the Amalfi Coast boasts of having a quantity of walls equivalent, in total length, to the Great Wall of China (8000 km). These data are only estimated, but they do not appear very far from reality. Assuming a retaining wall average of 1 km/ha (an average calculated from heterogeneous sample areas scattered in the Alps), we can reasonably state that at

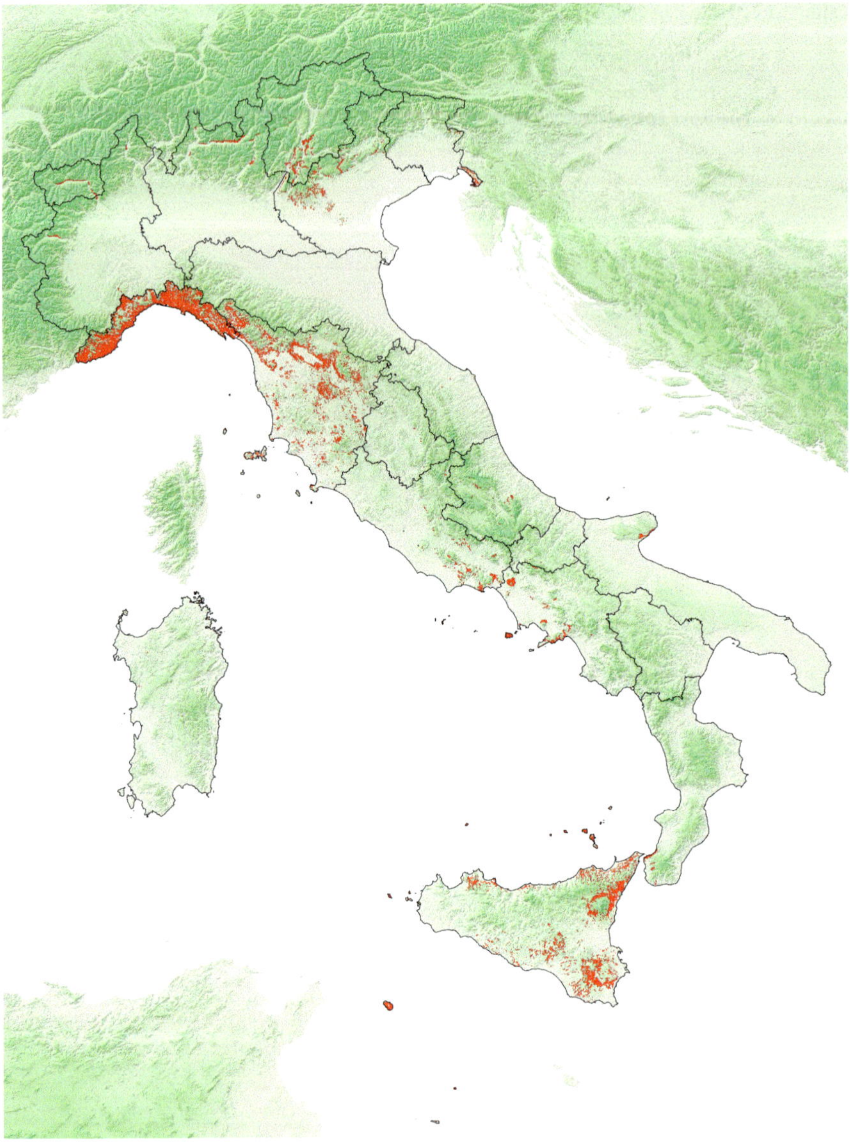

Fig. 3.1 Map of the national distribution of the main terraced areas. *Source* MAPTER project, 2016; elaboration: Francesco Ferrarese, Salvatore Eugenio Pappalardo, Mauro Varotto

least 170,000 km of dry-stone walls are still present in the peninsula. These walls are, on average, less than 2 m high. However, they can reach cyclopean heights of 8–10 m in areas such as the Amalfi Coast or the Brenta Valley, due to steep slopes, favorable lithological conditions and materials, and a high cultivation profitability to compensate construction investment.

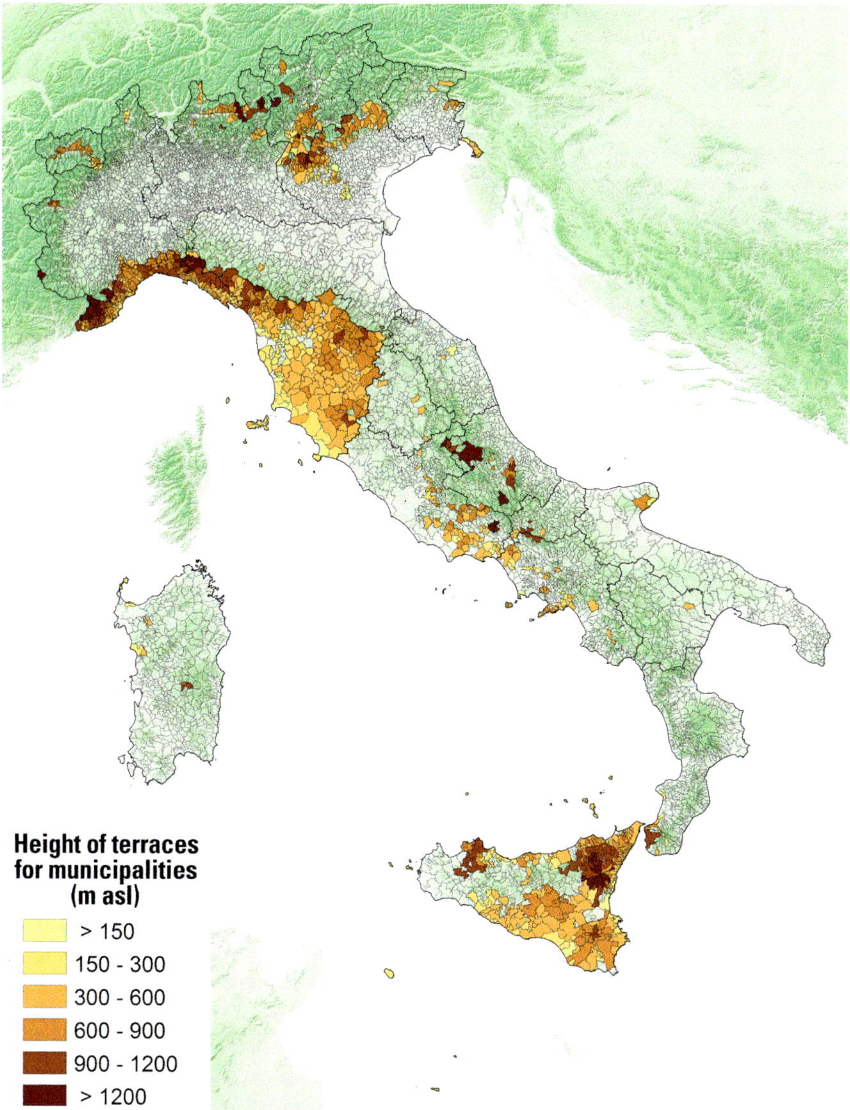

Fig. 3.2 Map of the altimetric distribution of the main Italian terraced areas. *Source* MAPTER project, 2016; elaboration: Francesco Ferrarese, Salvatore Eugenio Pappalardo, Mauro Varotto

Analyzing municipal terracing, the data show another peculiar aspect of Italian terraces—their strong character of insularity, with high concentrations in islands' municipalities. For example, in absolute values, Italy's most terraced municipality is Pantelleria Island (5500 ha), followed by other Sicilian municipalities (Genoa, the first non-Sicilian municipality, is in fifth place). However, in percentage terms, signifying terrace impact on individual municipal landscapes, the ranking sees other island territories in the first places, some with approximately 80% of their territory filled with terraces, as in the Aeolian Island municipalities of Leni and Santa Maria di Salina (Tables 3.2 and 3.3).

Regarding land use (Fig. 3.3), applying data from the Corine Land Cover 2012 to the approximately 170,000 ha of documented terraced areas shows that over 30% of this heritage is now unused and has been reconquered by natural afforestation and shrubs. Additionally, "pastures and grasslands" could be defined as substantial under-utilization. Thirty-two percentage of terraced areas is occupied by arable land, 19% by olive groves, 3% by orchards (lemon groves, chestnut woods), and only 3% by vineyards, while the advance of urbanization on agricultural terraces is

Tables 3.2 Ranking of the first 10 Italian municipalities for absolute width of terraced areas

	Municipality	Province	Terraced area (ha)
1	Pantelleria	TP	5503.1
2	Modica	RG	4267.7
3	Ragusa	RG	3825.0
4	Lipari	ME	3130.1
5	Genova	GE	2304.2
6	Noto	SR	2199.9
7	Roccamonfina	CE	1812.2
8	Castiglione di Sicilia	CT	1649.3
9	Mazzarino	CL	1606.9
10	Adrano	CT	1540.7

Source MAPTER project, 2016; elaboration: Francesco Ferrarese, Salvatore Eugenio Pappalardo

Tables 3.3 Ranking of the first 10 Italian municipalities for percentage incidence of terraced areas

	Municipality	Province	Terraced area (ha)	% Municipality area
1	Leni	ME	717.8	81.7
2	Santa Marina Salina	ME	701.3	79.9
3	Roccafiorita	ME	79.7	68.3
4	Pantelleria	TP	5503.1	65.1
5	Villa Faraldi	IM	605.0	63.6
6	Cervo	IM	227.3	63.4
7	Roccamonfina	CE	1812.2	58.4
8	Malfa	ME	491.7	56.3
9	Gaggi	ME	414.7	54.2
10	Graniti	ME	504.8	50.2

Source MAPTER project, 2016; elaboration: Francesco Ferrarese, Salvatore Eugenio Pappalardo

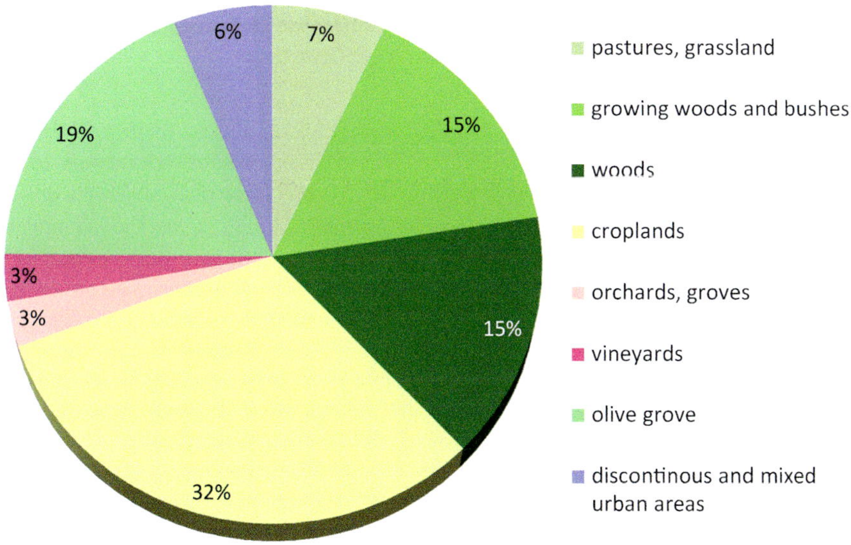

Fig. 3.3 Land use classification of the Italian terraced areas. *Source* MAPTER project—Corine land cover 2012; elaboration: Francesco Ferrarese, Mauro Varotto

estimated at around 6%. This is certainly rough data, calculated on a minimum cell area of 25 ha, which does not reflect the originally promiscuous nature of these plots. Still, it is useful at a small scale to capture the great variety of terrace use in Italy, where terracing has been wrongly associated in the contemporary image with the profitable and officially celebrated vineyard, which, in reality, concerns an extremely limited portion of terrace extent. This leads to some perplexity on whether such minimal land use, thanks to its economic profitability, can really affect the fortune and maintenance of such a vast terraced heritage.

An aspect of further interest is the remarkable development of terraced areas around the active or extinct volcanic cones of the Tyrrhenian coast (Etna, Somma and Vesuvius, Roccamonfina, and the Latium Volcanoes), particularly sought after for the peculiar soil minerality, used for orchards, chestnut groves, citrus groves, olive groves, and vineyards.

The data thus highlight the main threat Italian terraced landscapes face—namely abandonment and the advance of spontaneous reforestation. These are opposed, albeit less widespread, by the opposite trend toward urbanization and productive intensification, which transforms these complex landscapes into simplified, high-yield agrarian systems, often breaking down retaining walls or replacing them with gabions or concrete walls, compromising a valuable historical and cultural landscape. This is the “extremization” of habitats that characterized the Italian mountains during the second half of the twentieth century (Bätzing 2005).

3.3 A Second Survey: Toward a "Third Way" Between Abandonment and Agribusiness

In the last decades, in Italy and in other European countries, there has been a tendency to come back to abandoned lands and to the mountains (Dematteis 2011; Varotto 2013; Corrado et al. 2014; Varotto 2015). We have also observed this in some terraced landscapes (Bonardi and Varotto 2016: 115), as the Manifesto "Terraced landscapes: choosing the future" at the back of this book underlines. These new paths aim to overcome the profound dichotomy that characterized the mountains of the twentieth century, the gap between development processes and protection claims; they attempt to reconcile environmental, historical, and economic values, overcoming and reducing the fractures and poor territorial outcomes of twentieth century specialization (Du Guerny and Hsu 2012). Modern technologies and materials (concrete, plastics) and heavy agricultural mechanization are caesuras with a past considered outdated, and they have led to the gap between useful and useless, separating functional spaces from wasteland (Gri and Pascolini 2005; Bätzing 2005).

"Every stone is good", the dry-stone craftsmen say. Every stone, sooner or later, finds its place in building a wall, and there is no room for the "culture of waste." The stones are, at the same time, "alive" for those who cultivate around them, for the animal and plant microcosms that inhabit the interstices, for those who gaze on them from the outside as tourists or visitors. They welcome modernity, not as opposition, but as an integration of the plots inherited from the past. They live in the present, without erasing the past and without blocking the future. In this effort to reunite different functions, sometimes precarious and not always with a happy ending, we should forge a new "pact" (Camanni 2002; Donadieu 2006) between the urban and the rural (Guiseppelli 2005, 2006; Luginbuhl 2007), between mountains and plains, between local microcosms and the macrocosm crossed by globalization challenges. In fact, for many trajectories, the fruition, proximity, and relationship between the outside world and the terracing, the slope, the valley, and, more generally, the Alpine and Apennine mountains, are not oppositional but are proposed as a new alliance promoting development to enhance the starting context.

Despite diverse declinations; historical and geographical areas; individual or collective initiatives; spontaneous or structured efforts; or social, agronomic, and leisure values, these paths are crossed by common denominators outlining a "third way" for terraced landscapes. This "third way" escapes today's dominant tendencies: abandonment and degradation on the one hand, and the opposing intensification of production that stiffens, waterproofs, and trivializes the functional complexity of the dry-stone walls, of the crops, and of the people who work them.

3.3.1 The New Value of the "Longue Durée"

The first element linking these trajectories is an awareness of terraces' historical value. Far from making it appear an unused heritage of the past, the landscape's *longue durée* represents added value in an era of rapid change—a stability that balances mobility with the frenzy of everyday life. *Longue durée* can also play a role in symbolic marketing because returning to terraces, for many, means reconnecting themselves to history and geology, through lithological autochthony or the old cultivars adapted to the environment with time (see also Marson, in this book). In some of the experiences identified in the Livingstones Project (Varotto 2016), promoters of return are associations that, following the example of the pioneering Société Pierre Sèche established in France in 1997, recognize multiple values in the art of working with stone, or ecomuseums working to protect memory from the historical–ethnographic perspective.

Identity plays a part in reestablishing a link with the "land of the fathers." Many cases are of young people who return to forgotten farms inherited from their parents. However, this link does not exclusively mean possession but can instead delineate belonging earned through caring for the land. This search for identity does not translate into simple conservation. It is the product of an intergenerational dialogue. The vertical extension of temporality is associated with the horizontal, geographical extension of multiscalarity. This continuity exists thanks to an *in–out* dialectic, between insider and outsider, as in the case of Brenta Valley's small terraces (Varotto 2006, 2013; Lodatti 2013; Varotto and Lodatti 2014). It is, therefore, an identity combining roots and routes, re-fertilizing the mountains with contributions from outside the local reality, without imposing exogenous models of development.

3.3.2 Quality Turn: Local, Artisanal, Different

Quality is a nuanced and ambiguous term (Jackson 2013). It is cloaked in multiple values: the production reflecting characteristics of a place, the craftsmanship of work, the small production size and its sustainability, far beyond the simple *taste* sought and praised by *Masterchef* TV programs.

Quality means passion for the earth and for the stone, which translates into the manual skills of some processes—favoring adaptive techniques (see Murtas 2013, 2015) against the standardized intervention of an industrial matrix. It can be said that a "terroir" is built here starting from its minimum cell: the terraced field and the man who works there. This assumes a declination beginning with the biodiversity of a few square meters of cultivation and extending to the slopes. The slopes, in turn, compose a mosaic extending to the entire valley, often fighting against increasingly standardized and bureaucratized rules of production.

Quality is not necessarily synonymous with a pauperistic perspective or lack of profit. It means a market for niche products recognized all over the world, as is the case with the fine wines of Ischia. Defending quality means abandoning or overshadowing profit, high yield, interest, and immediate gain. Here, the concept of economy is linked with its original etymology of managing a house that is simultaneously environment, place, and landscape. Quality means knowing how to wait, dilating the return time of an investment. For this reason, in the era of high yields per hectare and maximum efficiency, the defense of quality seems a luxury possible only in situations of part-time or hobby farmers with access to multiple incomes. Far from being a point of weakness, this part-time farming is a strength if recognized and oriented to terrace maintenance toward functional plurality.

3.3.3 "Leaving Room for Others": The Ecological and Social Sustainability

Many returns to terraces avoid agronomic investment situations oriented to maximum yield, to the exploitation of every available space regardless of the impact on biodiversity, on the ecological and social balance, and on the landscape ecosystems (Plieninger et al. 2006; Haubenhofer et al. 2010a, b).

The terraced landscape, left at the margins by a certain model of agricultural development, becomes an opportunity to express alternative routes, to enhance old crops and methods of organic cultivation (Nossiter 2004). In this way, organic wine is opposed to Amarone's "desert of vineyards," and ancient varieties of tomatoes are recovered on the Amalfi Coast. The terraced landscape's biodiversity exists on at least three levels: the diversity of the species cultivated in the field; the "third landscape" represented by interstices and margins near the walls, in which rich microflora and microfauna flourish (Sarzo 2009); and the remote, abandoned spaces left to natural evolution because they are difficult to recover and cultivate. Cultivating *in* nature and not *against* nature is the median way these new farmers (Van der Ploeg 2009) pursue, debunking the "uncontaminated" concept—a spy of a modernist approach to nature—which fraudulently means man only "impacts" nature negatively.

When cultivated *in* rather than *against*, nature is not only preserved, but is even encouraged by the creation of specific anthropogenic ecologies: the Mediterranean habitats of the "masiere" or "parracine" (local dry-stone wall names) with incredible niches of microbotanical richness; the dry-stone wall flora seen as an ethnobotanical heritage to be used medicinally (Sarzo 2004); and, together with averting slope failure, preventing hydraulic risks by choosing the philosophy of protective care over resolving emergencies.

Looking closely at terrace sustainability, the answer is here—reconciling human and environmental needs by overcoming Man-Nature Manichaeism. Terracing is opposed to industrial modernity, but not to "modernity" *tout court*, which is always redefined with respect to terracing's past. In this sense, environmental sustainability does not mean going back. It can marry forms of high technological innovation, with low-impact research and production, with phytodepuration plants and renewable energy production, while reusing traditional housing and rural landscapes.

3.3.4 Sociability to Tame Verticality

Another common feature of return trajectories is rediscovering the economy-sharing and cooperation that has historically characterized the impervious conditions of life and work in the mountains. It is true that the terraced areas are, in general, privately owned, but this legal characteristic is grafted into the sociability of the complex "proximity networks" that have guaranteed wise system maintenance (Acovitsioti Hameau 2008).

In the experiences collected in the Livingstones Project (Varotto 2016), at least three levels of shared management can be distinguished. The first sharing dimension is at the family scale, played out on an intergenerational level, which proposes updating the traditional family farming model (Crowley 2013; Wymann von Dach et al. 2014; Varotto and Lodatti 2014). The second level is determined by the presence of broader, associative structures, which encourage participatory and inclusive recovery forms, such as the Committee for the Adoption of Abandoned Land in Valstagna or the project for social inclusion managed by the social cooperatives of Sondrio (Bonardi and Varotto 2016). In the most fortunate and structured outcomes, which are generally not so frequent, these association networks are also supported by the public administration.

From these projects—be they familiar, associative, or supported by public bodies—often emerges a "third level" of sharing with a wider user audience, which is played out, not only through web visibility, but also in concrete initiative promotion to raise awareness and share the landscape as a common good. There are no barbed wires, video surveillance systems, or threatening signs, even if the properties are private. Behind this lies an idea of hospitality oriented in two directions, toward the earth and toward humanity (Bonesio 2003). It seems that there cannot be one without the other, because the idea of common good and collective pleasure extends to embrace the field, humankind, and the planet together (Francesco 2015). Once again, this perspective marks its distance from solitary trajectories, which are more the result of desperate and losing attempts, as evidenced by the contrast between the farmers of Giuseppe Taffarel's "Fazzoletti di terra" and the new forms of adoption in the documentary "Small land" (Piccola terra: see Trentini and Romano 2012; Varotto and Rossetto 2016). The sharing of fatigue, a newfound sociality, and a sense of community that breaks away from the shortness of individual self-interest are aspects that meet the idea of landscape as a common good expressed by the European Landscape Convention and by the International Terraced Landscapes Alliance—a source of identity and well-being for all its inhabitants (see Peters and Junchao 2012; Tillmann and Bueno de Mesquita 2015; Alberti et al. 2018).

3.3.5 Landscape as a Theater: Aesthetic and Educational Values

In many cases, the places of return are chosen also for their aesthetic value—a harmoniousness of forms emanating from the dialogue between stones and landscape, from the adaptability and sweetness of the wall's lines supporting the relief, from the mastery and perfection that comes from processing boulders handed down by generations of craftsmen and farmers.

The terraced areas are, by nature, panoramic. They are balconies on the surrounding world, natural theaters, and, as such, they invite people to reflection and contemplation, to be, at the same time, "actors" and "spectators" in the landscape (Turri 2006). This requirement of beauty naturally transforms these places into educational workshops *en plein air*, an opportunity for experiences such as educational gardens and food courses—each one inviting people to re-appropriate knowledge and skills to guide behavior models toward local development.

This movement of experiences opens terraced spaces to the world and allows them to return to what they have always been—spaces born out of necessity and from the dynamics that postponed local microcosms, born for demographic or commercial reasons in past centuries, and continuing for the existential motivations of identity, quality, and sustainability today.

Most of the Italian terraces have been abandoned because society and the economy have turned their backs on them, thus condemning them to collapse. It is not an ineluctable destiny, if we create conditions so the terraces can exercise a plural function today: productive, environmental, economic, and social. The strength of these return trajectories, together with their virtuosity, is not given by this or that element taken singularly, nor by the conservation of the terrain's morphology and the walls themselves. The vision of the future these practices promise comes from being able to respond to many needs together. What, on the contrary, determines its fragility is the lack of political and economic guidelines capable of rewarding (not necessarily supporting) a multi-functional and multi-level perspective. The recognition of a historical rural landscape is not enough. The UNESCO World Heritage List, the European incentives, and policies for ecosystem services are not enough, nor is the recognition of a PDO or PGI specification, nor even restrictions for protected natural areas or museums (see Puleo 2012). These are beneficial, but they must be less specialized and more open to an overview and a shared project.

In this sense, the terraced landscapes that come back to life reflect the difficulties of our time—an era of specialization and efficiency incapable of building a unitary vision that holds the demanding judgment of conscious citizens and the future. But these terraces are also the seeds of a new vision of agriculture, of life, and of the world struggling to be something that is different and apparently, but not actually, impossible.

Acknowledgements A heartfelt thanks goes to all the collaborators of the MAPTER project, of the Livingstones Project carried out by the "Terre Alte" Group of the Italian Alpine Club, and to the friends of the Italian section of the International Terraced Landscapes Alliance for their support and active collaboration in these years of ambitious goals.

References

Acovitsioti Hameau A (2008) Territori terrazzati: atto tecnico e fatto sociale. In: Scaramellini G, Varotto M (eds) Paesaggi terrazzati dell'arco alpino—atlante. Marsilio, pp 19–27

Agnoletti M, Conti L, Frezza L, Santoro A (2015) Territorial analysis of the agricultural terraced landscapes of Tuscany (Italy): preliminary results. Sustainability 7:4564–4581

Alberti F, Lodatti L (eds) (2016) Venetian terraced landscapes. Regione del Veneto

Alberti F, Murtas D, Dal Pozzo A, Tillmann T (eds) (2018) Paesaggi terrazzati.Scelte per il futuro. Regione del Veneto

Asins Velis S, Romero González J (2015) Développement de politiques publiques pour la préservation de l'agriculture en terrasses dans l'Union Européenne. In: Harfouche R, Poupet Ph (eds) Du Mont Liban aux Sierras d'Espagne. Les sociétés méditerranéennes et la gestion des sols et de l'eau en montagne. Autour du projet franco-libanais CEDRE «Nahr Ibrahim». Archaeopress, Royaume-Uni

Barbera G, Cullotta S, Rossi-Doria I, Ruhl J, Rossi-Doaria B (2010) I paesaggi a terrazze in Sicilia. Metodologie per l'analisi, la tutela e la valorizzazione. ARPA Sicilia

Bätzing W (2005) Le Alpi, una regione unica al centro dell'Europa. Bollati Boringhieri

Bonardi L (2008) Paesaggi terrazzati d'Italia. L'Universo 88:56–70

Bonardi L (2010) Elementi per una geografia dei terrazzamenti agrari in Italia. Sul livello del Mare 36:16–21

Bonardi L (2011) "Salvare il salvabile". L'incerto futuro dei paesaggi terrazzati delle Alpi. In: Ferrata C, Crivelli P (eds) Paesaggio senza memoria? Perché e come tutelare il patrimonio. Museo Etnografico Valle di Muggio. Quaderno n. 6—GEA Associazione dei geografi, pp 72–79

Bonardi L (2014) Paesaggi e peculiarità dei terrazzamenti vitati. In: Bonardi L, Caligari A, Foppoli D, Gadola L, Grossi D, Stangoni T, Vanoi G (eds) Paesaggi valtellinesi, trasformazione del territorio, cultura ed identità locale. Mimesis

Bonardi L, Varotto M (2016) Paesaggi terrazzati d'Italia: eredità storiche e nuove prospettive. Club Alpino Italiano

Bonesio L (ed) (2003) La montagna e l'ospitalità. Il mondo alpino tra selvatichezza e accoglienza. Arianna editrice

Brancucci G, Ghersi A, Ruggiero ME (2000) Paesaggi liguri a terrazze. Riflessioni per una metodologia di studio. Alinea

Brandolini P, Faccini F, Pescetto C (2008) I terrazzamenti della Liguria: un bene culturale e del paesaggio a rischio. L'Universo 88:204–221

Camanni E (2002) La nuova vita delle Alpi. Bollati Boringhieri

Conti L, Tei T, Sorbetti Guerri F, Barbari M (2011) Analisi territoriale dei sistemi terrazzati agricoli della Regione Toscana. AIIA, pp 4564–4581

Corrado F, Dematteis G, Di Gioia A (eds) (2014) Nuovi montanari. Abitare le Alpi nel XXI secolo. Franco Angeli

Crowley E (2013) Family farming—the backbone of sustainable rural development? Rural 21 Int J Rural Dev. www.rural21.com

Dematteis G (ed) (2011) Montanari per scelta. Indizi di rinascita nella montagna piemontese. Franco Angeli

Di Fazio S, Malaspina D, Modica G (2005) La gestione territoriale dei paesaggi agrari terrazzati tra conservazione e sviluppo. AIIA

Donadieu P (2006) Campagne urbane. Donzelli
Du Guerny J, Hsu L-N (2012) Terraced landscapes: meeting the challenge to sustainability. A northern Mediterranean agricultural perspective. In: Peters HA, Junchao S (eds) First terraced landscape conference (Honghe, China). Paper collection. Kunming-Yunnan People Publishing, pp 22–38
Francesco P (2015) Laudato sì. Enciclica sulla cura della casa comune. Edizioni Paoline. Rapporto sullo stato del Paesaggio 06. www.paesaggiotrentino.it
Fondazione Montagne Italia (2015) Rapporto Montagne Italia 2016. FMI-Caire
Gravagnuolo A (2016) Valutazione e mappatura dei servizi culturali nei paesaggi terrazzati. Il caso studio della Costiera Amalfitana. https://doi.org/10.14609/ti_1_16_5i
Gri GP, Pascolini M (eds) (2005) Multiverso. Scarti e abbandoni. Forum
Guimbatan R, Baguilat T (2006) Misunderstanding the notion of conservation in the Philippine rice terraces—cultural landscapes. Int Soc Sci J 58(187):59–67
Guiseppelli E (2005) L'agriculture dans les zones périurbaines du sillon alpin: une figure de la nature pour les habitants? Rev Géographie Alp 4:59–70
Guiseppelli E (2006) Place et fonctions de l'agriculture en zones périurbaines de montagne: modes d'habiter et représentations du rural. L'Espace Géographique 35(2):133–147
Haubenhofer DK, Elings M, Hassink J, Hine RE (2010a) The development of green care in western European countries. Explore 6(2):106–111. https://doi.org/10.1016/j.explore.2009.12.002
Haubenhofer DK, Blom-Zandstra M, Kattenbroek I, Brandenburg W (2010b) Green Care as opportunity for knowledge systems, learning and collective action across Europe. In: Proceedings of the 9th European IFSA symposium, Wien Universität für Bodenkultur, 04–07 July 2011, pp 315–320
Italia Nostra (2004) La cultura dei terrazzamenti per la salvaguardia del paesaggio. Tecniche, risorse, strumenti. Menabò
Jackson P (ed) (2013) Foodwords. Essays in culinary culture. Bloomsbury, pp 166–168
Lodatti L (2013) Ritorno alle "terre alte": l'adozione di terrazzamenti in abbandono nel Canale di Brenta. Agriregionieuropa 9(33). Online available: www.agriregionieuropa.univpm.it
Luginbuhl Y (2007) Nouvelle urbanités nuovelle ruralites en Europe. Peter Lang
Murtas D (2013) Paesaggi terrazzati. Il valore delle persone. In: Varotto M (ed) La montagna che torna a vivere. Testimonianze e progetti per la rinascita delle Terre Alte. Nuova Dimensione, pp 69–80
Murtas D (2015) Pietra su pietra. Costruire, mantenere, recuperare i muri in pietra a secco. Pentagora
Nossiter J (2004) Mondovino. La guerra del gusto. Feltrinelli
Osservatorio del Paesaggio Trentino (2017) Atlante dei paesaggi terrazzati del Trentino. Trentino meridionale
Peters HA, Junchao S (eds) (2012) First terraced landscape conference (Honghe, China). Paper collection. Kunming-Yunnan People Publishing
Plieninger T, Hocthtl F, Spek Th (2006) Traditional land-use and nature conservation in European rural landscapes. Environ Sci Policy 9:317–321
Puleo T (2012) Parasitizing lanscape for UNESCO world heritage. Geoforum. http://dx.doi.org/10.1016/j.geoforum.2012.11.024
Sarzo A (2004) Mondi di pietra, per mano dell'uomo. Aspetti geografici e floristici dei muri a secco in Trentino. Natura Alpina 56(1-2):5–26
Sarzo A (2009) La flora dei muretti del Trentino. Cento fiori di campagna e di città: descrizione, distribuzione, ecologia, curiosità. Curcu & Genovese
Scaramellini G (2005) Il paesaggio agrario e il paesaggio culturale dei terrazzamenti artificiali nelle Alpi. In: Trischitta D (ed) Il paesaggio terrazzato. Un patrimonio geografico, antropologico, architettonico, agrario, ambientale. Città del Sole, pp 101–141
Scaramellini G (2010) I paesaggi terrazzati a fini agrari e la ricerca geografica in Italia. Sul livello del Mare 36:9–15

Scaramellini G, Varotto M (eds) (2008) Terraced Landscapes of the Alps—Atlas. Marsilio. Online available at: www.alpter.net

Scheytt S (2007) Cooperare o fallire. In: Noi Alpi! Uomini e donne costruiscono il futuro. 3° Rapporto sullo stato delle Alpi. CIPRA. CDA&Vivalda, pp 220–231

Tarolli P, Preti F, Romano N (2014) Terraced landscapes: from an old best practice to a potential hazard for soil degradation due to land abandonment. Anthropocene 6:10–25. http://dx.doi.org/10.1016/j.ancene.2014.03.002

Terranova R (1989) Il paesaggio costiero terrazzato delle Cinque Terre in Liguria. Studi e Ricerche di Geografia 12:1–58

Tillmann T, Bueno de Mesquita M (eds) (2015) II Congreso Internacional de Terrazas. Encuentro de culturas y saberes de terrazas del mundo. JICA-CBC

Trentini M, Romano M (2012) Piccola terra. Film DVD. Cierre-Antersass. Available on youtube: https://www.youtube.com/watch?v=yLeQeClpC-s

Turri E (2006) Il paesaggio come teatro. Marsilio

Van der Ploeg JD (2009) I nuovi contadini. Le campagne e le risposte alla globalizzazione. Donzelli

Varotto M (2006) Le "masiere" del Canale di Brenta: origine, declino e rilancio di un paesaggio culturale. Geotema 29:170–179

Varotto M (2008) Conclusions. Towards the rediscovery of the middle landscapes. In: Scaramellini G, Varotto M (eds) Terraced landscapes of the Alps—Atlas. Marsilio, pp 112–118. Online available at: www.alpter.net

Varotto M (ed) (2013) La montagna che torna a vivere. Testimonianze e progetti per la rinascita delle Terre Alte. Nuova Dimensione

Varotto M (2013) Nuove mani, vecchie pietre: recupero materiale e valori immateriali del ritorno alla terra in una località delle Prealpi venete. In: Bonato L, Viazzo PP (eds) Antropologia e beni culturali nelle Alpi. Studiare, valorizzare, restituire. Edizioni dell'Orso, pp 219–229

Varotto M (2015) Terraced landscapes of the Alps: decay, rediscovery, revitalization. In: Tillmann T, Bueno de Mesquita M (eds) II Congreso Internacional de Terrazas. Encuentro de culturas y saberes de terrazas del mundo. JICA-CBC, pp 38–48

Varotto M, Ferrarese F (2008) Mapping and geographical classification of terraced landscapes: problems and proposals. In: Scaramellini G, Varotto M (eds) Terraced landscapes of the Alps—Atlas. Marsilio, pp 38–45. Online available at: www.alpter.net

Varotto M, Lodatti L (2014) New family farming for abandoned lands. The adoption of terraces in the Italian Alps. Mt Res Dev 34(4):315–325

Varotto M, Rossetto T (2016) Geographic film as public research. Re-visualizing/vitalizing a terraced landscape in the Italian Alps (Piccola terra/small land, 2012). Soc Cult Geogr 17 (8):1140–1164

Wymann von Dach W, Romeo R, Vita A, Wurzinger M, Kohler T (eds) (2014) Mountain Farming is Family Farming. A contribution from mountain areas to the International Year of Family Farming 2014. FAO, CDE, BOKU. http://www.fao.org/docrep/019/i3480e/i3480e.pdf

Chapter 4
Slovenian Terraced Landscapes

Lučka Ažman Momirski

Abstract This review of Slovenian terraced landscapes presents some archaeological details, a detailed survey of published research on terraced landscapes, and the state of terraced landscapes in Slovenia. The study concludes with the inclusion of terraced landscapes in spatial planning and in protected cultural landscapes. Slovenia has hilly terrain that is not favorable for dense settlement. Exceptional archaeological findings on Mount Donatus (Donačka Gora) confirm that people already transformed slopes and settled on terraced platforms during the transition between the Bronze and Iron Ages. There have been two peak periods of publishing about Slovenia's terraced landscapes: the first one in 2007 and 2008 and the second one in 2015 and 2016. Terraced areas can be found in more than 90% of Slovenia's municipalities, but the presence of terraced landscapes in the municipalities is not uniform. Various types of terraces exist in Slovenia, and they can be defined according to the use or function of the terrace slope and terrace platform, the form of the terrace slope and terrace platform, and the construction of the terrace slope. Grassed terrace slopes are much more common in the Slovenian cultural landscape than dry stone wall construction, and they can be found in all Slovenian regions. Slovenian spatial planning has not recognized terraced regions as a landscape system sui generis and needs to adopt a procedure enabling institutions at the national and local levels to acknowledge their existence. In some cases, terraced landscapes are protected as a part of cultural landscapes and are recognized as cultural heritage sites.

4.1 Introduction

Slovenia has exceptionally diverse landscapes within its small territory of 20,273 km^2, which is inhabited by 2,063,371 people. The country does not have a terrain favorable for dense settlement, and the population density of Slovenia is

L. Ažman Momirski (✉)
Faculty of Architecture, University of Ljubljana, Zoisova 12, 1000 Ljubljana, Slovenia
e-mail: lucija.azman@fa.uni-lj.si

M. Varotto et al. (eds.), *World Terraced Landscapes: History, Environment, Quality of Life*, Environmental History 9,
https://doi.org/10.1007/978-3-319-96815-5_4

101.8 inhabitants/km^2. Urban areas in Slovenia cover less than five percent of its land. The present settlement structure is also a result of implementing a polycentric urban system, the main goal of long-term spatial planning in Slovenia, promoted by a two-tiered structured network of centers of national and regional importance, to which the network of other centers connects based on considering the proper division of functions and interconnection of transport links (OdSPRS 2004). The capital, Ljubljana, is the only Slovenian city with a population over 100,000 (Republika Slovenija Statistični urad 2016). Slovenia's wealth of diversity also results from the fact that the country lies at the intersection of four major European regions (the Alpine, Pannonian, Dinaric, and Mediterranean regions) and four different cultural spaces (German, Romance, Hungarian, and Slavic; Perko 1998a, b). This variety and the transitional nature of Slovenia's regions constitute its main geographic characteristic and are important elements of its identity.

The major part of Slovenian territory is wooded (about three-fifths of the territory), and the rest is cultural landscape (over one-third of the territory). The plains, fields, basins, and valleys have slope inclinations between 0° and 2°, which accounts for 14.3% of the country's territory (Perko 2001). The rest of the territory has a hilly configuration that required the people that settled it—who were more aware of their direct dependence on nature in the past and had to adapt to it—to grow their crops on the sides of hills or mountains by planting on graduated level areas built into the slope.

Farmers were particularly careful to retain and maintain their fertile soil (Ažman Momirski and Radikon 2008). Terraces maximized arable land in variable terrains and for various crops, reducing soil erosion and water loss at the same time. These terraces made it possible to create appropriate growing conditions for cultivating fruit trees and grapevines and to produce high-quality crops.

There is some archaeological evidence for the creation of terraced areas in Slovenia. Hillside terraces carved into rock are an exceptional archaeological discovery on the steep slope of Mount Donatus (*Donačka Gora*) at an elevation of 750 m in the Municipality of Rogatec. A study published in the early 1990s states that the archaeological site consists of terraces that were formed when collecting stone for querns (Ciglenečki 1992). After computer conversion of Lidar data for the terrain in 2015, archaeologists confirmed the existence of terraces that were clearly manmade. An archaeological field survey was conducted at the site in the fall of 2016. Remains of settlements from the transition of the Copper Age to the Iron Age (approximately when civilization was flourishing in the New Kingdom of Egypt, about 2800 years ago) were found on terraces south of the steep slope; these terraces enjoy excellent natural protection by cliffs on all sides. Houses and shacks stood on some terraces, and there were probably gardens and barns on other terraces. The terraces were about five meters wide and twenty or more meters long. Even after the Bronze Age, the terraces were occasionally in use (as is evident from pottery found at the site) in Late Antiquity and possibly in the Early Middle Ages (Odar 2016).

Other evidence that people were already building platforms and terraces reinforced with stone walls in pre-Roman times to erect houses on slopes has been found in Croatian Istria, south of the Slovenian border. Many prehistoric hillforts with fortified settlements at strategic locations from the Bronze Age and Iron Age in Istria are located there (Likar 2017). At the beginning of the second millennium BC, around 1800 BC, a newly settled community began extensive work to reshape the peak of the hill at the Monkodonja hillfort near Rovinj (Ivetić 2009). The site is laid out in oval concentric plateaus, the smallest one being the highest. The wider and more extended the platform, the lower it is. The material remains found in situ are evidence that the inhabitants were already able to build dry stone walls and earthen terraces for the village foundations (houses, platforms, and communication routes) and enclosure walls (Likar 2017). Research also indicates that people were compelled to build terraces for agricultural production in order to produce a sufficient quantity of food for a relatively large population living in the hilly Istrian environment. Studies of the finds at hillforts in Istria have shifted the boundary of features that until recently were ascribed to the Romans: Grape seeds, which were found during the last excavations at the Monkodonja hillfort in 1997, indicate that grapevines were already present before the arrival of the Romans (Ivetić 2009).

4.2 Terraced Landscape Research in Slovenia

The partnership between University of Ljubljana's Faculty of Architecture and the ALPTER project led to a decisive turning point in research on terraced landscapes in Slovenia. The ALPTER project (The Terraced Landscapes of the Alpine Arc, 2005–2008) reviewed the degradation of agricultural terraces caused by agricultural abandonment in the Alps and the possibilities for rehabilitating these terraced spaces (ALPTER 2015a, b). At least, 18 articles were published on terraced landscapes in Slovenia in connection with the project. The years 2007 and 2008 were one of the two peak periods of publishing about Slovenia's terraced landscapes.

The development of agriculture in Slovenia (as a Yugoslav Republic) was highly ideologically laden after 1945 and remained so at least until 1971 (Petek 2005). Most published studies on terraces before the ALPTER project dealt with terraced vineyards. Vrišer (1954) described the abandonment of cultivated terraces in the Upper and Central Gorizia Hills and the transformation of terraced plantations in the Lower Gorizia Hills in his study on the Gorizia Hills. Melik (1960) mentioned the same process for the Koper Hills. A year before the conference of winegrowers from Istria, the Kvarner Gulf, and the Slovenian coast was held in Portorož (1960), one of its initiators, Simčič (1959), presented an initiative for a plan to renovate terraced vineyards. Around the same time, some investment programs envisioned renovation of vineyards and orchards on terraces in eastern Slovenia (Tomaž 1959; Jeruzlem-Ormož 1960; Ljutomer 1962). Terracing was obviously an important theme at the beginning of the 1960s; the archives contain bachelor's theses on terraces in vineyards at the time (Kociper 1962) and on the renewal of vineyards on steep slopes (Ražem 1964). Addressing the social and geographical problems of the

Koper countryside in the light of historical development and economic changes, Titl (1965) studied cultivated terraces in Istria in southwest Slovenia in detail, including the typology of terraces, the causes for their abandonment, and a topographic map defining cultivated terraces, partly cultivated terraces, abandoned terraces, and destroyed terraces. In eastern Slovenia, after a major wine crisis between the two world wars and no systematic renewal of vineyards, which were also devastated and exhausted as a result of artillery bombardment during the war and as a result of labor shortages, new terraces for vineyards were constructed under communism (Bračič 1967). This transformation has completely altered the appearance of the cultural landscape in the Slovenian Hills (Belec 1968), where terrace renovation carried out in the national, public, or social sector during this period encompassed 80–90% of all areas (Simonič 2014). A practical guide to setting up vineyards (Colnarič 1971), in which the introductory part offers guidelines for various types of construction of terraced vineyards using agricultural machinery, is later summarized in two further books by the same author (Colnarič et al. 1985; Colnarič and Vrabl 1991). Colnarič's work was the basis for delivering practical information to farmers at agricultural lectures in the Littoral Region, which promoted knowledge as a key factor in agricultural development (Škvarč 1999, 2001; Kodrič 1999). Vršič and Lešnik (2005) also refer to Colnarič when discussing terracing of slopes, but they add new graphic material and some new interpretations of terracing. Colnarič served as the advisor for a 1993 thesis on three-row terraces and their impact on the growth, development, and fertility of grapevines depending on the types of distance from the edge of the terrace (Ozmec 1993). Among these comprehensive studies on terraced vineyards, one can find a rare contribution on how to handle orchards on terraces (Modic 1979). In 1990, a conference was held at which two papers were presented on terraced landscapes: The first dealt with the physical geographical significance of cultivated terraces and typology (Drobnjak 1990), and the second discussed opportunities to intensify their production (Kladnik 1990). A study of viticulture and terminology connected with terraces and old viticultural techniques in villages near Koper from the mid-nineteenth century to the 1950s (Presl 1995) presents important historical data. Some years later, agricultural advisers pointed out (Škvarč et al. 2002) that terraces had already been constructed on slope inclinations below 15% and that changing terraced vineyards into vertical plantations was unjustified because this would accelerate erosion and drought. Agricultural advisers in the Littoral Region also continually emphasized that constructing terraced vineyards is the best method for managing agricultural land on steep slopes (Škvarč and Kodrič 2007).

At the initiative of the ALPTER project, a recovery project for terrace construction in the village of Medana was prepared and carried out (Ažman Momirski and Berčič 2007a). A detailed process with the result of the recovery and also including general recommendations of terrace construction techniques for farmers and builders was published in a manual (Ažman Momirski et al. 2007). The Slovenian contribution to the two publications of ALPTER project, the atlas and the manual, was a description of the terraced vineyards of the Gorizia Hills (Ažman Momirski 2008a, 2014), a risk assessment of the slopes of the Gorizia Hills

(Komac and Zorn 2008), and a detailed study of the preparation, construction, and construction assessment of the terraced slope in Medana (Ažman Momirski et al. 2008). The results of Slovenian research on terraced landscapes, terraced landscapes in the southern and northern Gorizia Hills, and in Medana and Kožbana were published in parallel with Slovenian (Ažman Momirski 2008b, c, d). The International Conference Living Terraced Landscapes, held in Ljubljana at the conclusion of the ALPTER project, demonstrated that more experts in Slovenia deal with terraced landscapes than is immediately apparent (Living Terraced Landscapes 2008). Attentive planning of terraced landscapes is based not only on geotechnical terrain characteristics, but also on principles of sustainable development—which take into account the ecological aspects (design that complies with the local flora, fauna, terrain, bedrock, materials, climate conditions, and topography, as well as use of native plants and conserving existing flora and fauna), cultural aspects (respecting the local cultural landscape), and visual sustainability (Ribičič and Musek 2008). Vineyards on slopes are among the land-use types most exposed to slope mass movements in Slovenia (Jemec and Komac 2008), and landslides present a constant problem in the Gorizia Hills, affecting viticulture on slopes (Komac and Zorn 2008). In this respect, it was essential to show the positive results of the parametric stability studies carried out to prove the correctness of the plan for the recovery of terraced vineyards in settlement of Medana (Petkovšek et al. 2008). The first attempt to review all terraced landscapes in Slovenia, including their typologies, was introduced during a conference (Ažman Momirski and Kladnik 2008, 2009, 2012), as was an overview of vineyards and orchards on terraces in the Littoral Region (Škvarč and Kodrič 2008). Hauptman speaks about the Drava Valley wine region, where terraces were built between 1960 and 1990. After that time, wine production on terraces decreased, even though the humid climate has changed into an arid one due to climate change and terraces are the only reasonable option for revitalizing wine production (Štabuc and Hauptman 2008). Rebernišek (2008) presented the possibility of maintaining vineyards on steep slopes with mini-terraces. Hudoklin (2008) recognized terraced landscapes as areas of landscape identity that are changing significantly in some part of Slovenia. Prosen and Lisec (2008) emphasized a holistic approach toward the conservation of terraced landscapes, taking into account social, economic, and environmentally sustainable development.

After attempts to review all terraced landscapes in Slovenia, including their typologies (Ažman Momirski and Kladnik 2008, 2009, 2012), such studies expanded (Križaj Smrdel 2010a, b). The project to revitalize vineyards in the terraced landscape on the Karst ridge above Trieste aimed to recover the functional features of existing routes to improve access to agricultural land and thereby enhance cultivation on the terraces (Kačič and Lidén 2011).

The second peak period of publishing about Slovenia's terraced landscapes followed two world conferences on terraced landscapes: the Second World Conference on Terraced Landscapes (Peru 2014) and the Third World Conference on Terraced Landscapes (Italy 2016). A national research project focusing entirely on terraced landscapes in Slovenia concluded in 2014 and was then extended until the end of 2015 (Terraced Landscapes in Slovenia as Cultural Values 2016). Terraced slopes and special factors of land-use changes related to terraces are typical of the Mediterranean world in general, including the Slovenian sub-Mediterranean area (Ažman Momirski and Gabrovec 2014). A volume on terraced landscapes at the regional scale of sub-Mediterranean Slovenia (Ažman Momirski 2014) contains a survey of historical sources of terraces, geological data, physiognomy of the terrain and terraced landscapes, and use/functions of terraces in various areas at the level of region, municipality, cadastral municipality, settlement, and plot. In some cases, preventing landslides on slopes is not necessary because terraces and their slopes have preserved the same form for almost two hundred years. However, planning and constructing new terraces demand a contemporary methodological approach (Ažman Momirski 2015b). In the Mediterranean landscape, terraces are a traditional element, whereas in the Pannonian landscapes, terraces are modern phenomena, created mostly in the 1960s and 1970s to facilitate intensive mechanized farming (Ažman Momirski and Kladnik 2015a). After reviewing documents at the national level (laws and strategies) and local level (spatial documents) to determine whether and how they refer to terraced landscapes, it is possible to propose a procedure to allow institutions at the national and local levels to map and evaluate terraced landscapes (Ažman Momirski and Berčič 2016). Nowadays, Lidar data offer an unprecedented accurate new interpretation tool for detecting terraced landscapes. The most significant feature of this new method is its reliability for detecting the exact boundaries of terraced areas (Berčič 2016). This new methodology will probably challenge the findings of previous studies presenting the diversity of Slovenian terraced landscapes (Kladnik et al. 2016a, b). The same authors contributed to a publication for the general public presenting terraced landscapes of the world and Slovenia (Kladnik et al. 2016a, b). Recently, Likar (2017) called attention to the fact that terracing was not only carried out on slopes to prevent erosion and mitigate the effects of droughts and flooding on farmland, but was also used in excavations for house foundations in settlements, for example, in northern Istria.

4.3 State of Terraced Landscapes in Slovenia

Reviewing all Slovenian territory, only 19 municipalities (out of 211) have no terraced landscapes, which altogether represent 3.3% of the country's entire territory. In other words, terraced areas that form part of the cultural landscapes in Slovenia can be found in more than 90% of the country's municipalities, corresponding to a little less than 97% of Slovenian territory. The presence of terraced landscapes in the municipalities is not uniform: In some municipalities, where most of the territory is flat, there may be only a few terraces on slopes that are not very steep at the edge of the municipality's territory (borderline cases were included in the category of municipalities with terraced landscapes). In other municipalities, terraced landscapes may be the dominant landscape feature. Both active (i.e., cultivated) and abandoned terraces were considered in the review (Ažman Momirski and Berčič 2016).

Various types of terraces exist within Slovenia's terraced landscapes, but they are all made up of two basic formal elements: a terraced platform (or tread) and a terraced slope (or riser; Ažman Momirski 2008a, b, c, d). The criteria for determining the types of terraced landscapes are defined according to:

- The use or function of the terrace slope and terrace platform;
- The form of the terrace slope and terrace platform; and
- The construction of the terrace slope.

Land use relates to the exploitation of land through human activity in the landscape, and it is one of the best indicators of landscape structure and processes. Within Slovenia's cultural landscapes, three main types of terraces can be distinguished based on land use (Ažman Momirski and Kladnik 2008, 2009):

- Agricultural terraces;
- Viticultural terraces; and
- Fruit-growing terraces.

Mixed types of terraces can be also found. Vegetables are grown below olive groves on terraces, fruit trees are planted on the slopes of agricultural terraces, and so on.

Agricultural terraces can be found all over Slovenia, in all of its major regions. Agricultural terraces are mostly meadows and pastures, and sometimes they are cultivated more intensively with crops or vegetables. Agricultural terraces in Slovenia differ extensively in their forms: Terraces in the high mountains just below the ridges mainly have wide terrace platforms with a steep gradient, an irregular plan, and extremely low terrace slopes; however, because of the high terrain inclination, these terraces are extremely high. In contrast, terraces at the foot of the hills, which are often low, have medium-wide terrace platforms and an almost equal proportion between terrace slopes and terrace platforms. There are also uniform, regular, higher terraces with only a few centimeters of the gradient of the terrace slope, and therefore again an identical proportion between the terrace slope and terrace platform (Ažman Momirski and Berčič 2016).

In many Slovenian landscapes, the development of cultural landscapes is inseparably linked to the development of viticulture. In the Middle Ages, farmers started to cultivate unworked land, and in the Gorizia region, for example, they planted vines on flat fields. After 1574, this type of grape cultivation was sometimes banned because it began to threaten cereal production. Prohibitions on planting grapes in fields and pastures, and a general prohibition on planting vineyards, appeared until the eighteenth century. Since the middle of the nineteenth century, the development of viticulture has been significantly affected by pests and diseases, such as gray mold of grape, peronospora, and grape phylloxera. Because phylloxera destroyed many vineyards, they had to be reconstructed, which prompted the renewed progress of viticulture. The farmers dug deep into vineyards, planting vines in rows and at proper distances. The new arrangement also improved the production options. During the First and Second World Wars, many vineyards were destroyed and farmers started to abandon old vineyards and vineyards in less favorable areas because of labor shortages (Ažman Momirski 2008b).

Due to the specific features of production and favorable conditions for cultivating grapes, most vineyards in Slovenia are located on steep terrain, and one-fifth are terrace plantations (21%), among which 71% are arranged on slopes with a gradient over 9.1° (16%). Approximately, half of the vineyards on terraces can be found in the gradient class of 9.1°–17.0° (16–30%), and just under one-third are on slopes from 17.2° to 26.6° (31–50%). In 2015, 959 vineyards (229 ha) were planted on terraces with inclinations of 24.4°–31° (45–60%) and 101 vineyards (9 ha) on terraces with inclinations over 31° (60%) (Vineyard Census 2015). Viticulture on vineyard sites at elevations over 500 m, on slopes greater than 17.0° (30%), on terraces or embankments, and on small islands under difficult growing conditions is classified as "heroic viticulture." This is a marginal form of viticulture, which is widespread in less than a tenth of wine-growing areas in Europe, and for which the criteria have been defined by CERVIM (2017). Since 1987, the International Organisation of Vine and Wine (OIV) has paid special attention to it.

It is also important to emphasize that an average winegrower in Slovenia cultivates only a small area, about half a hectare of vineyard. In 2015, over 90% of winegrowers were cultivating less than one hectare of vineyard, and the average vineyard size was 0.3 ha. For comparison, an average winegrower was cultivating a much smaller area (0.1 ha less) than in 2009 (Vineyard Census 2015). The study further showed that 104 municipalities in Slovenia have terraced vineyards (Fig. 4.1).

The wine-growing area of Slovenia is divided into three wine-growing regions (the Drava Region, which includes the Drava Valley in the northeast; the Lower Sava Region, which includes the Lower Sava Valley, White Carniola, and Lower Carniola in the southeast; and the Littoral Region, covering the traditional region of the Littoral in the southwest) and nine wine districts (Prekmurje, Styria, Bizeljsko-Sremič, Lower Carniola, White Carniola, the Gorizia Hills, the Vipava Valley, the Karst Plateau, and Slovenian Istria) as a consequence of the great variety of the landscape, different climate and soil conditions, and different natural conditions (Pravilnik 2003).

According to data from orthophotograph maps, there is almost 19,300 ha of vineyards in Slovenia, although the Register of Grape and Wine Producers (RPGV) lists only about 16,000 ha (Vinogradništvo in vinarstvo 2016). Data show that vineyards are increasingly shrinking at the beginning of the twenty-first century. Based on data from orthophotograph maps, vineyards have decreased by 3651 ha in the last 10 years; based on data from the RPGV, vineyards have decreased by 1192 ha in the last 10 years (Table 4.1).

In 2007 in Slovenia, 37% of vineyards were located on terraces (in 2011 only 34.9% and in 2016 only 31%): in the Drava Valley 24% (1757 ha), in the Sava Valley 27% (824 ha), and in the Littoral 55% (3759 ha; Štabuc and Hauptman 2008; RPGV 2016) (Tables 4.2 and 4.3).

In some areas in Slovenia, such as the Brkini Hills, the majority of orchards were planted on terraces or terrace slopes by the beginning of the nineteenth century.

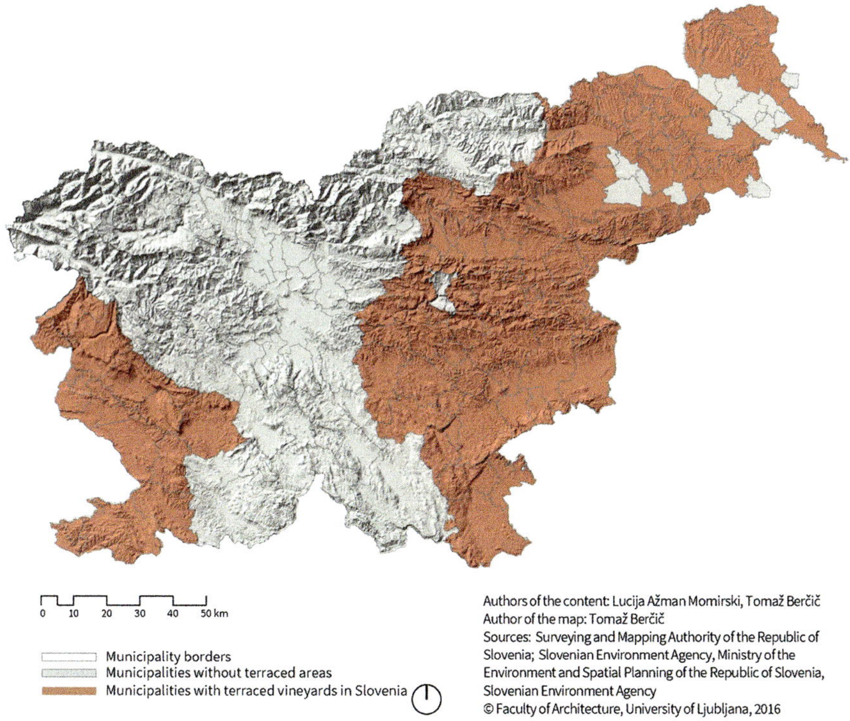

Fig. 4.1 Municipalities in Slovenia with terraced vineyards

Table 4.1 Comparison of vineyard land in Slovenia according to orthophotograph maps and according to the register of grape and wine producers (RPGV)

Years	Orthophotograph maps (ha)	Register of grape and wine producers data (RPGV) (ha)	Difference between data (ha)
2007	22,951	17,192	5759
2011	21,265	15,973	5292
2016	19,300	16,000	3300

Table 4.2 Comparison of vineyards located on terraces in the three wine-growing regions in Slovenia

Wine-growing regions	2007 (%)	2007 (ha)	2011 (%)	2011 (ha)	2016 (%)	2016 (ha)
Drava Valley	24	1757	22.7	1659	20	1254
Sava Valley	27	824	25.5	769	20	533
Littoral	55	3759	54.3	3571	49	3035
Slovenia	37	6340	34.9	5980	31	4822

Sources Štabuc and Hauptman (2008), RPGV (2016)

Table 4.3 Comparison of vineyards located on terraces in nine wine districts in Slovenia

Wine districts	2007 (ha)	2011 (ha)	2016 (ha)
Prekmurje	4	4	2
Styria	35	24	21
Bizeljsko-Sremič	42	34	27
Lower Carniola	27	18	15
White Carniola	27	23	20
Gorizia Hills	81	80	80
Vipava Valley	66	65	59
Karst Plateau	14	13	10
Slovenian Istria	30	21	20

Sources Škvarč and Kodrič (2008), Mavrič Štrukelj et al. (2012), RPGV (2016)

Here, orchard cultivation goes back to the late eighteenth century, when it was primarily promoted by teachers and parish priests (Volk et al. 2011).

A total of 8900 ha of orchards are found in Slovenia today, and intensive cultivation takes place in just over 4000 ha of orchards. There are two categories for orchards in the Slovenian land-use classification. The first is intensive orchards (category 1221), which are areas planted with only one type of fruit, except for mixed plantations of peaches and nectarines, and mixed plantations of hazelnuts, walnuts, and almonds. Farmers use modern intensive technologies to cultivate intensive orchards. The area of an intensive orchard plantation covers all plantation land together with turning areas, tracks, embankments, and other associated land. If there are more than fifty fruit trees per hectare on the land and if this plantation is not an intensive orchard, then this land is listed under land-use category 1222: extensive orchards. Natural conditions in Slovenia allow the production of mixed fruit, but apples are still the leading type of fruit (Ministry of Agriculture and Environment 2016). Only a small share of apples is grown on terraces (15.7%). The share of terraced orchards in Slovenia is higher for figs (58.1%), apricots (47.7%), cherries (47.6%), persimmons (40.7%), and chestnuts (23.7%; Škvarč and Kodrič 2007).

According to agricultural recommendations, grapevines in a terraced vineyard may be arranged as follows (terrace cross section; Ažman Momirski et al. 2007):

- Single-row terraces (the terrace platform width is 230–280 cm);
- Double-row terraces (the terrace platform width is 300–360 cm);
- Double-row terraces with a passage for a tractor between the row and the slope (the terrace platform width is 430–530 cm); and
- Multi-row terraces (the terrace platform width is 500–620 cm).

The slope and width of the terrace platform affect the width of the terraces and the height and length of the terrace slopes. The terrace platforms have a longitudinal inclination from 0.5 to 5%, which allows excess rainwater to slowly drain from vineyard terraces along their length, preventing falling of slopes, landslides, erosion, and drifting of slopes. According to agricultural recommendations, the usual length of a vineyard terrace is from 80 to 150 m (Vršič and Lešnik 2005).

Orchards on terraces can be arranged as follows (terrace cross section; Ažman Momirski et al. 2007):

- Single-row terraces with trees planted at the edge of the terrace platform (the terrace platform width is 300–650 cm); and
- Single-row terraces with trees planted on the terrace slope (the terrace platform width is 200–250 cm).

In some areas in Slovenia, planting trees on the terrace slopes is usual because it increases their stability.

The construction of terrace slopes can be divided into three basic categories:

- Dry stone wall construction (traditional terraces);
- Slopes reinforced with stones that were grubbed out during cultivation of farmland, and later covered with earth and grassed over (traditional terraces); and
- Grassed terrace slopes (modern terraces based on traditional terraces; modern terraces).

In some areas, it is difficult to distinguish between dry stone walls that divide property and the slopes of the terraces. Dry stone wall construction can be found in the Koper Hills, the Karst Plateau, and some places in the Alpine region (Fig. 4.2).

Fig. 4.2 Overgrown and deteriorating dry stone terrace walls in Krkavče in the Koper Hills, and new terrace construction and grassed terrace slopes in Medana in the Gorizia Hills. *Photo* Lučka Ažman Momirski

Fig. 4.3 Grassed terrace slopes in vineyards in Medana in the Gorizia Hills (*Photo* Lučka Ažman Momirski), in an olive grove in Krkavče in the Koper Hills (*Photo* Lučka Ažman Momirski), and in meadows and pastures in Artviže in the Brkini Hills (*Photo* Kras in Brkini) and the Materija Lowland (*Photo* Matevž Lenarčič)

Grassed terrace slopes are much more common in the cultural Slovenian landscape than dry stone wall construction and can be found in all Slovenian regions (Fig. 4.3).

Terraces can be also categorized by cultivation extent or state of decay:

- Cultivated terraces;
- Partly abandoned terraces; and
- Abandoned terraces.

The terraced landscape contributes to the identity and profile of local cultures. It is an important part of the quality of people's lives, providing variety and making the region attractive, and in this way making possible the preservation of the settlements and vitality of rural areas (Ažman Momirski 2008a, b, c, d). In this connection, terraces are often recognized as significant landscape features. Terraces with minor or medium significance are a very common feature in Slovenia.

4.4 Integration of Terraced Landscapes into Spatial Planning and Cultural Heritage

A recent study (Ažman Momirski and Berčič 2016) reviewed documents at the national level (laws and strategies) and local level (spatial documents) to determine whether and how they refer to terraced landscapes. The selected documents under review can be divided into spatial and sectoral texts. The final findings have shown that sectoral legislation covers only a small portion of landscape topics, but it does contain an essential part regarding terraced landscapes. On the other hand, the Spatial Development Strategy of Slovenia (OdSPRS 2004) provides suitable—but very general—guidelines, which are often vague and with no requirements for how to deal with landscapes at the local level. Terraced landscapes are hidden within the term *cultural landscape* and are undeniably considered part of the cultural landscape; certain statements also apply to terraced regions. In addition, parts of the strategy are obsolete due to changes in data, new strategic global and EU orientations, and so on. National documents mainly specify general guidelines, and so it is necessary to emphasize that more accurate data are needed at the operational level in order to define landscape diversity and other landscape issues. The fact that terraced landscapes are built, constructed landscapes (Ažman Momirski et al. 2007; Ažman Momirski and Kladnik 2008, 2009, 2015a, b) and that precise rules and methods for their construction exist today has not been made evident. No national or municipal document includes the search phrase *terraced landscape*.

It is possible to find many examples of terraced landscapes in the Register of Slovenian Cultural Heritage, which is a central repository of data on heritage maintained by the Slovenia Ministry of Culture, which is responsible for heritage. The purpose of the register is to provide information support to implementing heritage protection. There are 220 immovable cultural heritage units registered as cultural landscapes in the category "type of unit," but not all of them are examples of terraced landscapes. Every entry contains basic data: the name of the monument or site, its number (used in all procedures related to heritage conservation), the type and sort of heritage (of local or national importance), the location, and a short description. These descriptions highlight individual properties of the registered[1] landscapes and make it evident whether the landscape contains terraces or not. The register includes terraced landscapes in all parts of Slovenia: for example, in the Mediterranean region the Boršt area (no. 15,089) in the Koper Hills and cultural terraces in Korte (no. 21,627) in the countryside outside Izola; in the Alpine region terraced agricultural land divided by hedges in Zgornja Javoršica (no. 16,695); in the Pannonian region the Jeruzalem Hills with their terraced vineyards (no. 7867); and in the Dinaric region the terraced landscapes of Knej (no. 18,465). The main

[1]"Registered heritage" refers to heritage that is entered in the register and is not a monument. Landscape heritage is defined as a heritage site that is an open space with natural and artificial components in its structure, the development and use of which are chiefly determined by human processes and activities.

fear of the people that work and live in the terraced landscapes defined as heritage sites is that the protection measures could pose many obstacles for their lives, work, and residence. Terraces will survive in their complexity and uniqueness only if they satisfy the needs of the people involved in terraced landscapes, and if these people decide to keep them, protect them, and continue working there.

4.5 Conclusion

General knowledge and acknowledgment of terraced landscapes have greatly expanded in Slovenia in the twenty-first century. The international development of the theme is stimulating further identification of terraced systems and their technologies, research on their social organization, studies of natural hazards and land degradation, conservation and development of terraced landscapes, their role in agricultural production and food security, and tourism development and promotion. Last but not least, it is important to pay attention to policies, regulations, and management for preserving terraced landscapes and their functions.

Bibliography

Ažman Momirski L (2008a) The terraced vineyards of Goriška Brda. In: Scaramelini G,Varotto M (eds) Terraced landscapes of the alps—atlas. Marsilio, Venice, pp 102–107

Ažman Momirski L (2008b) Terasirana pokrajina. In: Kladnik D, Perko D (eds) Terasirana pokrajina Goriških brd. Založba ZRC, Ljubljana, pp 98–118

Ažman Momirski L (2008c) Kulturne terase v južnih Goriških brdih na primeru katastrske občine in naselja Medana. In: Kladnik D, Perko D (eds) Terasirana pokrajina Goriških brd. Založba ZRC, Ljubljana, pp 119–138

Ažman Momirski L (2008d) Kulturne terase v severnih Goriških brdih na primeru katastrske občine in naselja Kožbana. In: Kladnik D, Perko D (eds) Terasirana pokrajina Goriških brd. Založba ZRC, Ljubljana, pp 139–155

Ažman Momirski L (2014) Terasirana pokrajina sredozemske makroregije v Sloveniji. Ljubljana, Fakulteta za arhitekturo

Ažman Momirski L (2015a) History, culture and current state of terraced landscapes in the Gorizia Hills, Slovenia. In: Tillmann T, Bueno de Mesquita M (eds) II Congreso Internacional de Terrazas: encuentro de culturas y saberes de terrazas del mundo. Cusco, Lima, JICA, CBC

Ažman Momirski L (2016a) Types of Slovenian terraced landscapes (In print)

Ažman Momirski L, Berčič T (2007) Terraced areas in Goriška Brda. In Architettura del paesaggio, pp 1–6

Ažman Momirski L, Berčič T (in print) A detailed systematic inventory of terraced landscapes in Slovenia: case studies, development of methodology and results

Ažman Momirski L, Berčič T (2016) Ignored regions: Slovenian terraced landscapes. Ann Ser Hist Sociol26-3:399–418. http://dx.doi.org/10.19233/ASHS.2016.37

Ažman Momirski L, Berčič T, Kodrič I, Škvarč A (2007a) Smernice za ureditev trajnih nasadov na terasah: primer ureditve vinograda v Goriških brdih. Univerza v Ljubljani, Fakulteta za arhitekturo, Ljubljana

Ažman Momirski L, Berčič T, Škvarč A, Kodrič I (2007b) Priporočila za izdelavo teras: primer obnove vinograda v Goriških brdih. Fakulteta za arhitekturo Univerze v Ljubljani, Ljubljana
Ažman Momirski L, Gabrovec M (2014) Changes in land use in the Mediterranean terraced landscapes between 1819 and 2012: the case of two selected villages in Slovenia. In: Himiyama Y, Feranec J, Bičík I, Kupková L (eds) Land use, cover changes in selected regions in the world 9. IGU-LUCC, Asahikawa, pp 33–42
Ažman Momirski L, Kladnik D (2008) Agricultural production: terraced landscapes in Slovenia. In: Ažman Momirski L, Černič Mali B (eds) Living terraced landscapes: perspectives and strategies to revitalise the abandoned regions, Ljubljana
Ažman Momirski L, Kladnik D (2009) Terraced landscapes in Slovenia. Acta Geogr Slov 49–1:7–37. http://dx.doi.org/10.3986/AGS.1627
Ažman Momirski L, Kladnik D (2012) Terraces and terraced landscapes in Slovenia. In: Peters HA, Junchao S (eds) First terraced landscapes conference (Honghe–China) paper collection. Yunnan People's Publishing House, Kunming, pp 129–148
Ažman Momirski L, Kladnik D (2015a) Factors in the conservation and decline of cultivated terraces in Slovenia. In: Tillmann T, Bueno de Mesquita M (eds) II Congreso Internacional de Terrazas: encuentro de culturas y saberes de terrazas del mundo. CBC; Lima: JICA, Cusco, pp 49–55
Ažman Momirski L, Kladnik D (2015b) The terraced landscape in the Brkini Hills. Acta Geogr Slov 55–1:29–58. http://dx.doi.org/10.3986/AGS49101
Ažman Momirski L, Škvarč, A, Kodrič I (2008) The terraces of Goriška Brda–case study of Medana. In: Fontanari F, Patassini D (eds) Terraced landscapes of the alps: projects in progress. Marsilio, Venezia, pp 40–46
Belec B (1968) Ljutomersko-ormoške gorice: agrarna geografija. Obzorja, Maribor
Berčič T (2016) Discovering terraced areas in Slovenia: reliable detection with Lidar. Ann Ser Hist Sociol 26-3. http://dx.doi.org/10.19233/ASHS.2016.35
Bračič V (1967) Vinorodne Haloze: socialno-geografski problemi s posebnim ozirom na viničarstvo. Obzorja, Maribor
Ciglenečki S (1992) Polis Noricum. Zavod Spominski park Trebče, Podsreda
Colnarič J (1971) Ureditev vinograda. Kmečki glas, Ljubljana
Colnarič J, Gregorič J, Hrček L, Korošec Koruza Z (1985) Posebno vinogradništvo. Biotehniška fakulteta, VTOZD za agronomijo, Ljubljana
Colnarič J, Vrabl S (1991) Vinogradništvo. Kmečki glas, Ljubljana
Drobnjak V (1990) Fizičnogeografski pomen kulturnih teras. In: Orožen Adamič M (ed) Primorje: zbornik 15. zborovanja slovenskih geografov. Zveza geografskih društev Slovenije, Ljubljana, pp 139–142
Hudoklin J (2008) Terraced landscapes: areas of landscape identity in changing processes. In: Ažman Momirski L, Černič Mali B (eds) Living terraced landscapes: perspectives and strategies to revitalise the abandoned regions, Ljubljana
Ivetić E (2009) Istra kroz vrijeme: pregled povijesti Istre sa osvrtom na grad Rijeku. Centar za povijesna istraživanja, Rovinj
Jemec M, Komac M (2008) Landslide susceptibility in Slovenia on general and vineyard related landslides. In: Ažman Momirski L, Černič Mali B (eds) Living terraced landscapes: perspectives and strategies to revitalise the abandoned regions, Ljubljana
Kačič R, Lidén M (2011) Revitalisation of vineyards in the terraced landscape on the Karts ridge of Trieste. Oživljanje vinogradov v terasasti pokrajini tržaškega kraškega roba. AR 3 (2011):63–70
Kladnik D (1990) Možnosti intenziviranja rabe kulturnih teras. In: Orožen Adamič M (ed) Primorje: zbornik 15. zborovanja slovenskih geografov. Zveza geografskih društev Slovenije, Ljubljana, pp 143–150
Kladnik D, Ciglič R, Geršič M, Komac B, Perko D, Zorn M (2016a) Diversity of terraced landscapes in Slovenia. Ann Ser Hist Sociol 26-3:469–486. http://dx.doi.org/10.19233/ASHS.2016.38
Kladnik D, Perko D, Ciglič R, Geršič M (2016b) Terasirane pokrajine. ZRC SAZU, Ljubljana

Komac B, Zorn M (2008) Landslide hazard in the Goriška Brda hills. In: Ažman Momirski L, Mali Černič B (eds) Living terraced landscapes: perspectives and strategies to revitalise the abandoned regions, Ljubljana
Križaj Smrdel H (2010a) Kulturne terase v slovenskih pokrajinah. Dela 34:39–60. http://dx.doi.org/10.4312/dela.34.3.39-60
Križaj Smrdel H (2010b) Kulturne terase v slovenskih pokrajinah. Oddelek za geografijo Filozofske fakultete Univerze v Ljubljani, Ljubljana
Likar D (2017) An architectural and urban-planning perspective on cultivated terraces in northern Istria. Ann Ser Hist Sociol 26-3:399–418
Mavrič Štrukelj M, Brdnik M, Hauptman S, Štabuc R, Novak E, Martinčič J, Škvarč A (2012) Vinogradniške razmere v Sloveniji danes. In: Rusjan D (ed) Zbornik referatov, Slovenski vinogradniško-vinarski kongres z mednarodno udeležbo. Univerza v Ljubljani, Biotehniška fakulteta, Ljubljana
Melik A (1960) Slovensko Primorje. Slovenska matica, Ljubljana
Perko D (1998a) Geografija, regija in regionalizacija. In: Perko D, Orožen Adamić M (eds) Slovenija - pokrajine in ljudje. Založba Mladinska knjiga, Ljubljana, pp 12–15
Perko D (1998b) The regionalization of Slovenia. Geogr Zb/Acta Geogr 38:11–57
Perko D (2001) Analiza površja Slovenije s stometrskim digitalnim modelom reliefa (Analysis of the surface of Slovenia using the 100-meter digital elevation model. ZRC SAZU, Ljubljana
Petek F (2005) Spremembe rabe tal v slovenskem alpskem svetu. ZRC SAZU, Ljubljana
Petkovšek A, Klopčič J, Majes B (2008) Terraced landscapes and their influence on the slope stability. In: Ažman Momirski L, Mali Černič B (eds) Living terraced landscapes: perspectives and strategies to revitalise the abandoned regions, Ljubljana
Pravilnik (2003) https://www.uradni-list.si/glasilo-uradni-list-rs/vsebina?urlurid=20033317. Accesses on 11 Feb 2017
Presl I (1995) Vinogradništvo v vaseh koprskega zaledja od sredine 19. stoletja do petdesetih let 20. stoletja. In: Darovec D (ed) Dežela refoška: vinogradništvo in vinarstvo slovenske Istre. Zgodovinsko društvo za južno Primorsko, Koper, pp 177–196
Prosen A, Lisec A (2008) Conservation of terraced landscape as protection of soil against erosion (by sustainable land management). In: Ažman Momirski L, Mali Černič B (eds) Living terraced landscapes: perspectives and strategies to revitalise the abandoned regions, Ljubljana
Rebernišek A (2008) Is it possible to maintain vineyards on steeper slopes with mini terraces? In: Ažman Momirski L, Mali Černič B (eds) Living terraced landscapes: perspectives and strategies to revitalise the abandoned regions, Ljubljana
Ribičič M, Musek A (2008) Sustainable slope and geotechnical works. In: Ažman Momirski L, Mali Černič B (eds) Living terraced landscapes: perspectives and strategies to revitalise the abandoned regions, Ljubljana
Register of Grape and Wine Producers (RPGV) (2016) https://podatki.gov.si/dataset/register-pridelovalcev-grozdja-in-vina?resource_id=6c32828c-90a6-48f0-b559-46dc7391542b. Accessed on 17 Feb 2017
Škvarč A, Kodrič I (2007) Nature and regulation: management of vineyards and orchards on terraces. Urbani izziv 17(1/2):78–84
Škvarč A, Kodrič I (2008) Vineyards and orchards on terraces in Primorska region. In: Ažman Momirski L, Mali Černič B (eds) Living terraced landscapes: perspectives and strategies to revitalise the abandoned regions, Ljubljana
Škvarč A, Ozimič D, Maljevič J, Štabuc R, Novak, E, Carlevaris B (2002) Vinogradi za tretje tisočletje/Vineyards for the third millenium. In: Puconja M (ed) Vinogradi in vina za tretje tisočletje?. Strokovno društvo vinogradnikov in vinarjev Slovenije, Ljubljana; Zveza društev vinogradnikov in vinarjev Slovenije, Ljutomer; Poslovna skupnost za vinogradništvo in vinarstvo Slovenije, Celje, pp 1–18
Štabuc R, Hauptman S (2008) The possibilities of revitalisation of vineyards on the terraces in Podravje. In: Ažman Momirski L, Mali Černič B (eds) Living terraced landscapes: perspectives and strategies to revitalise the abandoned regions, Ljubljana
Titl J (1965) Socialnogeografski problemi na koprskem podeželju. Lipa, Koper

Vrišer I (1954) Goriška brda: gospodarska geografija. Geografski zbornik 2:51–113
Vršič S, Lešnik M (2005) Vinogradništvo. Kmečki glas, Ljubljana

Sources

ALPTER (2015a) Terraced landscapes in the Alpine Arc. http://www.alpter.net/. Accessed on 20 Jan 2017

ALPTER (2015b) Terasirane pokrajine v alpskem svetu (Terraced Landscapes in the Alpine Arc). http://predmet.fa.uni-lj.si/alpter/index.html. Accessed on 20 Jan 2017

Ažman Momirski L (2015b) Preventing landslides on terraced slopes. General Assembly of European Geosciences Union, Vienna. http://meetingorganizer.copernicus.org/EGU2015/EGU2015-9680-1.pdf. Accessed on 10 Jan 2017

Ažman Momirski L (2016b) Mediterranean landscapes, Slovene Littoral: Goriška Brda; Vipavska dolina; Kras; Brkini and dolina Reke; Podgrajsko podolje; Koprska Brda. In: Alberti F, Mattiuzzo F (eds) Terraced landscapes choosing the future: Italy October 2016, Exhibition: a world of terraced landscapes, Venice, pp 58–61

Ažman Momirski L (2016c) Pannonian landscapes, Slovenia Hills (Slovenske gorice): Jeruzalem. In: Alberti F, Mattiuzzo F (eds) Terraced landscapes choosing the future: Italy October 2016, Exhibition: a world of terraced landscapes, Venice, pp 62–65

Ažman Momirski L, Radikon B (2008) Terase v Brdih: najpomembnejša prvina kulturne krajine: uvodnik. Briški časnik, Pomlad 2008

CERVIM (2017) Centro di Ricerche, Studi e Valorizzazione per la Viticoltura Montana. http://www.cervim.org/en/heroic-viticulture.aspx. Accessed on 21 Oct 2017

Jeruzalem–Ormož VG (1960) Investicijski program za obnovo 140 ha vinogradov. Kmetijski pospeševalni zavod Maribor, Maribor

Kociper M (1962) Terase v sodobnem vinogradništvu. Univerza v Ljubljani, Biotehniška fakulteta, Agronomski oddelek, Ljubljana

Kodrič, I (1999) Obnova vinogradov. In: Stanek K, Srebrenič M, Radikon B (eds) Znanje kot ključni dejavnik razvoja kmetijstva. Kmetijska svetovalna slu\ba KZ, Dobrovo, Občina Brda; Združenje vinogradnikov in vinarjev Brda, Goriška Brda, pp 6–7

Living Terrased Landscapes (2008) Perspectives and strategies to revitalise the abandoned regions. In: International conference, 14–15 Feb 2008, Ljubljana, Slovenia. http://predmet.fa.uni-lj.si/alpter/conference/contact.html. Accessed on 21 Jan 2017

Ljutomer VK (1962) Investicijski program za obnovo 115 ha vinogradov. Agrobiro, Ljubljana

Modic D (1979) Impianti di frutteti e terazze. Frutticoltura e di ortofloricoltura

Odar B (2016) Naselbina na Donački gori obljudena v času največjega razcveta Novega kraljestva v Egiptu. http://kozjansko.info/2016/10/naselbina-na-donacki-gori-obljudena-v-casu-najvecjega-razcveta-novega-kraljestva-v-egiptu/. Accessed 2 Mar 2017

OdSPRS (2004) Spatial development strategy of Slovenia. http://www.uradni-list.si/1/objava.jsp?urlid=200476&stevilka=3397. Accessed 23 Feb 2017

Ozmec VM (1993) Trivrstne terase in njihov vpliv na rast, razvoj in rodnost trte glede na oddaljenost vrst od roba terase. Univerza v Ljubljani, Biotehniška fakulteta, Agronomski oddelek, Ljubljana

Portorož (1960) Posvetovanje vinogradnikov Istre, Kvarnerja in Slovenskega Primorja

Ražem M (1964) Obnova vinogradov v strmih terenih – terasiranje. Univerza v Ljubljani, Biotehniška fakulteta, Agronomski oddelek, Ljubljana

Republika Slovenija Statistični urad (2016) http://www.stat.si/StatWeb/Field/Index/17. Accessed 11 Feb 2017

Simčič Z (1959) Terasiranje vinogradov na Primorskem. Sadjarstvo, vinogradništvo, vrtnarstvo

Simonič S (2014) Pogled na sodobni vinski turizem, terase kot posebnost in promocija, iskanje priložnosti za razvoj vinskega turizma v Prlekiji. Društvo Antona Puklavca, Jeruzalem (in press)

Škvarč A (1999) Obnova vinogradov. In: Stanek K, Srebrenič M, Radikon B (eds) Znanje kot ključni dejavnik razvoja kmetijstva. Občina Brda: Kmetijska svetovalna služba KZ, Dobrovo; Združenje vinogradnikov in vinarjev Brda, Goriška Brda, pp 1–5

Škvarč A (2001) Znanje kot ključni dejavnik razvoja kmetijstva. Kmetijsko veterinarski zavod, Nova Gorica

Tomaž KZ (1959) Investicijski program za obnovo 46 ha nasada jablan. Kmetijski pospeševalni zavod Maribor, Maribor

Terraced Landscapes in Slovenia as Cultural Values (2016) http://predmet.fa.uni-lj.si/terasirane%2Dpokrajine/terraced-landscapes-in-slovenia-as-a-cultural-value.html. Accessed on 15 Mar 2017

Vineyard Census (2015) http://www.stat.si/StatWeb/News/Index/5919. Accessed on 11 Feb 2017

Vinogradništvo in vinarstvo (2016) http://www.mkgp.gov.si/si/delovna_podrocja/kmetijstvo/kmetijski_trgi/vino/. Accessed on 15 Feb 2017

Volk D, Štolfa M, Renčelj S, Zadnik D, Sotlar M, Furlan Brec M (2011) Brkinska sadna cesta. LAS Društvo za razvoj podeželja med Snežnikom in Nanosom, Ilirska Bistrica. LAS Krasa in Brkinov, Sežana

Chapter 5
Landscape Typology of French Agrarian Terraces

Jean-François Blanc

Abstract The landscapes of agricultural terraces occupy in France a geographical area essentially localized at east of a Strasbourg/Biarritz line. Built mainly by the peasants who cultivated them, they can be classified into two main types connected to their agricultural purposes: "food terraces", the most numerous, and terraces of opportunity on very limited spaces. The geography of terraces on the scale of France is, however, more complex. Indeed, these arrangements join very different regional contexts which produced each of the specific agrosystems in which terraces are more or less present. So, according to the important size of occupied surfaces, four types can be distinguished: the "Cévenol" model, the "High Valleys" model, the "Provencal" model and the "Isolates" model. All these models worked until the middle of the nineteenth century, and then gradually declined for a century, challenged by the crises, the technical progress and the political choices that are almost a general abandonment of mountain agriculture. However, from the 1980s, the question of the reconquest of the slopes was posed, and gradually, projects of rehabilitation of the old terraces emerged. Even if the pioneers of this revival were especially the winegrowers, today, the reconquest concerns other sectors carried by the evolution of consumption patterns, the questioning of chemical agriculture and monoculture.

5.1 Introduction

The landscapes of agricultural terraces occupy in France an important geographical area which far exceeds the Mediterranean frontage of the country. The vast majority of the slopes are located to the south-east of a line running from Strasbourg on the Rhine to Biarritz, in the western part of the Pyrenees.

J.-F. Blanc (✉)
Member of the Scientific and Executive Committee of International Terraced Landscape Alliance, Lumbin, France
e-mail: jf15blanc@gmail.com

M. Varotto et al. (eds.), *World Terraced Landscapes: History, Environment, Quality of Life*, Environmental History 9,
https://doi.org/10.1007/978-3-319-96815-5_5

The terraces of France were built mainly by the peasants who cultivated them. Two types of terraces can be distinguished: terraces of need, the most numerous, and terraces of opportunity, on very limited spaces. The first type can be called "food terraces" because their primary purpose was to feed men. They received varied cultures associated with polybreeding, mainly sheep–goat. Unlike the "food terraces", the second one very early specialized in monoculture: olive tree, floral crops, citrus. But it is mainly the vine that has settled on the slopes best exposed. The prestigious vineyards of the Eastern Pyrenees (Banyuls crus) and the Rhone Valley (Cornas, Hermitage, Côte Rôtie) testify to this.

It is not easy to present the geography of terraces in France because we do not have a precise map of these developments such as that made by our Italian neighbours (Bonardi and Varotto 2016). Indeed, if many researchers are interested in terraces, they all have limited their field of study to a few valleys (Alcaraz 1996; Castex 1980) and even those who have tackled larger sets like the northern part of the Mediterranean and have always limited their investigations and comparative analyzes to a few territories (Desbordes 1999). These completely assumed choices do not detract from the value of their scientific work and allow rich comparative approaches.

In the absence of exhaustive cartographic data, we will therefore propose a landscape typology of French terraces according to the importance of the areas occupied in the agrarian space.

Thus, four landscape models can be distinguished:

- The "Cévenol" model: when terraces and cultivated spaces were blended together.
- The "High Valleys" model of the Southern Alps, Southern Massif Central, Castagniccia and Eastern Pyrenees: small terraced areas associated with immense spaces of extensive agriculture.
- The "Provençal" model: when terraces adjoin flat lands potentially cultivable.
- The "Isolates" model: unusual developments in agrarian spaces that are generally without terraces.

The first model concerns the regions where most of the agricultural activities were carried out on the terraces because of the ubiquity of the slope. The land is heavily landlocked, with terraces corresponding to almost all of the cultivated land. The second model is associated with the first but serves as a transition to mountain areas where terraces gradually disappear. In the "Provençal" model, the agrarian space is divided between declining or flat plots and thus produces agricultures with or without terraces, even on steep slopes, but in more open regions. Finally, in other rural territories where terraces are virtually absent, certain productions, particularly the vine, can justify the choice of terraced crops very localized on insignificant surfaces sometimes, which thus present themselves in the form of unusual Isolates.

Two great types of land structures match these four models of landscape. In the "fully terraced" communes, tiny properties of 2–5 ha strongly fragmented (de Reparaz 1990) stand out. In communes where terraces occupy only a small part of the agrarian space, the rural society is more hierarchized and properties are more differentiated. We are in a less democratic social system, composed of both very small- and medium-sized proprietors, and a few notables living in "bastides" (French mansions) situated on the outskirts of the villages perched on more favourable lands.

In the four types of modelled spaces mentioned above, the functioning of agrosystem terraces is at its peak in the mid-nineteenth century. Then, and until the middle of the twentieth century, a long period of decline in the agriculture of slope begins, which massively fuels the rural exodus. Faced with this situation, from the years 1975 to 1980, geographers begin to question the well-founded and the consequences of this generalized abandonment. Our geography thesis, the first one devoted to terraced landscapes in France, was supported in 1983 (Blanc), but its content is considered "outdated" by agricultural institutions, then totally engaged in the revolution of a chemical agriculture strongly mechanized, and reserved for the plains.

Yet, associations and researchers sensitive to the preservation of rural landscapes, the environment and ecology are progressively getting interest in terraced agriculture and show that other forms of agriculture are possible. They began to be heard in the 1990s, which foreshadowed the first initiatives to reconquer slopes, supported by elected officials and carried out by agricultural institutions, regions and the European Union.

Thus, from a few examples, we will also show that cultivation in terraces can no longer be considered as a form of obsolete agriculture and non-adapted to a globalized economy. In addition to the ecological landscape and cultural interest, the reconquest of these marginalized territories would enable the reintegration of farmers and the economic revitalization of the rural slopes (Fig. 5.1).

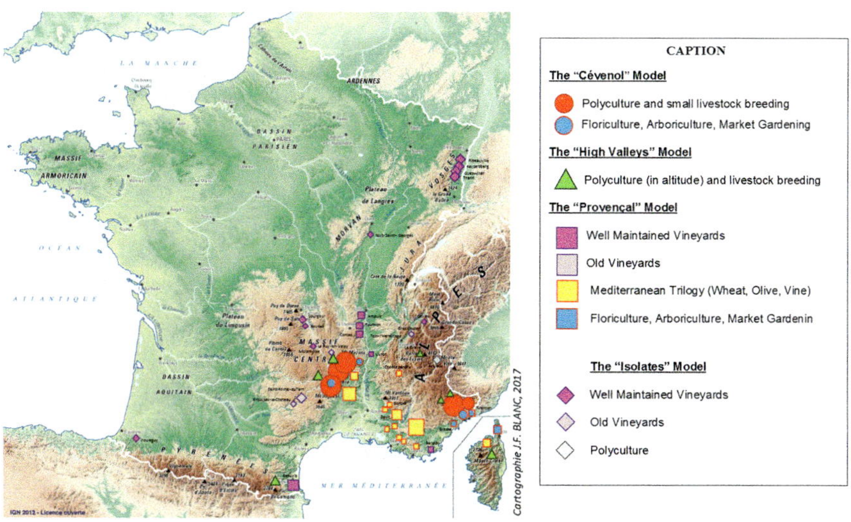

Fig. 5.1 Landscape typology of terraces in France

5.2 The Cévenol Model: When Terraces and Cultivated Areas Were Blended onto Enclosed Mountains

In absence of flat surfaces, in some French regions, agriculture could only be practiced on slopes. This first constraint, combined with a strong erosion linked to the Mediterranean climate, already justifies the generalization of the construction of terraces. Thus, the agriculture of the interfluves of the Southern Alps and southern part of the Massif Central would have been compromised without the use of the construction of walls initially intended to retain the land and facilitate the work of the men.

In the Southern Alps, the predominance of terraced landscapes in some valleys has been well analyzed in the Maritime Alps and in the Var (Castex 1996). Here, terraces dominate in the forms of practiced agriculture and cover between 30 and 50% of the surfaces. They are distributed between the sea and 1800 m above sea level, but most of them are located below 1300 m in the southern and eastern slopes, and even though they are at a low altitude, the slopes can be explained by a cooler microclimate (Castex 1983). In these hinterlands, the scarcity of available flat land and population growth are at the origin of the conquest of the slopes by a modest peasantry (de Reparaz 1990). The agriculture practiced on these "restanques" (local name for terraces) was initially based on the trilogy of Mediterranean cultures: wheat, olive and vine. This trilogy was part of a polycultural system involving a small sheep–goat farm.

The second French region that offers large terraces is situated in the south-east of the Massif Central, on three departments: Ardèche, Gard and Lozère. Often referred to as "Cévennes" (Martin 2006), it connects the uplands to more than 1000 m of altitude with the Rhone Valley and the hills of Languedoc, less than 300 m away. The slope here is still omnipresent "... between the valley of the Arre, in the south, and that of the Chassezac, in the north" (Usselmann 2006). In fact, this zone extends more to the north-east, in Ardèche, at least up to the catchment area of the Eyrieux. In these landlocked and isolated valleys, throughout history, men slowly and gradually settled, colonizing a constraining nature resulting from three geological components: shale, sandstone and granite that generated each type of specific terraces (Blanc 1983).

The landscape opposes a northerly forest, the shady side, to a south slope, the sunny side, where the peasants built their farms grouped in hamlets, surrounded by terraces. It is on this sunny slope that most of the "faïsses" or "traversiers" (local terms for terraces) are concentrated. But the terraces do not descend to the edge of the river which is reserved for a meadow irrigated by gravity. Above, from the habitat, the colonization of the slope was gradually carried out according to the demographic fluctuations. The oldest and most well-kept terraces adjoins the hamlets; the most recent climbs to the assault of the ridge lines, the summits, so characteristic of the Cévennes landscapes. Thus, over the centuries, peasant families have built the mountain to make arable lands that were not arable. With the sole objective of feeding, they created a unique landscape consisting of a mosaic of small properties of four or five hectares on average, which had to be rationalized to the extreme in order to optimize low yields. It was first of all a remarkable mastery of water and a complicated management of shortages and excesses. In Ardèche, for example, on the same slope, dry agriculture and irrigated agriculture were present. The latter has generated systems for the recovery and capture of water that is too scarce and associated with measures to limit the erosive effects of severe thunderstorms, very characteristic of the Mediterranean climate, known as "Cévenol periods", discharging up to 400 mm of precipitation per m^2 in 24 h.

This need to optimize the use of space has become more apparent with the demographic pressure exerted in the nineteenth century. With more than 100 inhabitants/km^2, the peasants settled on very small properties have rationalized the slopes by multiplying the facilities to make the land more profitable and try to feed their families. Thus, they have associated with the terraces many elements of vernacular architecture that make this "Cévenol" model an example, undoubtedly unique in the world. Indeed, there are no other forms of small rural architecture associated with terraces. The enormous work of inventory, carried out in Ardèche by Michel Rouvière, testifies to this richness (Blanc 1983). This diversity concerns firstly the circulation on the slopes which was made through narrow roads framed by walls, to the terraces in which one travelled by using ramps or stairs of various shapes, the most sophisticated using the cantilever technique. Water control has also produced multiple installations to capture, conduct, store or otherwise evacuate excess water in various ways, while also avoiding its erosive effects. The catalogue of these forms is rich: vaulted cisterns, basins, canals, wells, draining galleries, junctions, etc. To these elements of small rural architecture is added a great variety of shelters for the men and the harvests, far from the hamlets, often integrated into the retaining walls.

These terraces were the foundation of a self-sufficient agrarian system based on polyculture and sheep–goat farming, implemented by a poor peasantry. For the Ardèche, cereals occupied the most important areas (Bozon 1963). In the Gard, the olive tree covered many slopes. The vine was present everywhere, but on small surfaces, because everyone wanted to produce their wine, which explains why it could easily reach 800 m of altitude.

But the specificity of the terraced cultivation of these highly developed regions is also based on the "coltura promiscua" (Pinto-Correia and Vos 2005), which juxtaposed several crops on the same plot: vines and olive trees, chestnuts and cereals, mulberry trees and cereals, etc. This up-to-date farming, which had been brought back to life for its agro-ecological interest, was intensely practiced by peasants until the middle of the twentieth century.

This agrosystem reached its peak in the mid-nineteenth century. Thanks to the terraces, the communes of slopes were able to withstand demographic densities of 100 inhabitants/km^2, up to 150 inhabitants/km^2.

However, gradually, competing with the call of labour mines such as that of Alès and the booming industry that founded the rural exodus in France, terraced agriculture was the first affected by the abandonment of agricultural lands.

In these two highly developed areas, in the absence of flat land, no alternative was possible for crops without terracing systems. The omnipresence of steep slopes also made it impossible to envisage the practice of cattle breeding. The cow and the slope were difficult to associate with the interfluves of the Mediterranean mountains. However, between 600 and 800 m above sea level, in the areas where the slopes are softening a bit, another model of terraced landscape can be seen, characterized by fewer terraces, more meadows and extensive pastures.

5.3 The Model of the High Valleys of the Southern Massif Central, the Southern Alps, Castagniccia and the Pyrenees Orientals: Small Terraced Areas Associated with Immense Spaces of Extensive Agriculture

These landscapes of "confines" indicate regions of transition between slopes dominated by terraces, and highlands extensively exploited by livestock breeding which takes various forms, occupying the space permanently or seasonally, thanks to transhumance.

Here, the terraces are no longer so important. They are no longer adjacent to the farms, which are more often surrounded by meadows than "faïsses", as perfectly evidenced by the hamlet of Felgérolles in the upper valley of the Tarn towards Pont-de-Mauvert in Lozère. In addition to the smaller terraces, the architectural quality of drystone walls is not as neat as in the lower valleys. The walls are sometimes replaced by embankment, and it is difficult not to confuse them with another form of arrangement of the slope, "les rideaux"; unintentional arrangement of the slope (Fenelon 1956). Here, finally, the demographic pressure has always been less, and therefore, the need for rationalization of space is less necessary. Besides, there are very few ancillary constructions associated with the terraces, like the staircases which allowed to pass from one terrace to the other. There are also a few constructions for the recovery of water such as vaults or cisterns.

Finally, land tenure is larger, with averages exceeding 10/15 ha. Here too, the land structure is sometimes singular since it is a private property and community spaces sometimes juxtapose, as in the Maritime Alps or the south of the Massif Central with the communal system (Vivier 1998).

Rural architecture also illustrates these differences. Unlike the lower valleys, here, the basement is given less room; a larger volume is reserved for the barn above the stable. In steep slopes, the building inscribes volumes in height over three levels; in the high valleys, the buildings extend and have one or two levels maximum. Only perhaps the social organization of the habitat recalls the "Cévenol" model, with still a dominant of farms grouped in hamlets.

For the Massif Central, we had highlighted this type of landscape in the department of Ardèche. This model is very present in the high valleys of Chassezac, Drobie, Ardèche and Eyrieux, especially in the Hautes Boutières. In Saint Martial, for example, the "chambas" (local word for terraces) reach 1050 m, but adjoining immense pasture spaces reserved for cattle. The same model is found in Lozère, in the upper valley of the Tarn towards Pont-de-Mauvert for example, where terraces have been developed up to 1300 m above sea level with probably the primary purpose of cereal production.

In the high valleys of the Maritime Alps and the Var, especially in the valleys of the Roya and Tinée, modest peasants built "restanques" around the hamlets, housing subsistence crops and even irrigated meadows. Their surfaces do not exceed 10% of the communes because here we count on vast "communal", immense spaces used for an itinerant breeding characterized by a gradual climb towards the alpine pastures from spring to autumn (de Reparaz 1990). In this model, terraced cultivation only plays a secondary role in a self-sufficient economic system strongly marked by livestock breeding.

In Corsica, the terraces of the Castagniccia can be as a whole related to this model. In this region, which is strongly dominated by the slope, most of the chestnut grove is not found on the "ribades" (Corsican word for terraces) which only draw islets around the hamlets where gardening activities are concentrated. The numerous villages such as Piedicroce, Verdese, Piazzole, Carcheto Brustico, Valle d'Orezza, Monaccia d'Orezza and their hamlets, set up on ridges or staggered in the slopes, seem to protect themselves from the vegetable mass of the chestnut trees thanks to their belts of terraces more or less maintained (Garnier 2014).

Finally, in the Pyrenees Orientals, it is by leaving the coast and sinking in the valleys of the Têt or of the Tech and further in the high valley of the Ariège that one finds this type of landscape. Here, on the slopes close to the villages, the remains of the old buildings have been largely abandoned since the middle of the twentieth century. Indeed, the slopes exposed to the south called "soulanes" always show the traces of the old retaining walls, indispensable for an agriculture marked by traditional polyculture. In contrast, the northern slopes called "ombrées" are occupied by the forest as near the village of Albies in the upper valley of the Ariège between 500 and 1000 m of altitude. But, like other regions, managed slopes only represent a small part of the area devoted to all agricultural activities. For the other Pyrenees, it seems that we are here at the occidental limit of the terracing systems, the Upper Ariège making a transition between the climate of the Atlantic type and the Mediterranean climate of the Pyrenees Orientals.

However, even though polyculture remains largely dominant in this model, until the middle of the last century, some regions specialized very early in viticulture and later in other productions such as "cédrat" in Cape Corsica or floral culture, for example the carnation of Nice or the violet of Grasse. This is one of the characteristics of the "Provençal" model (Fig. 5.2).

Fig. 5.2 Evolution towards the land abandonment. Saint Andéol de Fourchade, Haute Boutière, Ardèche. *Photo* C. Blanc

5.4 The "Provençal" Model of Terraced Landscapes: When Terraces Adjoin Flat Lands that Are Potentially Cultivable

The third type of space that we distinguish concerns all the zones where sloping earth borders adjoin shallow surfaces. In these regions of France, mostly located in the great south-east, the slopes have been developed in terraces because of the flat surfaces, although close, presented constraints that were different depending on the eras.

The oldest is surely linked to History and insecurity. Indeed, after the Romanization of Gaul, which enabled numerous installations of "*villae*" in the plains, from the third century, the insecurity connected with the plundering of the invaders urged the peasants of the open plains to take refuge in height (Le Glay 1992). While these movements have marked the singular silhouette of the perched villages in the southern landscapes, they have not all translated into the generalization of terraces. It is likely that the great earthworks that produced the astonishing landscapes that characterize many of the slopes of the Southern Alps and the south of the Massif Central mostly date from after the Middle Ages (Carrier and Mouthon 2010). As time goes by, one will witness a back and forth between flat lands and slopes. Several articles from "L'histoire de la France Rurale" (Duby 1992) testify to this.

The second disadvantage of the plains was for a long time the difficulty of working on marshy flat land or frequently flooded. Although the monks developed the know-how related to drainage in limited spaces, it was not until the eighteenth and often the nineteenth century that, thanks to the mechanization and progress of civil engineering, the containment of rivers and the drainage are becoming widespread. For a long time, men preferred the slopes to the floodplains and the humid depressions, where only motorized mechanization allowed exploitation.

Until the beginning of the twentieth century, in these regions of France where most of the earth's work was done by hand, competition between plains and slopes did not really arise, even less with the demographic explosion of the eighteenth and nineteenth centuries, which generated a big labour. Thus, the few richest peasants appealed to the innumerable agricultural workers ready to get hired and to animal traction to relieve the painful tasks.

Although terraced agriculture specializes at a very early stage in certain soils, these lands are primarily subsistence crops (de Reparaz 1990) characterizing a small, planted polyculture in which cereals are ubiquitous. They mix here with potatoes or vines, olive trees, chestnut trees and other fruit trees. Here again, a small sheep–goat farm is almost always associated with these crops. It complements the resources needed for the self-sustaining survival of a poor population.

The "Provençal" model is well represented in the Massif Central, particularly in the departments of Ardèche, Gard, Lozère, Tarn and Aveyron.

For the department of Ardèche, we have highlighted these areas where plains, basins, depressions and slopes adjoin and produce strong oppositions, especially in Haut and Bas-Vivarais. The valley of the Doux perfectly illustrates this old duality between the slopes covered with terraces now strongly abandoned, and the plateau which has retained a dynamic polycultural agriculture, in which fruit production has long been illustrated. In the Bas-Vivarais around Privas or on the slopes surrounding the Chomérac basin, the cultivation of the vineyard pushed the peasants to build terraces that served more to take the stones out of the fields rather than to fight erosion. With the invasion of the phylloxera that ruined this vineyard from 1870, agriculture concentrated on the plain lands. Further to the north, the Rhodanian Coteaux also has the same morphological and architectural landscape characteristics. Between Saint-Péray and Vienne, men early made the choice of the wine specialization giving its nobility to the reputed "Northern Côtes du Rhône". In Aveyron, it is still the vine culture that has pushed the peasants towards the hillside. In Saint-Rome-du-Tarn or Brousses-le-Château in the Tarn Valley, or much further north in Estaing in the Gorges du Lot, the vineyards settle on terraced landscapes from the Middle Age. After being ruined by the phylloxera crisis, this vineyard reconquers the slopes: it has been classified as AOC (term of controlled origin) "Côtes-de-Millau" since 2012. In the Drôme, this model is rarer, but can be found in the east of the department, in the Baronnies, as in Villeperdrix where the olive trees on terraces did not withstand the frost of 1956.

In the Pyrenees, it is mainly the Eastern part of the massif which shelters the most emblematic slopes, those of the Vermeille coast in particular, also oriented from the Middle Ages to the production of wines. In Banyuls, the system of terraces is unique, in particular thanks to the arrangements put in place to channel and evacuate excess rainfall; the whole forms a singular geometric network called "*peu de gall*", literally: "pied de coq" (Olivier 2002). The vineyard rises from sea level to about 500 m of altitude, on steep slopes (sometimes over 45%), which allows an excellent sunlight exposure (325 days a year on average), and concerns the communes of Collioure, Port-Vendres, Banyuls-sur-Mer and Cerbère (Giorgis 2005).

In Corsica, two regions match this model. In Balagne, the terraces surround the villages perched or hung half-slope at about 400 m like in Sant'Antonino. They housed until the nineteenth century, olive groves, orchards, vineyards and cereal crops. In Eastern Corsica, the Eastern slopes are more austere because they are steeper, but original in the early specialization towards which they oriented themselves: the culture of the cedrat which is part of the Cap Corsica cultural heritage (northern part of Corsica). On the morphological and architectural level, these terraces are distinguished from the installations designed to protect the citrus orchards from the wind. Besides, near Nonza, the port remains still testify of the flourishing period of the export of cedrats to Italy and England (Garnier 2014). The cultivation of the Corsican cedrats peaked at the end of the nineteenth century. By the year 1920, annual production exceeded 8000 ton out of about 1000 ha. In the early 1950s, Corsica supplied one-third of the world cured brine market. Even if the walls that supported the ground sometimes descend to the sea like in Minerviu, most of the buildings are located around the hamlets, higher up the slopes, as the coastline was for a long time dangerous.

In the south of the Alps, terraces of the "Provençal" type also cover limited areas occupying the margins of the basins and limestone plateaus, as well as some interfluves, but are in the form of aureoles whose surfaces considerably vary from one department to another. Concerning the Bouches-du-Rhône, the communes of Velaux or Rognes can be retained. The terroirs of Goult or Gordes in the Vaucluse also perfectly illustrate this type of landscape where the proximity of flat mechanized land led very early to the abandonment of terraces, transformed today, at best, into a rural museum (Frapa 1999).

In the Maritime Alps and in the Var, mainly along the coast, terroirs in terraces located close to cities or traffic links developed speculative cultures from the mid-nineteenth century. Opportunity terraces, they have first specialized in the vineyard as evidenced by the vineyard of Cassis or the vineyard of Bandol. On the outskirts of Marseille, it is the culture of the capers which was enthroned on the sunny "restanques" of Gémenos, Cuges and Roquevaire. The specialization also concerned cherry, fig and almond orchards that were often planted with cereals such as in Sollies-Pont (Vaucluse) or in Ventabren and Velaux, which exported their almonds to Holland (de Reparaz 1990).

Finally, and later, the Riviera coast was marked by a conquest of floral crops that set out to attack the slopes between Grasse and Menton. But these terraced spaces represent only tiny islands adjoining lands that are easier to work and irrigate, highly coveted by an exponential urbanization from the middle of the twentieth century (Fig. 5.3).

Fig. 5.3 Vineyard of St-Joseph Hospice, Tournon. *Photo* E. Guigal

5.5 The "Isolates" Model: Unusual Developments in Agrarian Spaces Globally Without Terraces

The latter type of terraces concerns the geographical margins of terraced cultivation in France. Here, in fact, nothing apparently seems to justify the necessity of the development of the slopes by the construction of drystone walls. Climatic constraints, in particular linked to very high rainfall over a short period, do not concern these regions. Nor is it in the choice of crops that we can propose a justification of the terraces, since the same agriculture is practiced here with or without terraces, especially the vine. What are the reasons then for this unusual geography? Are the ecological advantages empirically apprehended by the peasants which justify such efforts of the development of the agrarian space? Is it necessary to go for pedological responses that would explain a very high exposure to erosive risks? Is it rather the excess of lithic material present on these slopes that requires more stone removal than the erection of retaining walls? Is it to facilitate the work of men on steep slopes? It is still very difficult to provide reasoned and scientifically credible answers to these questions. What we know is that almost everywhere the culture of the vine has occupied these terraces, carried by the immense desire to produce wine.

In the Massif Central, the setting up of a vineyard explains many small islets of terraces whose northernmost may be found in Courgoul in the Cantal, between Issoire and Puy de Sancy. In the same department, other terraced vineyards exist as in Boudes labels in the AOC Côtes d'Auvergne. In Cantal, it is the passion for wine that justifies the "palhàt" (local word for terraces) of Molompize installed between

600 and 750 m of altitude on the slopes of the valley of the Alagnon, between Massiac and Murat. In Haute Loire, the town of Puy-en-Velay also had an important vineyard whose traces are still clearly visible on the slopes of Vals-près-le-Puy. The very intensive development of these slopes coveted by the city's shopkeepers sharply contrasts with other far more extensive agricultural activities.

The northernmost terraces of France are lost in the immensity of the Alsatian vineyards between 200 and 400 m above sea level. These unusual terraces concentrated on a few slopes exposed to the south form rare islets, from Thann to Ribeauvillé, passing by Ammerschwihr, Kaysersberg and Guebwiller. It is undoubtedly the pedological constraints and a greater sensitivity to erosion that explain the construction of the walls on these plots of lands. But this constraint, combined with a remarkable exhibition, has become an asset, thus creating exceptional terroirs, as on the slope of "Kitterlé" in Guebwiller, which is renowned for a few hectares, facing the extent of the Alsatian vineyard without terraces. Here, the poor and sandy soil can never yield large quantities of grapes. But exposure to the sun from the east to the west fosters the concentration of aromas in the grapes and gives wines an exceptional longevity. "For the capacious force, the finesse of the bouquet, no other Alsace vintage surpasses the Kitterlé", (anonymous chronicler nineteenth century). This is why these plots managed by small peasants are now in the hands of the flags of the industrialists and vine growers of this region.

In the Alps, islet terraces are rare in the northern Alps. One can, however, mention the tiny vineyard of Cevins in Tarentaise, installed by the monks in the very early Middle Ages, abandoned after the invasion of the phylloxera and recently rehabilitated. But it is less than 10 ha. In the valley of "Moyen Grésivaudan", a few tiny vineyards were set up on the steep slopes facing south, at the foot of the "Balcons de Belledonne", between Grenoble and Montmélian. Further south in the Massif des Ecrins, remarkable terraces are always visible around La Grave, in the upper valley of the Romanche. They range from 1300 to 2400 m above the hamlets of the "Clots" or "Hières". The technique of grading is fostered here in view of the large quantities of earth available. Built in the eighteenth century, their only purpose was the production of cereals and potatoes. Continuing further east in the Massif du Queyras, some terracing systems exist, but still more limited like in Arvieux at 1600 m above sea level around the only hamlet of "Maison". Consisting of drystone walls or slopes, the fields are also framed at the right of the slope by "clapiers". Called elsewhere "murgers" or "clapas", they recovered the materials resulting from the stone. Food crops and hay meadows succeeded before the agricultural depression condemned these slopes where the terraces occupied only a few hectares.

5.6 What is the Reality of Terraced Agriculture in France in 2017?

Landscapes of poor peasants living on tiny, fragmented and marginalized farms, these islands of terraces were the first to be affected by agricultural decline and rural exodus. But everywhere in France the agricultural function of the terraces gradually declined from the middle of the nineteenth century, showing the rural exodus and the successive crisis that particularly affected this sloping agriculture: the Phylloxera which fell on the vine in 1863, the Ink which attacked the chestnut trees in 1860 and 1865, the Pebrine which ravaged the silkworm farms and the complacent worms to which they were associated. Other constraints will also contribute to emptying the mountains of their peasants, especially the effects of mechanization inappropriate to sloping spaces and land fragmentation. Finally, the political choices of the post-war period decided on the "end of the peasants" (Mendras 1967) and the disappearance of this mode of production characterized by a subsistence economy accompanied by a great autonomy in the organization of the process of production and of labour. According to Mendras, the French peasantry is then gradually replaced by agricultural professionals who organize their production according to a capitalist mode. The peasants of the terraces were the first victims. Since the 1950s, the main agricultural actors have opted for the abandonment of the mountains, fostering monoculture farming in the plain on large farms. Only a few terraced spaces have escaped this general decline, saved by productions with high added value, mainly winegrowing. But this survival has served as a model for those who, beginning in the 1980s, were wondering about a possible reconquest of certain terroirs of terraces in a perspective of sustainable development (Fig. 5.4).

Little by little, many reclamation projects were born in France. This reconquest was certainly carried out on very limited areas, but with promising results that demonstrate that terraced farming is not obsolescent. The pioneers of this revival are mainly winegrowers who set out again to attack the hillsides of the Northern

Fig. 5.4 Production of sweet onions. Cévennes. *Photo* J. du Boisberranger

Rhone Valley from the late sixties and on a smaller space in the Eastern Pyrenees. In the 1970s, other attempts were made. In Ardèche, for example, in the lower valley of the Eyrieux around new crops such as the feijoa, a fruit tree that could take the place of peach trees, or the crosne of japan, a small tubercle much appreciated by the gourmets. Other farmers set up medicinal and aromatic plants on their terraces, but always on isolated farms. Starting in the 1990s, vegetable crops also found terroirs with strong identity on terraces, as in the Cévennes with the cultivation of sweet onion. This production, which is now recognized by an AOC obtained in 2003 and a PDO awarded in 2008, concerns about thirty farms, forty hectares and a hundred members grouped together in a cooperative located in Saint-André de Majencoules in the Gard. In Ardèche, on smaller areas, the early potato "échamp de l'Eyrieux" tries to relaunch agriculture in terraces. At the level of France, we do not have an exhaustive census of all these projects, but their number probably exceeds a hundred. Through the renewal of the vineyard of Cevins in Savoie, the reconquest of the terraces in Boudes in the Puy de Dôme, the revival of the "Chatus" grape variety in Ardèche or the revaluation of the AOC "Côtes-de-Millau" measure the dynamics triggered.

Will the evolution of consumption patterns, the questioning of chemical agriculture and monoculture, offer terraces and those who cultivate them new prospects for the twenty-first century?[1]

References

Alcaraz F (1996) La Soulane du Haut-Conflent entre le XVIIIème et le début du XXème siècle. Revue Conflent n°202, pp 25–37

Blanc JF (1983) Un paysage en crise: les versants à terrasses en Ardèche. Université Lyon III, Thèse. Géographie

Bonardi L, Varotto M (2016) Paesaggi terrazzati d'Italia. Angeli

Bozon P (1963) La vie rurale en Vivarais. Imprimeries Réunies

Carrier N, Mouthon F (2010) Paysans des Alpes. Presses universitaires de Renne

Castex J-M (1980) L'aménagement des pentes et des sols dans les Alpes-Maritimes et le Var. Université de Nice, Thèse de Troisième Cycle

Castex J-M (1983) L'aménagement des pentes et des sols dans les Alpes-Maritimes et le Var. Méditerranée 1:3–15

Castex J-M (1996) Quelques exemples de cartographie des terrasses de culture dans les Alpes-Maritimes et le Var. In: ProTerra, Actes de l'atelier 1, "Le système terrasses: définitions, outils et méthodes d'approche"

de Reparaz A (1990) La culture en terrasses, expression de la petite paysannerie méditerranéenne traditionnelle. Méditerranée, Revue géographique des pays méditerranéens 71(3–4):23–29

Desbordes E (1999) Les paysages de terrasses agricoles dans le bassin méditerranéen septentrional. Thèse de géographie, Université de Limoges, Etude comparée des Alpes maritimes et de la Ligurie

Duby G (1992) Histoire de la France Rurale. Tomes 1-2-3-4. Editions du Seuil

[1]Translated from the French by Nicolas Di Rito, French Ministry of Agriculture, and Catherine Blanc, French Ministry of National Education.

Fenelon P (1956) Les «rideaux» de Picardie et de la Péninsule ibérique. Bull de l'Association de géographes français 33(255):2–9

Frapa P (1999) La négociation avec les propriétaires: un lourd préalable. Un exemple dans le Vaucluse: le conservatoire des terrasses de Goult. Forêt Méditerranéenne XII 4:547–549

Garnier F (2014) Atlas des paysages de la Région Corse. DREAL de Corse

Giorgis S (2005) Le paysage singulier du cru Banyuls dans les Pyrénées-Orientales. In: Les paysages culturels viticoles. www.icomos.org/studies/viticoles.htm

Le Glay M (1992) Histoire de la France Rurale. Tome 1. Editions du Seuil, pp 195–290

Martin C (2006) Espaces en terrasses et prévention de risques naturels en Cévennes. Publication de l'UMR 6012 «Espace» . Maison de la Géographie

Mendras H (1967) La fin des paysans, innovations et changement dans l'agriculture française. D.E. I.S, S.E

Olivier G (2002) Le paysage de terrasses du cru "Banyuls" (Pyrénées Orientales) et son évolution, Étude des pratiques d'entretien des terrasses de culture. Montagnes méditerranéennes 5:21–26

Pinto-Correia T, Vos W (2005) Multifonctionality in Mediterranean landscapes - past and future. ActesWagenigen. The New Dimensions of the European Landscapes, Netherlands, pp 135–164

Usselmann P (2006) Les systèmes de terrasses Cévenols: exemples de la vallée Obscure et le Vallon du Rouquet. *Supplément au n°* XXXIII UMR 6012 "ESPACE" - Équipe G.V.E., Département de Géographie–Nice, pp 15–21

Vivier N (1998) Propriété collective et identité communale. Les biens communaux en France 1750–1914. Publications de la Sorbonne

Chapter 6
Terraced Fields in Spain: Landscapes of Work and Beauty

Sabina Asins-Velis

Abstract The sheer diversity of approaches and the volume of scientific output underscore the growing social and policy interest in these landscapes. Recognition of their productive, environmental and cultural functions—and of the grave problems arising from their mismanagement or abandonment—shows the need to allocate resources to their study at both the national and European levels. Finally, the need for a Catalogue of European Union Terraced Landscapes is outlined as a preliminary step to considering these as EU Landscapes of Special Relevance. Such an initiative would be reinforced by activities being undertaken at the international level.

6.1 Introduction

Over the centuries, farmers in mountainous areas of Spain have used their knowledge of the environment to build a valuable landscape with a unique "character" (Countryside Agency and Scottish Natural Heritage 2001; Wascher 2005): the terraced landscape. In its integrated design, they adapted their methods to climatic, edaphic, and topographic circumstances and the crops they wished to grow. Moreover, they went about this task in a holistic, socially organised way. These terraces constitute a superb agronomic, cultural and environmental heritage that still covers thousands upon thousands of hectares in rural Spain. They are particularly common along Mediterranean regions, especially in the Valencian Community, Murcia, Catalonia and The Balearic Islands; in certain parts of Andalusia and Aragon, and in The Canary Islands (which has its own chapter in this book). The terraced landscapes of Spain are a magnificent example of the work of

S. Asins-Velis (✉)
Centro de Investigaciones Sobre Desertificación-CIDE, CSIC,
Universitat de València, Generalitat Valenciana, València, Spain
e-mail: sabina.asins@uv.es

M. Varotto et al. (eds.), *World Terraced Landscapes: History, Environment, Quality of Life*, Environmental History 9,
https://doi.org/10.1007/978-3-319-96815-5_6

farmers over the centuries and the beauty, uniqueness and intangible values that these "heritage landscapes" represent today.

At present, the analysis of these terraced landscapes is a complex matter given local differences in cause and effect. Broadly speaking, one can place these under three heads:

(1) Industrial–commercial terraced agriculture

- Those areas of terraces that survived the twentieth-century rural exodus because they were more suitable for mechanised agriculture. Examples are terraced piedmont or larger terraces (citrus trees in Murcia; vineyards, citrus and fruit trees in the Valencian Community; olive trees in Mallorca; vineyards in Ribeira Sacra-Galicia).
- New land terracing for highly profitable vineyards (in Catalonia, for example) or fruit trees (Valencian Community).

(2) Territorial agriculture: Small terraced fields kept up by their farmer-owners and of high ecological value (mainly in the smallest rural areas of the inland mountains of Mallorca, the Iberian Range and Pyrenees).
(3) Abandoned terraced fields (found throughout Spain).

(1) Industrial–commercial terraced agriculture shares similar opportunities and problems with farming in flat areas. In Spain, industrial capital-intensive agriculture (which is highly mechanised and competitive) tends to be quick to rise new challenges. The new opportunities arising in these mechanised, terraced landscapes stem from the world market and world demographic trends. The world's population is forecast to reach 8.5 billion by 2030 (United Nations 2015: 1) but more significant than total population growth are the market opportunities created by the burgeoning middle classes in developing countries. Over the last two decades, domestic material consumption has soared in Asia-Pacific, eastern Europe and Central Asia and even in Africa. This sharp rise in consumption is attributable to the rapid growth in the middle classes (UNEP 2016). It is envisaged that the growth of the middle classes will continue apace. This new middle class is demanding quality food products, of proven origin and that meets the highest standards for agricultural produce. Such rigorous control is what sets wine and cava from the terraced areas of Catalonia, Valencian Community or Galicia. Then there is the olive oil and a great variety of fruits from Valencian Community, Murcia, Andalusia and Mallorca. Spanish products from terraced landscapes meet these quality criteria (Fig. 6.1).

The problems arising from this kind of agriculture fall, among others, under four heads: (i) how to overcome the environmental impact of emissions from the transport needed to serve faraway markets (Verburg et al. 2013); (ii) greenhouse gas emissions from agriculture (Sanz-Cobena et al. 2017); (iii) expansion of irrigation into former rain-dependent uphill areas or a shift to higher-value monoculture fruit crops with greater water requirements, etc. (Sese-Mínguez et al. 2017; MAAMA 2016); (iv) how to deal with land grabbing and land concentration (European Parliament 2015; Soler and Fernández 2015). Growing concentration of

Fig. 6.1 New land terracing for highly profitable vineyards in Priorat (Catalonia). *Photo* J.A. Martínez-Casasnovas

landholdings (land over > 100 ha) means that, in Spain, 55.5% of the Utilised Agricultural Area belongs to just 5.4% of the total country holdings, and that 1.3% of Common Agricultural Policy (CAP) direct beneficiaries received, in 2013, 23.4% of CAP direct payments (Kay 2016: 12, 16). In Priorat (Catalonia), this key issue already affects the new mechanised, terraced vineyards: 12% of the farmers owned 61% of the new vineyards, 42% of the total agricultural land and received 68% of CAP subsidies assigned to the region (Cots-Folch et al. 2009). Highly profitable terraced fields also raise environmental issues, such as soil erosion and the constant cost of maintaining the terraces (Martínez-Casasnovas and Ramos 2006; Ramos et al. 2007; Ramos and Martínez-Casasnovas 2010).

(2) However, Spain's agriculture presents strong contrasts. While produce from industrial, terraced agriculture competes in the international market, scores of small hill farmers are struggling to make ends meet. These terraced fields are mainly sited in areas stricken by the rural exodus that took place mostly between 1950 and 1990 (Collantes 2004). Back then, those terraced fields farthest from the villages—on land that was either marginal or hard to mechanise (mainly in the Central Pyrenees and the Iberian Range)—were abandoned (Lasanta 1988; Arnáez-Vadillo et al. 1990; Rodríguez and Lasanta 1992; García-Ruiz et al. 1996). Moreover, between 1990 and 2016, farmland declined by 3,185,101 ha (MMAMRM 2010:18; MAPAMA 2016) and, between 1990 and 2013, 48% of the agricultural holdings under 10 ha were abandoned (Kay 2016:14).

Nevertheless, in some villages, fields near the settlement and with good transport links and/or that are more productive (terraced orchards, for example) have bucked the trend. Such land constitutes what has been termed "territorial agriculture" (González Regidor 2003). Here, the problems are different, for instance, rural ageing, middlemen and small holdings. The environmental and social functions of these terraced landscapes have been highlighted over the last few years (Asins-Velis 2007; Hernández Hernández 2009; Lasanta et al. 2013). It is precisely these values that offer such farming a second chance (Fig. 6.2). These new times coincide with the key debate on the direction we want our rural areas to follow, and the opportunities arising from the "circular economy in agriculture" (EIP-AGRI 2015).

(3) A third consideration is the present state of the terraced fields abandoned since 1950. Here, there are two processes at work. One is the spontaneous revegetation of scrubland and the arboreal stratum (Bonet and Pausas 2004; Nadal-Romero et al. 2016a). The other is soil erosion and the formation of gullies (Calvo-Cases et al. 2005; García-Ruiz and Lana-Renault 2011; Romero-Díaz 2016). Which process predominates varies from one plot of land to another. Fields on which scrubland/forest has sprung up face the risk of bush fires (due to lack of forest management) or the incision of rivers channel (Lasanta et al. 2017a). Land subject to erosion is losing topsoil through landslides and mass movements, among other risks (Fig. 6.3).

Fig. 6.2 Territorial agriculture. Castellfort (Castellón). *Photo* S. Asins-Velis

Fig. 6.3 Abandoned terraced fields. Puerto de la Cadena (Murcia) *Photo* A. Romero-Díaz

Currently, the study of terraced landscapes in Spain requires both a retrospective analysis and an assessment of future scenarios, linked to the regional, national and European policies focused on the management of rural areas. There is considerable scholarly activity in this field and a rich bibliography in both the human and environmental sciences. In this paper, we will try to sketch the main lines of work followed by various research groups and to summarise some of the main opportunities in the near future. Due to the considerable scientific output, we can only quote a tiny sample of the articles published per work teams; we redirect to their publications to expand the concepts treated.

6.2 Looking Back, Looking Forward

6.2.1 Geo-historical Research

The historical study of Spanish terraced landscapes has shifted from a sectoral approach, attempting to reconstruct ancient farming techniques, irrigation systems and the expansion of terraces, among others (Bazzana and Guichard 1981; Carbonero Gamundi 1984; Barceló 1989; Grimalt and Blázquez 1989), to a cross-cutting vision. These studies led to the formation of multidisciplinary teams.

Among other issues, the teams are interested in interpreting productive spaces (Orejas 2006) in relation to the settlement patterns, population trends, social organisation, climate and land management techniques, etc.; also terrace structures, location preferences, types, materials and functions; the continuous introduction of new crops both in dry lands and irrigated ones or the role of agriculture in landscape transition.

Thanks to these ground-breaking studies, we can trace the history of terraced landscapes in The Bronze Age in eastern Spain—at the beginning of the second Millennium BC in Valencia (De Pedro Michó 1998); in The Iron Age in Denia, Alicante (Gisbert Santonja 2001); or in Roman times in a mining area in the west of the Iberian Peninsula (Ruiz del Árbol et al. 2003) or in the mountainous East (Grau Mira and Pérez Rodríguez 2008). The building and functions of terraces in The Early Middle Ages are well documented in the rural communities in northern Spain, mainly in Galicia since the fifth century (Ballesteros Arias 2010); the sixth century in Bizkaia (Varón Hernández et al. 2012); the eighth century in Asturias (Fernández Mier et al. 2014) and since the tenth century in Álava (Quirós Castillo et al. 2014). The building of terraces has also been studied in al-Andalus (territory under Islamic control since the eighth century): Murcia (Puy and Balbo 2013); mountainous, inland Alicante (Torró 2007); in the "Orchard of Valencia" (Guinot 2008; Esquilache Martí 2015); The Balearic Islands (Glick and Kirchner 2000; Sitges 2006); Andalusia (Malpica Cuello 1996; Trillo San José 2010) and the Ebro valley, north-eastern Spain (Alonso et al. 2014). Terraces in feudal areas have also been studied in depth in Catalonia (Riera and Palet 2005; Retamero et al. 2015), in The Balearic Islands (Kirchner 2012) and the Valencian Community (Torró 2005; Guinot 2009).

Terraced fields spread fast with the dawning of The Modern Age and reached their greatest extension and the highest altitudes at the end of the nineteenth century as a result of population growth and the fact that the country's incipient industrialisation still required relatively few workers. Andalusia, the Iberian range, The Pyrenees, the western part of The Ebro Valley and, in general, mountainous areas saw farming creep up the hillsides (Cavanilles i Palop 1795; Garrabou 1985; García-Ruiz 1988; Hernández Hernández 1997; Giménez-Font 2008; Boixadera et al. 2016). However, the process reversed from the mid-twentieth century as rural abandonment affected these same areas. The emigration of the young was compounded by the difficulty of mechanising hill farming areas to offset the loss of rural labourers (García-Ruiz et al. 2013).

Looking forward, there is still much interdisciplinary work to do. Among other issues, this includes archaeological and archival research; analysing intensification and extensification in terraced areas in relation to demographic trends and land ownership models; the impact of technological innovations; the links between climate, crops and their diversification. Working together, archaeologists, historians, hydrologists and soil scientists may learn how agricultural techniques affect hydrological and geomorphological dynamics in a watershed over the centuries; the socio-economic and environmental problems derived from shrub clearing in order to enlarge grazing areas and to promote the development of livestock (as tested by

Lasanta et al. 2015); to analyse soil fertility and cultural practices (as tested by Ferro-Vázquez et al. 2015); the long-term assessment of the impact of soil and water conservation techniques on the soil's organic matter content, and so on. Speaking of techniques, great scope is offered by organic geochemical analyses and radiocarbon dating; to date terraces construction by combining archaeological survey with sampling for luminescence profiling and optically stimulated luminescence (as Kinnaird et al. 2017 have already demonstrated in Catalonia); and to apply statistical modelling techniques.

The cultural importance of these landscapes shows the need to draw up a Catalogue of Spanish Terraced Landscapes, identifying their historical context, the various types of terraces and related constructions, their state of conservation, etc. This task has become affordable because some regions have already published landscape atlases that cover terraced areas, including those for Mallorca (Colomar Mari 2002); Catalonia (Observatori del Paisatge 2008–2016); the Province of Granada (Junta de Andalucía 2014) and the Valencian Community (Generalitat Valenciana 2011). One should also consider drawing up "Best Practice" guidelines for farmers, landowners and decision-makers that take into account the European Union, and national and regional policies bearing on cultural and historical landscapes. This offers new opportunities for responsible tourism in recognising the value of local host communities. The final step should be incorporating terraced landscapes into regional government territorial plans and policies.

6.2.2 *Eco-geomorphological Studies*

There is a long tradition of studies in this field. Since the late 1980s, various teams from Spain's Scientific Research Council (CSIC) and Spanish universities have analysed the effects of terraces on: soil depth; organic matter content; aggregate stability (Sánchez-Marañón et al. 2002); water repellency (Bodí et al. 2012; Burguet et al. 2016) and bulk density (Seeger and Ries 2008); soil water erosion and run-off generation (Gallart et al. 1994; Durán Zuazo et al. 2005; Cerdà and Doerr 2007; Gaspar et al. 2013; Prosdocimi et al. 2016; García-Ruiz et al. 2017); mineral salts and nutrients; soil compaction and surficial crusts; hydromorphic conditions; the hydrological and geomorphological processes involved (Arnáez et al. 2015); soil infiltration capacity (Arnáez et al. 2017); the influence of soil's physical–chemical properties on the composition of microbial communities (Zornoza et al. 2009)—in short: soil properties, soil management and soil conservation.

Other issues of interest focus on the consequences of agricultural abandonment in The Pyrenees (Lasanta 1988; Ruiz-Flaño et al. 1992; López-Vicente et al. 2016), Rioja (Arnáez et al. 2011; Lasanta 2014), Catalonia (Dunjó et al. 2003; Poyatos et al. 2003; Olarieta et al. 2008), Mallorca (Reynés et al. 2007), Valencian Community (Ruecker et al. 1998), Andalusia (Douglas et al. 1994; Jiménez Olivencia 2010) and Murcia (Romero-Díaz et al. 2007; Alonso-Sarría et al. 2016; Romero-Díaz 2016).

These researchers work both on erosion modelling and on the analysis of new environmental dynamics when farming stops: their effects in soil organic carbon dynamics (Emran et al. 2012; Nadal-Romero et al. 2016b; Boix-Fayos et al. 2017); vegetation recolonisation; species composition; species diversity; vegetation cover and their effects in soil properties (Marco-Molina et al. 1996; Padilla Blanco 1998; Gallego-Fernández et al. 2004; Pueyo and Beguería 2007; Gispert et al. 2017; Pardini et al. 2017; Romero-Díaz et al. 2017). In connection with this revegetation and the lack of forest management, one should also take into account the greater risk of forest fires and their impact on plant and ecosystem recovery (Vicente-Serrano et al. 2000; Llovet et al. 2009). In some areas, abandonment of farming can lead to the recovery of the geo-ecosystem, with improvements in vegetation and soil quality (Romero-Díaz 2016). These studies have yielded a great deal of valuable knowledge.

Looking forward: Interesting research areas are: (a) the links between terrace location and agricultural abandonment; (b) sediment yield in abandoned areas and long-term soil erosion; (c) tillage erosion, water erosion and soil quality on farmed terraces; (d) the combination of qualitative and quantitative methods to gauge soil erosion in terraced fields; (e) the application of large-scale soil erosion modelling in conservation plans for terraced landscapes; (f) the response of terraced slopes to torrential rain events; (g) analysis of the effects of building terraces on soil properties by crop types; (h) the effectiveness of abandoned terraces as part of afforestation schemes to tackle erosion, etc, last but not least, those studies that focus on soil and water conservation under global change (the first World Conference on the subject took place in Lleida in June 2017).

6.2.3 Natural and Cultural Drivers Influencing Mediterranean Landscapes

This section analyses land-use change and societal transformations arising from the appraisal of agricultural land uses, urban expansion, tourism and leisure uses (Plieninger et al. 2016: 210). Driving forces cover both natural and socio-economic issues and include: decreasing attractiveness of farming as a way to make a living in comparison with service industries; the rising impact of second-holiday homes; the job opportunities offered by cities; the low profitability and productivity of farming; the rise of tourism as an economic activity; national and international markets (Lasanta et al. 2017a). In these areas, researchers try to identify primary, secondary and tertiary driving forces, whether intrinsic (influenced by local actions and people) or extrinsic (covering broad processes, such as globalisation, climate change, urbanisation and, of course, European Union policies—especially those bearing on agriculture, the environment, rural development and culture).

Specialists are focusing on past and current land management systems and the processes of change in Mediterranean rural landscapes—intensification

(Vanwalleghem et al. 2017); erosion (García-Ruiz 2010); the geo-ecological factors controlling the location of field types (Lasanta et al. 2017b); the introduction of drip irrigation in terraced fields (Puy et al. 2016; Sese-Mínguez et al. 2017); the ecosystems services (Quintas-Soriano et al. 2016); the maintenance of "agricultural terraced islands" and pasturelands to lessen the risk and severity of forest fires (Asins-Velis et al. 2016; Lasanta et al. 2016; Nadal-Romero et al. 2016c); the links between tourist-urban-industrial sprawl and rural abandonment. The impact of changing land management on landscape character and diversity, experimental socioecology in landscape dynamics (Barton et al. 2016); the participation of the population in landscape evaluation (López Martínez et al. 2016) and stakeholders perception (Vila Subirós et al. 2016) are also research subjects.

Looking forward: It is desirable to design scenarios anticipating the possible trajectories of terraced landscape changes and their environmental impacts. One needs to: (1) pay attention to the social demand for quality landscapes (including peri-urban areas); (2) promote the values of the landscape and share collective knowledge; (3) promote the co-operative identification of the elements diminishing the quality of the local landscape and its dysfunctions; (4) design methodologies for the shared identification and evaluation of regional terraced landscapes through mapping; (5) analyse strategies for the community-based rehabilitation of terraced landscapes and get farmers' views. This will facilitate heritage-led rural regeneration, helping agricultural landscapes, farm products, the environment and tourism become factors boosting competitiveness.

6.2.4 New Technologies

Díaz-Varela et al. (2014) pointed out that in implementing and monitoring EU policies, there is a need to develop robust, repeatable, cost-effective methodologies for automatically identifying and monitoring terraced landscape features at the farm scale. At the European level, various teams are using different methods to yield this information. An excellent example of this work is furnished by Díaz-Varela et al. (2014), who tested imaging by low-cost drones in Cordoba. In this connection, Spanish regions need draw on new technologies to meet this EU need. For example, at the national level, one needs to delimit the total terraced areas (in a way similar to the Italian "Progetto MAPTER: mappatura dei paesaggi terrazzati italiani").

The international teams working on terraces are also interested in modelling hillslope sediment delivery using the WATEM/SEDEM model (Quijano et al. 2013); in techniques such as detailed geomorphological mapping and geo-radar profiles using RAM.

AC/GPR; to parametrise the effect of terraces on run-off and sediment yield by Soil Water Assessment Tool (SWAT) modelling (Ben Khelifa et al. 2016) and in the new opportunities offered by LiDAR [Light Direction And Ranging] data, such as those developed for vineyards in Italy (Tarolli et al. 2015).

Looking forward: There is a real need for Spain's central government and regional governments to spend more on innovation and to incorporate new techniques for mapping and characterising terraced landscapes.

6.2.5 Policy Needs

All the foregoing fields of knowledge have practical application to regional, national or EU policies—especially those covering rural development in mountainous areas and other areas facing natural constraints. Sectoral policies, such as those on agriculture, the environment, the economy, tourism, employment, benefit from them. Moreover, other lines of work that might be of interest to policy-makers include analysis of the financial cost-benefit for farmers keeping terraced landscapes (Martínez-Casasnovas and Ramos 2006) and the problems that may arise from applying given criteria when defining terraced land (Asins-Velis et al. 2016).

Looking forward, policy-makers need more information on the integration of socio-economic and biophysical aspects in soil conservation when considering the sustainability of terraced landscapes. To this end, territorial managers and planners need to foster innovative solutions that boost farmers' trust and well-being. On the other hand, more research is needed on (among other issues): high-quality crops, ecological agriculture, grazing circuits and innovative cropping systems on terraced landscapes. The aim here should be to achieve long-term environmental quality. Regional, national and EU policies require detailed studies for making EU agricultural policy better-adapted to the special features of terraced landscapes and to foster public participation in their design and implementation.

6.3 Conclusions

Studies by various research groups on terraced agricultural areas in Spain underscore the need to give these landscapes the recognition they so richly deserve. Here, the aim should be to conserve the productive function, environmental and socio-cultural attributes of these terraced areas. Such recognition will bring security that will benefit investment in these territories.

Spanish researchers collaborate with scientists from other European and non-European countries to identify both fitting opportunities for these historical landscapes and the risks they face. Drawing up a Catalogue of European Union Terraced Landscapes would help to identify their values and provide a basis for their interpretation. This could be the first step in beginning a campaign for their consideration as EU Landscapes of Special Relevance. Conferring this status on them would do much to prevent their future mismanagement or abandonment. This initiative would tie in nicely with the interest that these landscapes are arousing internationally, as evidenced by the creation of the International Terraced Landscapes Alliance (ITLA), which forges and strengthens links among terraced landscapes around the world.

Acknowledgements We wish to congratulate the organisers of the Third World Meeting of ITLA "Terraced Landscapes: Choosing the Future" for the interesting programme and experiences they prepared. Our gratitude also to the Editors of this book for their always intense and continuous work in defence of terraced landscapes. Part of this work has been carried out under the Spanish National Project I+D+I (CGL2013-44917-R).

References

Alonso N, Ferran A, Kirchner H (2014) Novelties and legacies in crops of the Islamic period in the northeast Iberian Peninsula: The archaeobotanical evidence in Madîna Balagî, Madîna Lârida, and Madîna Turṭûsa. Quatern Int 346:149–161

Alonso-Sarría F, Martínez-Hernández C, Romero-Díaz A, Cánovas-García F, Gomariz-Castillo F (2016) Main environmental features leading to recent land abandonment in Murcia region (Southeast Spain). Land Degrad Dev 27:654–670

Arnáez-Vadillo J, Lasanta-Martínez T, Ortigosa-Izquierdo L, Ruiz-Flaño P (1990) L'abandon de l'espace agricole dans la montagne submediterranéenne espagnole: Pyrénées et Système Iberique. Revue Géographique des Pyrénées et du Sud-Ouest 61(2):237–253

Arnáez J, Lasanta T, Errea MP, Ortigosa L (2011) Land abandonment, landscape evolution, and soil erosion in a Spanish Mediterranean mountain region: The case of Camero Viejo. Land Degrad Dev 22:537–550

Arnáez J, Lana-Renault N, Lasanta T, Ruiz-Flaño P, Castroviejo J (2015) Effects of farming terraces on hydrological and geomorphological processes. A review. Catena 128:122–134

Arnáez J, Lana-Renault N, Ruiz-Flaño P, Pascual N, Lasanta T (2017) Mass soil movement on terraced landscapes of the Mediterranean mountain areas: a case study of the Iberian Range, Spain. Cuadernos de Investigación Geográfica 43(1):83–100

Asins-Velis S (2007) Los aterrazamientos mediterráneos. Paradigma ambiental-agro-cultural. In: El paisaje mediterráneo. Opciones de multi-funcionalidad. Cuadernos de Sostenibilidad y Patrimonio Natural 11. Madrid, Fundación Santander Central Hispano, pp 81–91

Asins-Velis S, Arnau-Rosalén E, Romero-González J, Calvo-Cases A (2016) Analysis of the consequences of the European Union criteria on slope gradient for the delimitation of "areas facing natural constraints" with agricultural terraces. Annales, Series Historia et Sociologia 26(3):433–449

Ballesteros Arias P (2010) La Arqueología Rural y la construcción de un paisaje agrario medieval: el caso de Galicia. In: Kirchner H (ed) Por una arqueología agraria. Perspectivas de investigación sobre espacios de cultivo en las sociedades medievales hispánicas. BAR International Series 2062, Oxford, pp 25–40

Barberá GG, Mosch W, Navarro Cano JA, Gaspar L, Navas A (2017) Carbon redistribution by erosion processes in an intensively disturbed catchment. Catena 149(3):799–809

Barceló M (1989) El diseño de espacios irrigados en al-Andalus: un enunciado de principios generales. In: El agua en zonas áridas: arqueología e historia. I Coloquio de Historia y Medio Físico, vol I, XIII-L. Diputación Provincial, Almería

Barton CM, Ullah IIT, Bergin SM, Sarjoughian HS, Mayer GR, Bernabeu-Auban JE, Heimsath AM, Acevedo MF, Riel-Salvatore JR, Arrowsmith JR (2016) Experimental socioecology: Integrative science for Anthropocene landscape dynamics. Anthropocene 13:34–45

Bazzana A, Guichard P (1981) Irrigation et société dans l'Espagne Orientale au Moyen Age. In: L'Homme et l'Eau en Mediterranée et au Proche Orient. GIS-Maison de l'Orient. Presses Universitaires de Lyon, pp 115–140

Ben Khelifa W, Hermassi T, Strohmeier S, Zucca C, Ziadat F, Boufaroua M, Habaieb H (2016) Parameterization of the effect of bench terraces on runoff and sediment yield by SWAT modelling in a small semi-arid watershed in Northern Tunisia. Land Degrad Dev 28(5): 1568–1578

Bodí MB, Cerdà A, Mataix-Solera J, Doerr SH (2012) Repelencia al agua en suelos forestales afectados por incendios y en suelos agrícolas bajo distintos manejos y abandono. Cuaderno de Investigaciones Geográficas 38(2):53–74

Boixadera J, Riera S, Vila S, Esteban I, Albert RM, Llop JM, Poch RM (2016) Buried A horizons in old bench terraces in Les Garrigues (Catalonia). CATENA 137:635–650

Boix-Fayos C, Martínez-Mena M, Pérez Cutillas P, de Vente J, Barberá GG, Mosch W, Navarro Cano JA, Gaspar L, Navas A (2017) Carbon redistribution by erosion processes in an intensively disturbed catchment. Catena 149(3):799–809

Bonet A, Pausas JG (2004) Species richness and cover along a 60-year chronosequence in old-fields of southeastern Spain. Plant Ecol 174(2):257–270

Burguet M, Taguas EV, Cerdà A, Gómez JA (2016) Soil water repellency assessment in olive groves in Southern and Eastern Spain. CATENA 147:187–195

Calvo-Cases A, Boix-Fayos C, Arnau-Rosalén E (2005) Patterns and thresholds for runoff generation and sediment transport on some Mediterranean hillslopes. In: García C, Batalla RJ (eds) Catchment dynamics and river processes: Mediterranean and other climate regions. Elsevier, Amsterdam, pp 31–51

Carbonero Gamundi MA (1984) L'origen i morfologia de les terrasses de cultiu a Mallorca. Bolletí de la Societat Arqueològica Lul·liana 40:91–100

Cavanilles i Palop AJ (1795–1797) Observaciones sobre la Historia Natural, Geografía, Agricultura, Población y Frutos del Reyno de Valencia. Tomos I y II. Available via Biblioteca Digital Real Jardín Botánico CSIC. http://bibdigital.rjb.csic.es/spa/Libro.php?Libro=244. Accessed 10 Feb 2017

Cerdà A, Doerr SH (2007) Soil wettability, runoff and erodibility of major dry-Mediterranean land use types on calcareous soils. Hydrol Process 21:2325–2336

Collantes Gutiérrez F (2004) La evolución de la actividad agrícola en las áreas de montaña españolas (1860–2000). Revista Española de Estudios Agrosociales y Pesqueros 201:79–104

Colomar Marí A (ed) (2002) Patrimoni de marjades a la Mediterrània Occidental. Una proposta de catalogació. Consell de Mallorca. Departament de Promoció i Ocupació. FODESMA, Mallorca

Cots-Folch R, Martínez-Casasnovas JM, Ramos MC (2009) Agricultural trajectories in a Mediterranean mountain region (Priorat, NE Spain) as a consequence of vineyard conversion plans. Land Degrad Dev 20(1):1–13

Countryside Agency and Scottish Natural Heritage (2001) Landscape character assessment—guidance for England and Scotland. Countryside Agency Publications, Cheltenham

De Pedro Michó MJ (1998) La Lloma de Betxí (Paterna, Valencia) un poblado de la Edad del Bronce. Diputación Provincial de Valencia, Servicio de Investigación Prehistórica, Serie de Trabajos Varios 94, Valencia

Díaz-Varela RA, Zarco-Tejada PJ, Angileri V, Loudjani P (2014) Automatic identification of agricultural terraces through object oriented analysis of very high resolution DSMs and multispectral imagery obtained from an unmanned aerial vehicle. J Environ Manage 134: 117–126

Douglas TD, Kirkby SJ, Critchley RW, Park GJ (1994) Agricultural terrace abandonment in the Alpujarra, Andalucía, Spain. Land Degrad Dev 5(4):281–291

Dunjó G, Pardini G, Gispert M (2003) Land use change effects on abandoned terraced soils in a Mediterranean catchment, NE Spain. CATENA 32:23–37

Durán Zuazo VH, Aguilar Ruiz J, Martínez Raya A, Franco Tarifa D (2005) Impact of erosion in the taluses of subtropical orchard terraces. Agric Ecosyst Environ 107(2–3):199–210

Emran M, Gispert M, Pardini G (2012) Patterns of soil organic carbon, glomalin and structural stability in abandoned Mediterranean terraced lands. Eur J Soil Sci 63:637–649
EIP-AGRI (European Innovation Partnership for Agricultural Productivity and Sustainability), European Commission (2015) Workshop opportunities for agriculture and forestry in the circular economy. Workshop report, 28–29 Oct 2015
Esquilache Martí F (2015) Els espais agraris i l'estructura social d'una gran horta fluvial andalusina. La construcció i evolució de l'Horta de València entre els segles VIII i XIII. University of Valencia, Valencia. Available via RODERIC. http://roderic.uv.es/handle/10550/50815. Accessed 15 Feb 2017
European Parliament (2015) Directorate-general for internal policies. Policy Department B: structural and cohesion policies. Agriculture and Rural Development. In: Kay S, Peuch J, Franco J (eds) Extent of Farmland Grabbing in the EU. Transnational Institute. IP/B/AGRI/IC/2014–069, May 2015
Fernández Mier M, Fernández Fernández J, Alonso González P, López Sáez JA, Pérez Díaz S, Hernández Beloqui B (2014) The investigation of currently inhabited villages of medieval origin: Agrarian archaeology in Asturias (Spain). Quat Int 346:41–55
Ferro-Vázquez C, González Prieto S, Martínez Cortizas A, Criado Boado F (2015) Deciphering the evolution of agrarian technologies during the last $\sim$1600 years using the isotopic fingerprint ($\delta^{13}C$, $\delta^{15}N$) of a polycyclic terraced soil. Estudos do Quaternário 12, APEQ, Braga:39–53
Gallart F, Llorens P, Latrón J (1994) Studying the role of old agricultural terraces on runoff generation in a Mediterranean small mountainous basin. J Hydrol 159:291–303
Gallego-Fernández JB, García-Mora MR, García-Novo F (2004) Vegetation dynamics of Mediterranean shrublands in former cultural landscape at Grazalema mountains, South Spain. Plant Ecol 172:83–94
García-Ruiz JM (1988) La evolución de la agricultura de montaña y sus efectos sobre la dinámica del paisaje. Revista de Estudios Agro-Sociales 146:7–37
García-Ruiz JM (2010) The effects of land uses on soil erosion in Spain: a review. CATENA 81: 1–11
García-Ruiz JM, Lasanta T, Ruiz Flaño P, Ortigosa L, White S, González C, Martí C (1996) Land-use changes and sustainable development in mountain areas: a case study in the Spanish Pyrenees. Landsc Ecol 11(5):267–277
García-Ruiz JM, Lana-Renault N (2011) Hydrological and erosive consequences of farmland abandonment in Europe, with a special reference to the Mediterranean region—a review. Agr Ecosyst Environ 140(3–4):317–338
García-Ruiz JM, Nadal-Romero E, Lana-Renault N, Beguería S (2013) Erosion in Mediterranean landscapes: changes and future challenges. Geomorphology 198:20–36
García-Ruiz JM, Beguería S, Lana-Renault N, Nadal-Romero E, Cerdà A (2017) Ongoing and emerging questions in water erosion studies. Land Degrad Dev 28:5–21
Garrabou R (1985) Un fals dilema: modernitat o endarreriment de l'agricultura valenciana (1850–1900). Institució Alfons el Magnànim. Institució Valenciana d'Estudis i Investigació
Gaspar L, Navas A, Walling DE, Machín J, Gómez Arozamena J (2013) Using ^{137}Cs and $^{210}Pb_{ex}$ to assess soil redistribution on slopes at different temporal scales. CATENA 102:46–54
Generalitat Valenciana (2011) Catálogo de los paisajes de relevancia regional. Propuesta de Plan. http://www.upv.es/contenidos/CAMUNISO/info/U0549937.pdf. Accessed 26 Feb 2017
Giménez-Font P. 2008. *Las transformaciones del paisaje valenciano en el siglo XVIII. Una perspectiva geográfica.* Institució Alfons el Magnànim, Diputació de València
Gisbert Santonja J (2001) Muros y bancales de la Marina Alta en el Castell de Dénia. Ajuntament de Dénia
Gispert M, Pardini G, Colldecarrera M, Emran M, Doni S (2017) Water erosion and soil properties patterns along selected rainfall events in cultivated and abandoned terraced fields under renaturalisation. CATENA 155:114–126

Glick T, Kirchner H (2000) Hydraulic systems and technologies of Islamic Spain: history and archaeology. In: Squatriti P (ed) Working with water in Medieval Europe. Technology and resource-use. Technology and change in history, vol 3. Brill Academic Publishers, pp 267–330

González-Regidor J (2003) Territorial agriculture and rural development: from agricultural support to territorial policies. In: The future of rural policy: from sectoral to place-based policies in rural areas. Organisation for Economic Co-operation and Development, pp 175–187

Grau Mira I, Pérez Rodríguez V (2008) Aproximación arqueológica al estudio de los paisajes aterrazados en el área central valenciana. Lucentum XXVII:33–50

Grimalt Gelabert M, Blázquez Salom M (1989) El mapa de marjades de la Serra de Tramuntana de Mallorca. Treballs de Geografía 42:43–47

Guinot E (2008) El paisaje de la Huerta de Valencia. Elementos de interpretación de su morfología espacial de origen medieval. In: Historia de la ciudad. V. Tradición y progreso. Icaro-Colegio Territorial de Arquitectos de Valencia, Universidad Politécnica de Valencia, pp 116–129

Guinot E (2009) El regadío histórico en el mundo mudéjar-morisco (siglos XIII–XVII). In: Entre tierra y fe. Los musulmanes en el reino cristiano de Valencia (1238–1609). Publicacions de la Universitat de València, pp 219–228

Hernández Hernández M (1997) Paisajes agrarios y medio ambiente en Alicante. Evolución e impactos medioambientales en los paisajes agrarios alicantinos: 1950–1995. Publicaciones de la Universidad de Alicante

Hernández Hernández M (2009) El paisaje como seña de identidad territorial: valorización social y factor de desarrollo, ¿utopía o realidad? Boletín de la Asociación de Geógrafos Españoles 49:169–183

Jiménez Olivencia Y (2010) Consecuencias del abandono del regadío en la montaña mediterránea. In: Guzmán Álvarez JR, Navarro Cerrillo R (eds) El agua domesticada: Los paisajes de los regadíos de montaña en Andalucía. Consejería de Medio Ambiente de la Junta de Andalucía, pp 508–513

Junta de Andalucía (2014) Catálogo de paisajes de la provincia de Granada. Consejería de Medio Ambiente y Ordenación del Territorio; Universidad de Sevilla, Centro de Estudios Paisaje y Territorio

Kay S (2016) Land grabbing and land concentration in Europe. A research brief. Transnational Institute for HOTL

Kinnaird T, Bolòs J, Turner A, Turner S (2017) Optically-stimulated luminescence profiling and dating of historic agricultural terraces in Catalonia (Spain). J Archaeol Sci 78:66–77

Kirchner H (2012) Conquista y colonización feudal: arqueología de los cambios producidos en los espacios irrigados de origen andalusí. El caso de las Islas Baleares. In: Eiroa Rodríguez J (ed) La conquista de al-Andalus en el siglo XIII. Servicio de Publicaciones de la Universidad de Murcia, pp 41–63

Lasanta T (1988) The process of desertion of cultivated areas in the Central Spanish Pyrenees. Pirineos 132:15–36

Lasanta T (2014) El paisaje de campos abandonados en Cameros Viejo (Sistema Ibérico, La Rioja). Gobierno de la Rioja; Instituto de Estudios Riojanos

Lasanta T, Arnáez J, Ruiz Flaño P, Lana-Renault Monreal N (2013) Los bancales en las montañas españolas: un paisaje abandonado y un recurso potencial. Boletín de la Asociación de Geógrafos Españoles 63:301–322

Lasanta T, Nadal-Romero E, Arnáez J (2015) Managing abandoned farmland to control the impact of re-vegetation on the environment. The state of the art in Europe. Environ Sci Policy 52: 99–109

Lasanta T, Nadal-Romero E, Errea MP, Arnáez J (2016) The effect of landscape conservation measures in changing landscape patterns: a case study in Mediterranean mountains. Land Degrad Dev 27:373–386

Lasanta T, Arnáez J, Pascual N, Ruiz-Flaño P, Errea MP, Lana-Renault N (2017a) Space-time process and drivers of land abandonment in Europe. CATENA 149:810–823

Lasanta T, Errea MP, Nadal-Romero E (2017b) Traditional agrarian landscape in the Mediterranean mountains. A regional and local factor analysis in the Central Spanish Pyrenees. Land Degrad Dev 28(5):1626–1640

Llovet J, Ruiz-Valera M, Josa R, Vallejo R (2009) Soil responses to fire in Mediterranean forest landscapes in relation to the previous stage of land abandonment. Int J Wildland Fire 18 (2):222–232

López Martínez F, Pérez Morales A, Gil Guirado S (2016) In landscape management all of us have something to say. A holistic method for landscape preservability evaluation in a Mediterranean region. Land Use Policy 51:172–183

López-Vicente M, Nadal-Romero E, Cammeraat ELH (2016) Hydrological connectivity does change over 70 years of abandonment and afforestation in the Spanish Pyrenees. Land Degrad Dev 28(4):1298–1310

MAAMA (Ministerio de Agricultura, Alimentación y Medio Ambiente) (2016) Encuesta sobre superficies y rendimientos de cultivos: Informe sobre regadíos en España 2015. ESYRCE, Secretaría General Técnica, Subdirección General de Estadística

Malpica Cuello A (1996) Análisis arqueológico de las terrazas de cultivo. El ejemplo de los Albercones de la Alhambra de Granada. In: Cara Barrionuevo L, Malpica Cuello A (eds) Agricultura y regadío en al-Andalus. Síntesis y problemas. Actas del II Coloquio de Historia y Medio físico. Instituto de Estudios Almerienses, pp 409–424

MAPAMA (Ministerio de Agricultura, Pesca, Alimentación y Medio Ambiente) (2016) Encuesta sobre superficies y rendimientos de cultivos: Resultados nacionales y autonómicos. ESYRCE, Secretaría General Técnica, Subdirección General de Estadística

Marco-Molina JA, Olcina Cantos J, Padilla Blanco A, Rico Amorós AM (1996) Abandono de terrazas de cultivo: recolonización vegetal y erosión en el sureste peninsular. In: Grandal d'Anglade A, Pagés Valcarlos JR (eds) IV Reunión de Geomorfología. Sociedad Española de Geomorfología. O Castro, pp 133–146

MMAMRM (Ministerio de Medio Ambiente y Medio Rural y Marino) (2010) Anuario de Estadística. Ministerio de Medio Ambiente y Medio Rural y Marino 2009. Subdirección General de Estadística

Martínez-Casasnovas JA, Ramos MC (2006) The cost of soil erosion in vineyard fields in the Penedès-Anoina Region (NE Spain). CATENA 68:194–199

Nadal-Romero E, Cammeraat ELH, Pérez-Cardiel E, Lasanta T (2016a) Effects of secondary succession and afforestation practices on soil properties after cropland abandonment in humid Mediterranean mountain areas. Agr Ecosyst Environ 228:91–100

Nadal-Romero E, Cammeraat ELH, Pérez-Cardiel E, Lasanta T (2016b) How do soil organic carbon stocks change after cropland abandonment in Mediterranean humid mountain areas? Sci Total Environ 566–567:741–752

Nadal-Romero E, Lasanta T, Cerdà A (2016c) Integrating extensive livestock and soil conservation policies in Mediterranean mountain areas for recovery of abandoned lands in the Central Spanish Pyrenees. A long-term research assessment. Land Degrad Dev. https://doi.org/10.1002/ldr.2542

Observatori del Paisatge (2008–2016). Catàlegs de Paisatge de Catalunya. http://www.catpaisatge.net/esp/catalegs.php. Accessed 22 Feb 2017

Olarieta JR, Rodríguez-Valle FL, Tello E (2008) Preserving and destroying soils, transforming landscapes: soils and land-use changes in the Vallès County (Catalunya, Spain) 1853–2004. Land Use Policy 25:474–484

Orejas A (ed) (2006) Arqueología Espacial: espacios agrarios. Arqueología Espacial 26. Seminario de Arqueología y Etnología Turolense. Teruel

Padilla Blanco A (1998) Colonización vegetal en campos de cultivo abandonados en la provincia de Alicante. Universidad de Alicante

Pardini G, Gispert M, Emran M, Doni S (2017) Rainfall/Runoff/Erosion relationship and soil properties survey in abandoned shallow soils of NE Spain. J Soils Sedim 17(2):499–514

Plieninger T, Draux H, Fagerholm N, Bieling C, Bürgi M, Kizos T, Kuemmerle T, Primdahl J, Verburg PH (2016) The driving forces of landscape change in Europe: a systematic review of the evidence. Land Use Policy 57:204–214
Poyatos R, Latrón J, Llorens P (2003) Land-use and land cover change after agricultural abandonment. The case of a Mediterranean Mountain Area (Catalan Pre-Pyrenees). Mt Res Dev 23(4):52–58
Prosdocimi M, Cerdà A, Tarolli P (2016) Soil water erosion on Mediterranean vineyards: a review. CATENA 141:1–21
Pueyo Y, Beguería S (2007) Modelling the rate of secondary succession after farmland abandonment in a Mediterranean mountain area. Landsc Urban Plann 83:245–254
Puy A, Balbo A (2013) The genesis of irrigated terraces in al-Andalus. A geoarchaeological perspective on intensive agriculture in semi-arid environments (Ricote, Murcia, Spain). J Arid Environ 89:45–56
Puy A, García Avilés JM, Balbo A, Keller M, Riedesel S, Blum D, Bubenzer O (2016) Drip irrigation uptake in traditional irrigated fields: the edaphological impact. J Environ Manage 202 (3):550–561
Quijano L, López-Vicente M, Gaspar L, Machín J, Navas A (2013) Modelling soil redistribution in a hydrologically defined crop field with WATEM/SEDEM. In: Geophysical research abstracts, vol 15, EGU2013–5413
Quintas-Soriano C, Castro AJ, Castro H, García-Llorente M (2016) Impacts of land use change on ecosystem services and implications for human well-being in Spanish drylands. Land Use Policy 54:534–548
Quirós Castillo JA, Nicosia C, Polo-Díaz A, Ruiz del Árbol M (2014) Agrarian archaeology in northern Iberia: geoarchaeology and early medieval land use. Quat Int 346:56–68
Ramos MC, Cots-Folch R, Martínez-Casasnovas JA (2007) Sustainability of modern land terracing for vineyard plantation in a Mediterranean mountain environment -the case of the Priorat region (NE Spain). Geomorphology 86:1–11
Ramos MC, Martínez-Casasnovas JA (2010) Effects of field reorganisation on the spatial variability of runoff and erosion rates in vineyards of Northeastern Spain. Land Degrad Dev 21:1–15
Retamero F, Kirchner H, Oliver J, Vila JM (2015) Agricultura en terrasses en espais feudals: els casos del mas l'Agustí (Tagamanent, Vallès oriental) i del solar dels Centelles (Sant Martí de Centelles, Osona). In: Actes de III Jornades d'arqueologia de la Catalunya central, 188–196. Departament de Cultura, Generalitat de Catalunya
Reynés A, Alvaro P, Alomar G, Vadell J (2007) Marjades i prevenció de riscs naturals a la serra de Tramuntana. Departament de Medi Ambient i Natura, Consell de Mallorca, Palma de Mallorca
Riera S, Palet JM (2005) Aportaciones de la Palinología a la historia del paisaje mediterráneo: estudio de los sistemas de terrazas en las Sierras Litorales Catalanas desde la perspectiva de la Arqueología Ambiental y del Paisaje. In: Riera Mora S, Julià Brugués R (eds) Una aproximació transdisciplinar a 8.000 anys d'història dels usos del sòl. Seminari d'Estudis i Recerques Prehistòriques, Universitat de Barcelona, Barcelona, pp 55–74
Rodríguez A, Lasanta Martínez T (1992) Los bancales abandonados en la montaña mediterránea: una revisión bibliográfica. Pirineos 139:105–123
Romero-Díaz A (ed) (2016) Abandono de cultivos en la Región de Murcia. Consecuencias ecogeomorfológicas. Universidad de Murcia.
Romero-Díaz A, Marín Sanleandro P, Sánchez Soriano A, Belmonte Serrato F, Faulkner H (2007) The causes of piping in a set of abandoned agricultural terraces in southeast Spain. CATENA 69:282–293
Romero-Díaz A, Ruiz-Sinoga JD, Robledano-Aymerich F, Brevik EC, Cerdà A (2017) Ecosystem responses to land abandonment in Western Mediterranean Mountains. CATENA 149:824–835
Ruecker G, Schad P, Alcubilla MM, Ferrer C (1998) Natural regeneration of degraded soils and site changes on abandoned agricultural terraces in Mediterranean Spain. Land Degrad Dev 9 (2):179–188

Ruiz del Árbol M, Sánchez-Palencia FJ, López Sáez JA, López P, Macías R, López O (2003) A geoarchaeological approach to the study of Roman terraces: landscape transformations in a mining area in the north-western Iberian Peninsula. In: Fouache E (ed) The Mediterranean world environment and history. Elsevier, Amsterdam, pp 331–339

Ruiz-Flaño P, García-Ruiz JM, Ortigosa L (1992) Geomorphologial evolution of abandoned fields. A case study in the Central Pyrenees. CATENA 19:301–308

Sánchez-Marañón M, Soriano M, Delgado G, Delgado R (2002) Soil quality in Mediterranean mountain environments. Effects of land use change. Soil Sci Soc Am J 66(3):948–958

Sanz-Cobena A, Lassaletta L, Aguilera E, del Prado A, Garnier J, Billen G, Iglesias A, Sánchez B, Guardia G, Abalos D, Plaza-Bonilla D, Puigdueta-Bartolomé I, Moral R, Galán E, Arriaga H, Merino P, Infante-Amate J, Meijide A, Pardo G, Álvaro-Fuentes J, Gilsanz C, Báez D, Doltra J, González-Ubierna S, Cayuela ML, Menéndez S, Díaz-Pinés E, Le-Noë J, Quemada M, Estellés F, Calvet S, van Grinsven HJM, Westhoek H, Sanz MJ, Gimeno BS, Vallejo A, Smith P (2017) Strategies for greenhouse gas emissions mitigation in Mediterranean agriculture: a review. Agr Ecosyst Environ 238:5–24

Seeger M, Ries JB (2008) Soil degradation and soil surface process intensities on abandoned fields in Mediterranean mountain environments. Land Degrad Dev 19:488–501

Sese-Mínguez S, Boesveld H, Asins-Velis S, van der Kooij S, Maroulis J (2017) Transformations accompanying a shift from surface to drip irrigation in the Cànyoles Watershed, Valencia, Spain. Water Altern 10(1):81–99

Sitges E (2006) Inventario y tipología de sistemas hidráulicos de Al-Andalus. In: Arqueología Espacial: espacios agrarios. Arqueología Espacial 26, Seminario de Arqueología y Etnología Turolense. Teruel, pp 263–291

Soler C, Fernández F (2015) Estructura de la propiedad de tierras en España. Concentración y acaparamiento. Fundación Mundubat, Revista Soberanía Alimentaria, Biodiversidad y Culturas

Tarolli P, Sofia G, Calligaro S, Prosdocimi M, Preti M, dalla Fontana G (2015) Vineyards in terraced landscapes: new opportunities from Lidar data. Land Degrad Dev 26(1):92–102

Torró J (2005) Terrasses irrigades a les muntanyes valencianes. Les transformacions de la colonització cristiana. Afers 51:301–356

Torró J (2007) Vall de Laguar. Asentamientos, terrazas de cultivo e irrigación en las montañas de Šarq al-Andalus: un estudio local. Recerques del Museu d'Alcoi 16:151–182

Trillo San José C (2010) Paisajes, cultivos y culturas. In: Guzmán Álvarez JR, Navarro Cerrillo RM (eds) El agua domesticada: Los paisajes de los regadíos de montaña en Andalucía. Consejería de Medio Ambiente de la Junta de Andalucía, Sevilla, pp 103–119

United Nations, Department of Economic and Social Affairs, Population Division (2015) World population prospects: the 2015 revision, key findings and advance tables. Working Paper No. ESA/P/WP.241

UNEP, Schandl H, Fischer-Kowalski M, West J, Giljum S, Dittrich M, Eisenmenger N, Geschke A, Lieber M, Wieland HP, Schaffartzik A, Krausmann F, Gierlinger S, Hosking K, Lenzen M, Tanikawa H, Miatto A, Fishman T (2016) Global material flows and resource productivity, an assessment study of the UNEP International Resource Panel. United Nations Environment Programme, Paris

Varón Hernández FR, Hernández-Beloqui B, Sopelana Salcedo I, Fernández Carvajal JA (2012) Las terrazas de Abanto. Nuevas aportaciones desde la Arqueobotánica a las cronologías de la Alta Edad Media vizcaína. Munibe (Antropologia-Arkeologia) 63:293–303

Vanwalleghem T, Gómez JA, Infante Amate J, González de Molina M, Vanderlinden K, Guzmán G, Laguna A, Giráldez JV (2017) Impact of historical land use and soil management change on soil erosion and agricultural sustainability during the Anthropocene. Anthropocene 17:13–29

Verburg PH, Mertz O, Erb K-H, Haberl H, Wu W (2013) Land system change and food security: towards multi-scale land system solutions. Curr Opin Environ Sustain 5(5):494–502

Vicente-Serrano SM, Lasanta T, Cuadrat JM (2000) Influencia de la ganadería en la evolución del riesgo de incendio en función de la vegetación en un área de montaña: el ejemplo del valle de Borau (Pirineo aragonés). Geographicalia 38:33–58

Vila Subirós J, Rodríguez-Carreras R, Varga D, Ribas A, Úbeda X, Asperó F, Llausàs A, Outeiro L (2016) Stakeholders perceptions of landscape changes in the Mediterranean mountains of the North-Eastern Iberian Peninsula. Land Degrad Dev 27:1354–1365

Wascher DM (ed) (2005) European landscape character areas—typologies, cartography and indicators for the assessment of sustainable landscapes. Final Project Report as deliverable from the EU's Accompanying Measure project "European Landscape Character Assessment Initiative" (ELCAI), funded under the 5th Framework Programme on Energy, Environment and Sustainable Development (4.2.2), X

Zornoza R, Guerrero C, Mataix-Solera J, Scow KM, Arcenegui V, Mataix-Beneyto J (2009) Changes in soil microbial community structure following the abandonment of agricultural terraces in mountainous areas of Eastern Spain. Appl Soil Ecol 42(3):315–323

Chapter 7
Terraced Landscapes in the Canary Islands: La Gomera, "The Terrace Island"

Lidia Esther Romero Martín, Antonio Ignacio Hernández Cordero, Aarón Santana Cordero, Carla Vargas Negrín and Juan Manuel Palerm Salazar

Abstract The aim of this work is to map and characterize the maximum surface area of terraced slopes on La Gomera Island. To do so, two digital orthophotos, from 1964 and 2016, were used and analyzed with GIS. These sources represent two key moments in the recent history of the island, corresponding in the first instance to a peak in population and agricultural activity and in the second to the fall into disuse of the terraces. In this study, terrace characterization was based on a map-based analysis of environmental and human factors which influence this kind of landscape. The environmental factors considered were topographically—(slope, elevation, landforms) and geologically-based, while the human factors analyzed were land use, land cover, and protected natural areas. The results show the

L. E. Romero Martín (✉) · A. I. Hernández Cordero
Grupo de Investigación Geografía Física y Medio Ambiente,
Instituto Oceanográfico y Cambio Global, IOCAG,
Universidad de Las Palmas de Gran Canaria, Las Palmas, Spain
e-mail: lidia.romero@ulpgc.es

A. I. Hernández Cordero
e-mail: hernandez.cordero@ulpgc.es

A. Santana Cordero
Department for Management of Science and Technology Development,
Ton Duc Thang University, Ho Chi Minh City, Vietnam
e-mail: aaron.santana@tdt.edu.vn

A. Santana Cordero
Faculty of Environment and Labour Safety, Ton Duc Thang University,
Ho Chi Minh City, Vietnam

C. Vargas Negrín
Palerm & Tabares de Navas, Santa Cruz, Spain
e-mail: carlavargasnegrin@gmail.com

J. M. Palerm Salazar
Department of Graphic Expression and Architectural Projects,
Universidad de Las Palmas de Gran Canaria, Las Palmas, Spain
e-mail: jm.palerm@ulpgc.es

M. Varotto et al. (eds.), *World Terraced Landscapes: History, Environment, Quality of Life*, Environmental History 9,
https://doi.org/10.1007/978-3-319-96815-5_7

enormous, diverse, and vulnerable heritage of the terraces, which are an example of a cultural landscape whose history reveals the isolation, scarcity, subsistence, and identity of La Gomera islanders, as well as the terraces' current state of abandonment and degradation. This research constitutes the beginning of an inventory process (identification, location, and description) which will serve as a prior step to classification and evaluation, necessary to protect this landscape. It is also the first contribution from an Atlantic-Macaronesian island to the atlas of world terraces, as promoted by the International Terraced Landscapes Alliance (ITLA).

7.1 Introduction

Terraces with traditional walls are fine examples of locally developed architectural engineering. They are the most abundant type of dry stone construction in the world, and their study, protection and conservation are the main objective of the International Scientific Society for the interdisciplinary study of dry stone (French initials: SPS), founded in 1997. Additionally, thanks to the initiatives of public[1] institutions and private associations,[2] the traditional mode of construction with dry stone is in the process of obtaining recognition as a *Bien de Interés Cultural* by the Government of Spain and inclusion in UNESCO's List of the Intangible Cultural Heritage of Humanity.

The terraces constitute cultural landscapes[3], the result of the interaction of people and their environment over time and more specifically of the dynamic relationship between environmental limitations and socioeconomic opportunities. These landscapes belong to the "organically evolved" category (World Heritage Committee 1992) comprised of two subdivisions, relic and continuing, depending on the extent of the activity of their evolutionary process. The risk of fossilization is present in many territories where terraces have ceased fulfilling the active social function for which they were built. This last situation is particularly noticeable in islands, as in the Atlantic archipelagos, where there has been a rapid explosion over recent years in the service economy and tourism.

The cultural character of these landscapes is related to the concept of "heritage," as manifested through the perception and evaluation that the corresponding society

---

[1]Resolution of September 2, 2016, of the Island Council of Mallorca (Balearic Islands), referring to the declaration as intangible cultural interest good of the traditional technique of construction with dry stone. Spanish Official State Gazette No. 224, September 16, 2016.

[2]The necessary legal steps to request the inclusion in the Intangible Cultural Heritage list by UNESCO were initiated by the French Federation of Dry Stone Professionals (FFPPS), who proposed to the SPS the request as a transnational nomination, in September 2016 (at the XV International Conference of Dry Stone, Greece). Eight countries have since been included in this initiative (Greece, France, Switzerland, Spain, Italy, Croatia, Slovenia, and Cyprus).

[3]As defined in the World Heritage Convention by UNESCO (1992).

has of their qualities, converting them into an element of identity[4] (Hernández 2009). However, not all terraced landscapes are in the same phase of "social appropriation of the landscape" and, therefore, they have not all attained the same value in terms of heritage. They can be considered "heritage" once they fulfill three basic conceptual premises: when the corresponding society accepts the idea of collective ownership, when it acknowledges the material and cultural values of these landscapes, and when it ensures their values are preserved and handed down to subsequent generations (Silva 2009).

Some 22% of the cultural landscapes included in UNESCO's Heritage of Humanity List (23 of 103) are terraced agricultural landscapes. Of these, ten are located in Europe, with four in Italy and one in Spain: the Cultural Landscape of Sierra de Tramuntana (Mallorca, Balearic Islands) incorporated in the list in 2011. In the Canary Archipelago, in only 7.5% of its Protected Nature Areas (11 of 146) are terraces considered basic elements for landscape protection and so included among essential and/or priority objectives (Romero et al. 2016).

An increased recognition of the value of these landscapes could potentially trigger their inclusion in the World Heritage list of terraced landscapes and, following the European Landscape Convention (Article 2), they would thus become "quotidian" landscapes, with acceptance of their heritage-based value and multi-functional character with resources that favor economic activity and sustainable development.

A heritage-centered analysis of these cultural terraced landscapes needs to incorporate their origin, historical evolution and social value, as well as a description of the resulting land structure (Silva 2009). It should begin with the identification and characterization of the area they occupy.[5] Such identification includes their location and a delimitation of the geographical area, while their characterization comprises a description, analysis, and interpretation of the components of the landscape (physical environment and human action) and their interrelationships or processes.

Peru was one of the first countries to draw up a national inventory of terraces, thanks to the initiative of its National Office of Natural Resources Assessment (Spanish initials: ONERN) started in 1987 (Masson 1994). In Europe, through various projects that have considered terraced landscapes as an object of study, including MEDSTONE (Alomar et al. 2002), PATTER (Colomar 2002), TERRISC (Reynés 2007), and ALPTER (Fontanari and Patassini 2008), important advances have been made regarding their identification, cataloging, classification, and diagnostics in several countries in the southern and central parts of the continent.

The knowledge we have about the extension of terraced landscapes in Spain remains deficient despite studies undertaken by, for example, Grimalt and Blázquez (1989) in Sierra de Tramuntana (Mallorca), Rodríguez-Aizpeolea (1992) in Vall

[4]Idea extracted from Spain's Plan Nacional de Paisaje Cultural (2012).

[5]Area identification and characterization form part of the commitment acquired by Spain when ratifying the European Landscape Convention (in force since 1st of March, 2008).

d'Ebo municipality (Alicante), Asins (2009) in Petrer municipality (Alicante), and Lasanta (1989), Maiso and Lasanta (1990) and Lasanta et al. (2009) in the Aragonese Pyrenees and the Iberian System of La Rioja. In the Canary Islands, only the mapping of the area and typology of the Guiniguada basin (Gran Canaria) (Romero 2015) and of La Gomera Island have been performed (Plan Insular de Ordenación de La Gomera).

The terraces of La Gomera can be considered as an example of the expression of human adaptation to adverse topographical conditions. In this small island, the enormous amount of terraced landscapes results in a diverse landscape[6] and toponymy (Perera 2015). The steep terraced slopes of Lepe (Agulo), Erque, and Erquito (Vallehermoso) are particularly noteworthy for their complicated environmental conditions. They have also been given other names, such as "Giant's Stairway" (Sabaté 2012) and the "Machu Picchu of La Gomera" (Rodríguez 2012).

The aim of this work is to begin a heritage-based analysis of the cultural terraced landscapes of La Gomera Island. To this end, we have identified, mapped, and characterized the total terraced surface area in the island.

7.2 Study Area

La Gomera is one of the smallest islands in the Canary archipelago (Spain), which is comprised of seven islands and is located in the NE Atlantic Ocean, opposite the northwest coastline of Africa (Fig. 7.1). La Gomera is the sixth largest island (369.8 km^2) and the fourth highest (1487 m, Garajonay).

La Gomera is a volcanic island, although there has been no volcanic activity during the Quaternary (the last 2 million years). As a result, the island constitutes an "authentic museum of erosive volcanic landforms" (Carracedo 2008) including, most notably, sub-volcanic morphologies or sub-volcanic bodies in the form of abundant and spectacular dikes, stocks, and domes.

The island has an approximate age of 10–10.5 million years (Ancochea et al. 2006). It was constructed in three volcanic phases separated by two erosive periods, during which various volcanic structures overlapped and eroded. The oldest is the Miocene basaltic shield which experienced a giant landslide on its northern flank, followed by a trachyphonolitic stratovolcano that subsequently collapsed. Finally, from the most recent volcanic phase only Pliocene basalt lava traps have been preserved.

The island is an old massif with an elliptical plant shield form. Mid-range heights dominate (73.4% of its surface is between 200 and 1000 m above sea level), along with steep slopes (with 61.2% of them >12°), while flat surfaces are notable for their scarcity (Santana and Villalba 2008).

[6]Typology of terraced landscapes in La Gomera (AIDER La Gomera). Project "Custody of the territory."

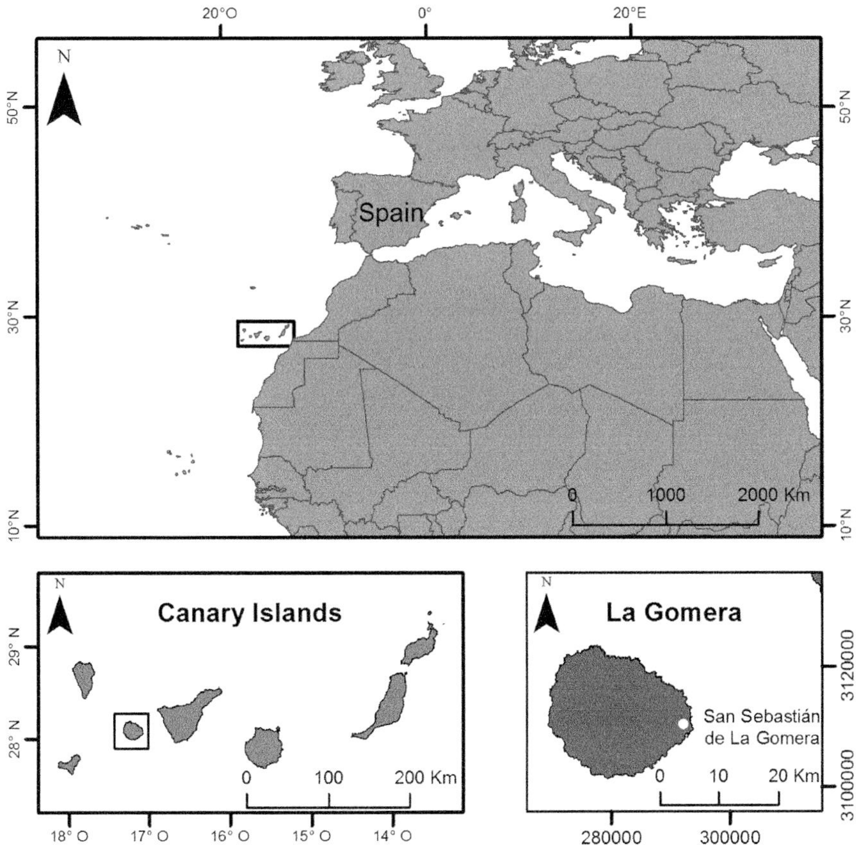

Fig. 7.1 Study area. Island of La Gomera

Its cliffed coast, which is surrounded by a shallow platform, the central high plateau and a radial network of incised ravines are its most significant topographical elements. Cliff height and coastal platform width attain their maximum values in the NNE section, due to greater exposure and the antiquity of the geological materials. The central plateau, located above 1000 m, constitutes the origin of most of the island's numerous ravines.

The ravines and their water-divider lines present an unequal degree of incision and narrowness, respectively, according to the age of the materials and the characteristics of the rainfall. There are four types of ravine: mature wet, mature dry, young, and forest (Llánes et al. 2009). Ravines of the N and NE are the most incised, and their interfluves are narrow, while in the S and SW the ravines have wider riverbeds and their interfluves are *quasi*-horizontal.

Due to its subtropical Atlantic location (cold oceanic stream, Azores anticyclone, trade winds, and stratocumulus mantle) and its mountainous nature, there are significant climatic contrasts between different areas of La Gomera despite its reduced size. The annual mean temperature is 20 °C on the northern coast and 13 °C at the summits. Mean annual precipitation is low (450 mm, approximately), with average values for the southern coastal region of 200 mm/year and for the mid-northern mountainous region above 700 mm/year.

In accordance with climate variation with height, four types of zonal vegetation can be identified in the island (Del Arco et al. 2009): (1) coastal xerophilous shrubs formed of two plant communities: sweet spurge scrub (*Neochamaeleo pulverulentae-Euphorbietum balsamiferae*) and cardon scrub (*Euphorbietum berthelotii-canariensis*); (2) juniper woodland (*Brachypodio arbusculae-Juniperetum canariensis*); (3) laurel forest, formed of four plant communities: dry evergreen laurel forest (*Visneo mocanerae-Arbutetum canariensis*), humid evergreen laurel forest (*Launaeo novocanariensis-Perseetum indicae*), hygrophilous evergreen laurel forest (*Diplazio caudati-Ocoteo foetentis*), and cold evergreen laurel forest (*Violo rivinianae-Myricetum fayae*); and (4) pine woodland (*Cisto gomerae-Pinetum canariensis*). In addition, there are some azonal plant communities conditioned by substrate characteristics: leafless spurge scrub (*Euphorbietum aphyllae*), canary palm grove (*Periploco laevigatae-Phoenicetum canariensis*), hydrophilic vegetation (*Rubo-Salicetum canariensis*), *Plocama pendula* scrub (*Plocametum pendulae*), tamarisk thicket (*Atriplici ifniensis-Tamarici canariensis sigmetum*), and rupicolous vegetation complexes. Human activities have degraded part of the vegetation of the island, generating natural secondary plant communities, such as *Euphorbio berthelotii-Schizogynetum sericeae*, *Euphorbietum berthelotii*, *Launaeo arborescentis-Schizogynetum sericeae*, *Rhamno crenulatae-Hypericetum canariensis,* and *Myrico fayae-Ericetum arboreae.*

Natural and cultural values of the island have been acknowledged by a number of international, national and regional institutions. Spain declared Garajonay (a forest in the La Gomera summits) a National Park in 1981. The Canary Government has also established 17 Protected Nature Areas in the island (33% of its area). Likewise, at international level, UNESCO incorporated Garajonay as the first natural Spanish asset in its World Heritage List (1986) and La Gomera's whistle language as an Intangible Cultural Heritage of Humanity in 2009, while the island in its entirety was declared a Biosphere Reserve in 2012.

Human colonization of the island began around 500 BCE, with immigrants from Northern Africa. They were herders who also gathered seafood and vegetables, resulting in the first alteration to the island's vegetation. However, human impact on the island's natural resources increased radically after its conquest by the Spanish (1488). Over the subsequent centuries, the crops that were exported changed and were differentiated from the subsistence crops. Such changes were related to shifts in market demand, the adaptation of foreign species, and the gradual increase in population. The result of these changes is the current landscape, influenced by the intense establishment of terraces and crop spatial distribution (Fig. 7.2).

The biggest changes in La Gomera's history took place over the course of the past century. A rapid growth in agricultural activity (mainly involving the banana and tomato plant) resulted in an extensive sprawl of terraces and a rise in population that reached its zenith in 1960 (30,784 inhabitants). The end of the century saw a

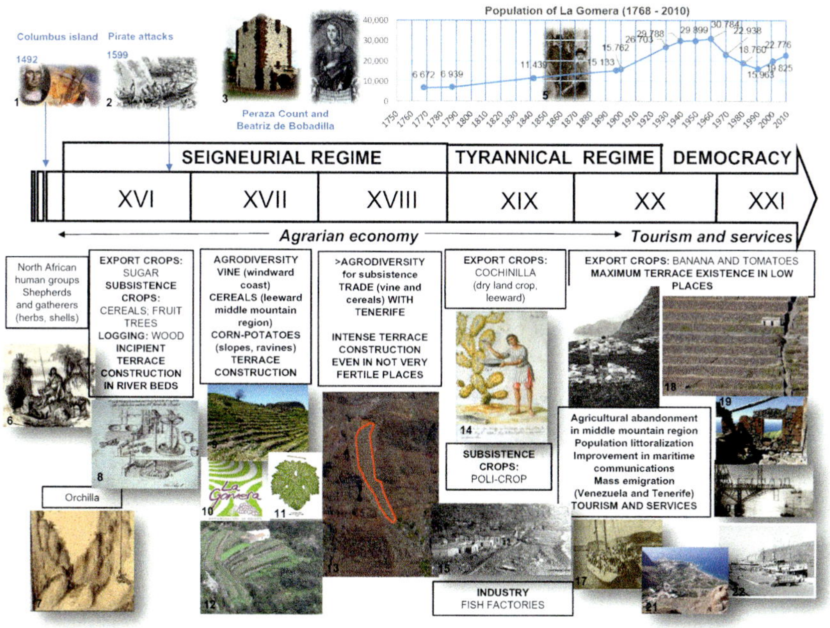

Fig. 7.2 Timeline of the socioeconomic history of La Gomera. (List of sources used for the timeline: 1. http://iberoamericasocial.com/los-viajes-de-cristobal-colon-a-america/ [08/04/2017]. 2. http://www.gomeratoday.com/wp-content/uploads/2015/01/piratas-holandeses-en-la-gomera.jpg [08/04/2017]. 3. https://commons.wikimedia.org/wiki/File:Torre_del_Conde,_San_Sebasti%C3%A1n_de_la_Gomera,_La_Gomera,_Espa%C3%B1a,_2012-12-14,_DD_02.jpg [08/04/2017]. 4. http://www.guanches.org/enciclopedia/index.php?title=Archivo:Beatriz_de_Bovadilla.JPG [08/04/2017]. 5. http://images.eldiario.es/lagomeraahora/centro-Leoncio-Bento_EDIIMA20150605_0893_5.jpg [08/04/2017]. 6. http://i.bssl.es/unomasenlafamilia/2013/03/nombres-guanches.jpg [08/04/2017]. 7. http://3.bp.blogspot.com/- [08/04/2017]. 8. http://www.revistacanarii.com/ext/img/molinotorriani.jpg [08/04/2017]. 9. Author: Juan Carlos Hernández Marrero. 10. http://vinosyaceites.com/base/ui/imagenest/DO/do-la-gomera.gif [08/04/2017]. 11. http://www.eltambor.es/wp-content/uploads/2016/03/FORSTESRA-BLANCA.jpg [08/04/2017]. 12. http://gomeraverde.es/upload/img/periodico/img_50431.jpg [08/04/2017]. 13. Author: Juan Carlos Hernández Marrero. 14. http://www.museosdetenerife.org/assets/images/events/event–c705f43153.jpg [08/04/2017]. 15. http://images.eldiario.es/lagomeraahora/Rajita_EDIIMA20150211_0205_5.jpg [08/04/2017]. 16. https://www.facebook.com/isladelagomera/photos/a.174249062602488.43727.154491724578222/524423120918412/?type=3&theater [08/04/2017]. 17. http://lh3.ggpht.com/-XFkNQA4LnR0/T5doCsBJd- [08/04/2017]. 18. Author: Juan Carlos Hernández Marrero. 19. Author: Concepción Fagundo García. 20. http://images.eldiario.es/lagomeraahora/Antigua-fotografia-pescante-Vallehermoso_EDIIMA20140503_0297_4.jpg [08/04/2017]. 21. http://vallegranrey.es/wp-content/uploads/2014/11/litoral-1080x675.jpg [08/04/2017]. 22. http://www.eltambor.es/wp-content/uploads/2014/09/Le%C3%B3n-y-Castillo-atracando-y-del-La-Palma-saliendo-en-junio-de-1958-en-el-puerto-de-San-Sebasti%C3%A1n-de-La-Gomera.jpg [08/04/2017])

rapid change on the island from an agricultural-based economic system to a service-based one (tourism and services), which in turn triggered a trend for agricultural abandonment. Currently, the island is a nature (hiking), cultural, and scientific tourist destination.

In addition to the impact of tourism, the abandonment of agricultural practices and an aging and falling population are endangering conservation of the highly valued cultural landscapes of the island, as well as its agrodiversity.

The terraced slopes are the most remarkable of the agrosystems in La Gomera, given their beauty, abundance, and magnitude. They constitute a human adaptation to the adverse conditions of the island's topography. Although terrace walls (Ecoplán de La Gomera, planes de Medianías) have been repaired and the landscaping idea for the walls or "paredones" has been restored with a new procedure put into practice, this custody of the land has not been enough to stop the present state of neglect of many of them.

Other elements of these landscapes that should not be separated from the terraces include the palm groves and hamlets. Examples of terraced landscapes with associated palm groves include Tazo, Alojera, Taguluche, Vallehermoso, and Valle Gran Rey.

7.3 Methods

This study was based on the use of different spatial sources (Table 7.1) and its analysis using geographic information systems (GIS).

Table 7.1 Spatial sources

	Year	Spatial resolution (m)	Source
Digital orthophoto	1964	0.80	IDE Canarias. GRAFCAN (Gobierno de Canarias)
Digital orthophoto	2016	0.25	IDE Canarias. GRAFCAN (Gobierno de Canarias)
DEM (digital elevation model)	2009	5	Instituto Geográfico Nacional (IGN)
Geology GIS layer	2010	–	Instituto Geológico y Minero de España (IGME)
Land cover GIS layer	2002	–	GRAFCAN (Gobierno de Canarias)
Crops GIS layer	2003	–	Consejería de Agricultura, Ganadería, Pesca y Alimentación (Gobierno de Canarias)
Protected Nature Areas GIS layer	–	–	Dirección General de Ordenación del Territorio (Gobierno de Canarias)

Digital orthophotos from 1964 and 2016 were photo-interpreted to identify the maximum surface area of terraced landscapes on La Gomera. The terraces were digitized through a GIS, obtaining polygon layers. The 1964 orthophoto allowed identification of the terraces existing at the time, when agricultural activity was at its peak. However, some of these terraces are currently in a highly deteriorated condition while others have even disappeared. Meanwhile, the 2016 orthophoto allowed identification of new terraces and it was possible to resolve some doubts which had arisen in relation to the first orthophoto due to shadows, spatial resolution limits, etc.

A digital elevation model (DEM) was subsequently used to create a digital slope model. This new raster was reclassified using the following intervals: 0°–5°, 5°–15°, 15°–25°, 25°–35°, 35°–45°, 45°–55°, 55°–65°, 65°–75°, and 75°–80.5°. The elevations from the DEM were then also reclassified according to the following intervals: 0–200, 200–400, 400–600, 600–800, 800–1000, 1000–1200, and >1200 m.

Various GIS analyses were performed to characterize the terraces and determine their current use. The terraced layer was overlaid over the following layers: slope, elevation, geology, protected areas (Protected Nature Areas and Special Areas of Conservation), and land cover.

To analyze terrace distribution in the watersheds of the island, a map was generated using the "watershed" GIS tool of the ArcGIS software. The resulting raster was converted into a layer of polygons. The smallest adjacent watersheds/basins were merged with their neighboring basins to generalize these spatial features according to the scale of the final map, thus obtaining the final layer of watersheds.

7.4 Characterization of the Terraces of La Gomera

7.4.1 Environmental Factors (Altitude, Slope, Lithology and Landforms)

The area occupied by terraces amounts to a total of 8233.1 ha, or 22.4% of the island. They are located in almost all the watersheds of the island (Fig. 7.3), from the coast to a height above sea level of 1362.6 m.

In terms of altitudinal distribution, 78.5% of the terraced areas are concentrated between 200 and 1000 m (Fig. 7.4). At lower and upper heights, terraced landscapes represent 13.6% (1114.7 ha) and 8% (571.7 ha), respectively. An unequal altitudinal distribution was observed according to the face of the island on which they are located. In the south, leeward of the trade winds and the sea of clouds, the terraces rise above 1000 m. By contrast, in the north, the almost permanent presence of the sea of clouds between 600 and 700 m has generally limited the construction of terraces and favored the recovery of laurel forest in areas that were briefly used for farming in the 1950s and 1960s (during periods of famine and when demographic pressure was greatest) and then abandoned.

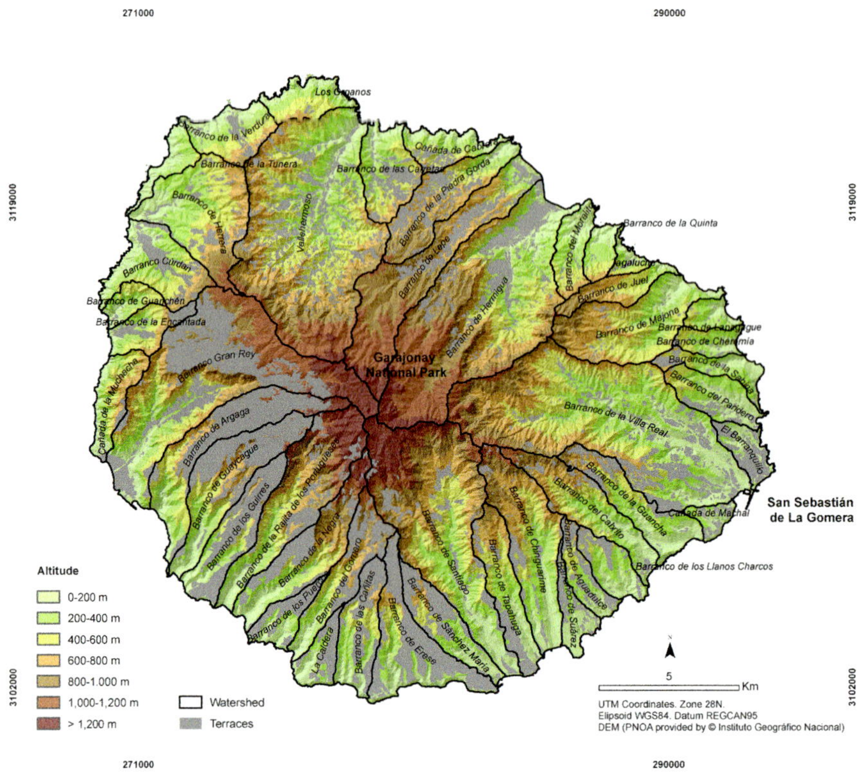

Fig. 7.3 Terraces of La Gomera

Fig. 7.4 Distribution of terraced area by altitude

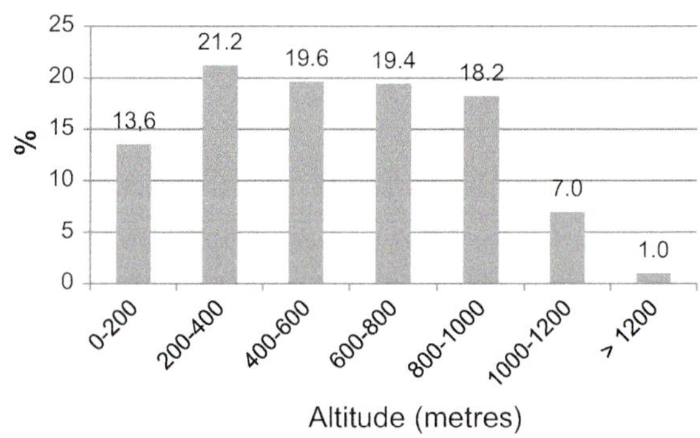

Some 39.3% (3226.7 ha) of the terraced area are located on slopes between 5° and 15°, while 30.7% (2524.1 ha) are located on slopes between 15° and 25°, where agricultural use is only recommended for terraces (Fig. 7.5). In addition, 24.3% of terraced areas have been constructed on slopes unsuitable for agricultural

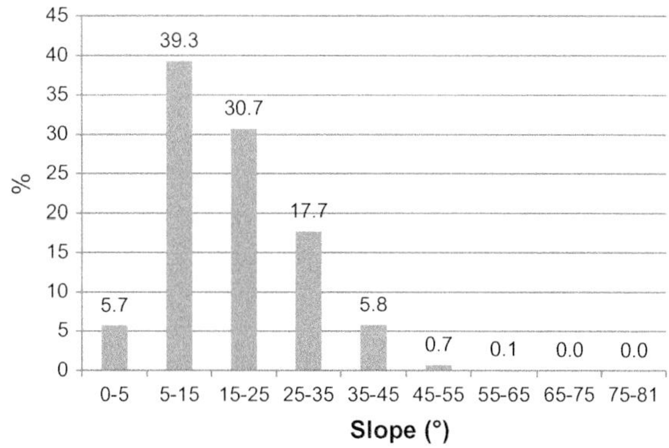

Fig. 7.5 Distribution of terraced areas by slope (°)

use (>25°), of which 6.6% are considered highly unstable and in serious danger of collapse. The largest slope where terraces are located is 76.9° in the watershed Valle Gran Rey (Fig. 7.8). In contrast, there are also terraces on slopes <5°, representing 5.7% (469.2 ha) of the total terraced area, located in the lower part of ravines and interfluves (Figs. 7.5, 7.6, and 7.7).

From a geological point of view (Table 7.2), most of the terraces are located on the largest-sized geological deposits of the island. Some 72.0% (5929.4 ha) are

Fig. 7.6 Mouth of Hermigua ravine. *Photo* M. C. Iglesias López

Fig. 7.7 El Cabrito ravine. *Photo* J. Izquierdo Valli

Table 7.2 Distribution of terraced areas by lithology

Lithology	Area (ha)	%
Basaltic and trachybasaltic lava flows	4543.0	55.2
Basaltic lava flows	613.5	7.5
Mafic trachyte lava flows	538.9	6.5
Basaltic lava flows and related breccias	233.9	2.8
Subtotal	5929.4	72.0
Basic and ultrabasic plutonic stocks	222.5	2.7
Felsic dike swarms and associated lava flows	184.2	2.2
Tephra cones	312.5	3.8
Subtotal	719.2	8.7
Colluvium and hillside deposits	353.7	4.3
Landslide deposits, debris flows, coastal rock falls	308.7	3.7
Alluvial deposits and ravines	217.5	2.6
Debris avalanche of Vallehermoso	110.9	1.3
Subtotal	990.7	12.0
Other	593.8	7.2
Total	8233.1	100.0

Fig. 7.8 Ravine of Valle Gran Rey. *Photo* L. E. Romero Martín

located over basaltic and trachybasaltic lava flows, which constitute the remains of Miocene and Pliocene shields, while 8.7% (719.2 ha) are found over other deposits of igneous origin, such as tephra cones, intrusion of felsic dikes and basic and ultrabasic plutonic stocks. Terraces are also found over types of sedimentary deposit (12%, 990.7 ha) as the result of slope erosive activity (colluvium and landslide deposits, coastal rock falls and a debris avalanche in Vallehermoso).

As ravines and watersheds are the dominant landforms on the island, differences are observable between the windward (N–NE) and leeward (S–SW) sides. In the S–SW, the watersheds have, on average, 30.1% terraced land, while in the N–NE side this value falls to 15.1% (Fig. 7.3). The watersheds with the highest presence of terraces are El Barranquillo (59%) in the N–NE and Barranco de Los Guirres (65.8%) in the S–SW. However, Valle Gran Rey watershed, located on the southern slope (Fig. 7.8), is the biggest contributor to the island (11.5% of the island's terraced slopes are located here). Among those located on the northern slope, Hermigua stands out for its profuse terraces and banana cultivation (Figs. 7.9 and 7.10).

The dotted vertical line shows mean percentage value for terraced land/watershed/island. The continuous vertical line shows mean percentage value of terraced land/watershed.

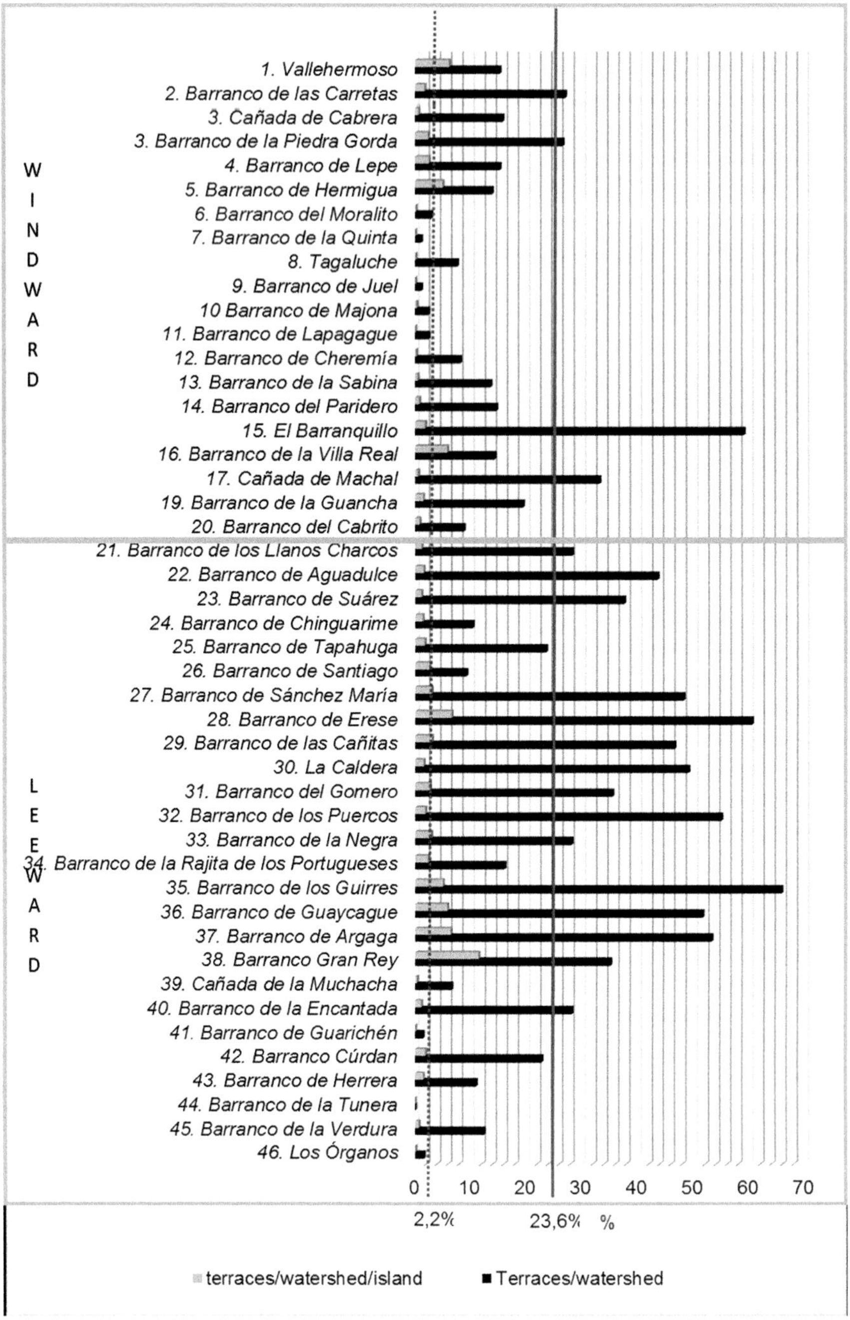

Fig. 7.9 Terraced surface area in each watershed (black) and as percentage of total terraced surface area on the island (grey).

Fig. 7.10 Hermigua ravine. *Photo* M. C. Iglesias López

7.4.2 Human Factors (Land Occupation and Protected Nature Areas)

Through an analysis of current land cover, it is possible to study the evolution of the terraced landscapes. Almost half of the terraced landscapes have been colonized by shrubby and herbaceous vegetation (48.6%), mainly by climax shrubs (*Neochamaeleo pulverulentae-Euphorbietum balsamiferae*, *Euphorbietum berthelotii-canariensis*, etc.) or natural secondary shrubs (Table 7.3). In the terraced areas that remain in cultivation today, the most notable crops are orchards (13.8%), wine grapes (8.1%) for local consumption and bananas for exportation (4.2%).

Furthermore, forests have colonized 12% of the terraced area. Evergreen forests (laurel forest and *Myrica faya-Erica arborea*) have colonized 8.3% of the terraced area,[7] thermophilous forests 2.7%, while reforestation with foreign species represents 0.1% (Fig. 7.11).

[7]No terrain surface data re-colonized by evergreen forest affected by the fire of 2012 were available for this study.

Table 7.3 Distribution of terraces by land cover

Group	Class	Area (ha)	%
Scrub and herbaceous vegetation	Coastal scrub	1622.4	19.7
	Secondary scrub	1390.8	16.9
	Pasture	983.7	11.9
	"Escobón" shrubland	4.9	0.1
Subtotal		4001.8	48.6
Agricultural	Herbaceous crops	1138.1	13.8
	Abandoned crops	701.4	8.5
	Vineyard	667.9	8.1
	Banana	349.8	4.2
	Greenhouse	16.7	0.2
	Tropical fruit trees	13.0	0.2
	Citrus fruit trees	3.4	0.0
	Ocean temperate fruit trees	2.0	0.0
Subtotal		2892.2	35.1
Forests and reforestation	Laurel forest	363.6	4.4
	Myrica faya-Erica arborea	321.2	3.9
	Thermophilous forests	226.2	2.7
	Ravine vegetation	49.0	0.6
	Pinus canariensis	13.7	0.2
	Pinus halepensis	7.6	0.1
	Pinus radiata	4.1	0.0
	Eucaliptus sp.	0.4	0.0
Subtotal		985.9	12.0
BARE SOIL	Extraction of materials, works	107.6	1.3
	Sparse vegetation	71.5	0.9
	Dumping site, graveyard	2.8	0.0
	Beach, dunes, and sandy area	1.3	0.0
Subtotal		183.2	2.2
Urbanizations and infrastructures	Dense housing developments	81.0	1.0
	Urbanization along roads	46.7	0.6
	Airports and infrastructures	13.8	0.2
	Industrial complexes	6.1	0.1
	Dispersed urbanization	5.3	0.1
	Sports, recreation, health, and education facilities	3.9	0.0
	Military installations	2.2	0.0
Subtotal		158.9	2.0
Water	Dam	8.9	0.1
	Pond	2.2	0.0
Subtotal		11.1	0.1
Total		8233.1	100.0

Fig. 7.11 Effects of the 2012 fire in Guadiana ravine. *Photo* J. C. Hernández Navarro

Terraces are present in 16 of the 17 Protected Nature Areas existing on the island, but only represent 8.7% (712.7 ha) of total terraced area. One of the goals of these Protected Nature Areas is the conservation of humanized landscapes. They include Valle Gran Rey Rural Park, Orone Protected Landscape (Fig. 7.12), and Barranco del Cabrito Natural Monument (Table 7.4).

Of the 25 terrestrial Special Areas of Conservation (SAC) existing in the island, 24 contain terraces, representing 20.4% (1683.3 ha) of total terraced surface area (Table 7.5). Most of the terraces are located in SAC Teselinde—Cabecera de Vallehermoso, Montaña del Cepo, Cuenca de Benchijigua—Guarimiar, Orone, Barranco del Charco Hondo, Valle Alto de Valle Gran Rey, Barranco del Cabrito, and Barranco del Cedro y Liria.

Fig. 7.12 Orone Protected Landscape. *Photo* J. C. Hernández Navarro

Table 7.4 Distribution of terraces in Protected Nature Areas

Protected Nature Areas	Area (ha)	Terraces of the island (%)
Valle de Gran Rey Rural Park	259.3	3.1
Orone Protected Landscape	163.9	2.0
Barranco del Cabrito Natural Monument	118.8	1.4
Majona Natural Park	44.7	0.5
Lomo del Carretón Natural Monument	42.2	0.5
La Fortaleza Natural Monument	16.6	0.2
Garajonay National Park	14.9	0.2
La Caldera Natural Monument	13.8	0.2
Roque Cano Natural Monument	11.7	0.1
Puntallana Special Nature Reserve	11.6	0.1
Charco del Conde Site of Scientific Interest	5.7	0.1
Roque Blanco Natural Monument	4.3	0.1
Acantilados de Alajeró Site of Scientific Interest	4.3	0.1
Benchijigua Integral Nature Reserve	0.6	0.0
Los Órganos Natural Monument	0.2	0.0
Charco de Cieno Site of Scientific Interest	0.0	0.0
Total	712.7	8.7

Table 7.5 Distribution of terraces in Special Areas of Conservation

Special Areas of Conservation (SAC)	Area (ha)	Terraces of the island (%)
Teselinde–Cabecera de Vallehermoso (ES7020097)	350.2	4.3
Montaña del Cepo (ES7020098)	200.2	2.4
Cuenca de Benchijigua–Guarimiar (ES7020107)	187.4	2.3
Orone (ES7020039)	164.0	2.0
Barranco del Charco Hondo (ES7020102)	160.2	1.9
Valle Alto de Valle Gran Rey (ES7020104)	131.4	1.6
Barranco del Cabrito (ES7020035)	118.8	1.4
Barranco del Cedro y Liria (ES7020109)	106.3	1.3
Majona (ES7020030)	44.7	0.5
Laderas de Enchereda (ES7020101)	43.3	0.5
Lomo del Carretón (ES7020037)	42.2	0.5
Taguluche (ES7020108)	37.7	0.5
Garajonay (ES0000044)	18.9	0.2
La Fortaleza (ES7020034)	16.6	0.2
Cabecera Barranco de Aguajilva (ES7020106)	12.5	0.2
Roque Cano (ES7020032)	11.7	0.1
Puntallana (ES7020029)	11.6	0.1

(continued)

Table 7.5 (continued)

Special Areas of Conservation (SAC)	Area (ha)	Terraces of the island (%)
Barranco del Águila (ES7020105)	10.4	0.1
Charco del Conde (ES7020041)	5.7	0.1
Barranco de Argaga (ES7020103)	4.3	0.1
Roque Blanco (ES7020033)	4.3	0.1
Benchijigua (ES7020028)	0.6	0.0
Los Organos (ES0000108)	0.2	0.0
Charco del Cieno (ES7020042)	0.0	0.0
Total	1683.3	20.4

7.5 Conclusions

The results obtained after identification and quantification of the terraced surface area of La Gomera Island are a clear reflection of what can easily be observed from the presence of these agricultural landscapes, namely the enormous human effort that has been carried out over five centuries, especially during the first half of the twentieth century.

The removal of stones and of tons of soil to level the land, the construction of stone walls of differing size, and the type of farming according to different environmental conditions have brought about a clear diversity in terrace landscape typology.

The wild character of the topography (hillsides with marked and deep ravines) makes access to many of these spaces difficult and increases the risk of erosion (mass movements) due to their instability.

Terrace loss affects up to 73.4% of the land of this island. The risk of erosion is very high and the danger posed by fires increases on a yearly basis. Both are processes that destroy landscapes which are part of the island's idiosyncrasy and uniqueness. With a view to ensuring their protection, further work is therefore needed to continue with the characterization, typology, and cataloguing of these terraced slopes, as well as their present state of conservation.

Acknowledgements The authors would like to acknowledge the help of José Mangas Viñuela (Chair in Geology, member of the "Geología Aplicada y Regional" research group (ULPGC)). Thanks are also due to Carolina Peña Alonso for her participation in the making of the cartography of the La Gomera basin slopes and, finally, to Juan Carlos Hernández Marrero, archeologist and head of the Archaeological Museum of La Gomera for sharing his knowledge, such as the history and heritage of the island, and for providing some of the photographic material.

References

Asins S (2009) El paisaje agrario aterrazado. Published by the University of Valencia, Diálogo entre el hombre y el medio en Petrer (Alicante)

Alomar G, Ferrer I, Grimalt M, Reynés A, Rodríguez R (2002) La pedra en sec. Materials, eines i tècniques tradicionals a les illes mediterrànies, vol 2. Officine Grafiche Riunite, Mallorca

Ancochea E, Hernán F, Huertas MJ, Brändle JL, Herrera R (2006) A new chronostratigraphical and evolutionary model for La Gomera: implications for the overall evolution of the Canarian Archipelago. J Volcanol Geoth Res 157:271–293

Carracedo JC (2008) Los volcanes de las Islas Canarias. IV La Palma, La Gomera y El Hierro. Editorial Rueda S.L

Colomar A (2002) Patrimoni de marjades a la mediterrania occidental. Una proposta de catalogacio. Publication in the framework of the PATTER project of the European programme RAPHAEL, DGX, European Commission. Palma de Mallorca. Published by the Consell de Mallorca

Del Arco MJ, Rodríguez-Delgado O, Acebes JR, García-Gallo A, Pérez De Paz PL, González Mancebo JM, González-González R, Garzón-Machado V (2009) Bioclimatology and climatophilous vegetation of Gomera (Canary Islands). Ann Botanici Fennici 46(3):161–191

Fontanari E, Patassini D (2008) Paesaggi terrazzati dell'arco alpino. Esperienze di progetto. Publication of the European project ALPTER part of the EU Programme Interreg IIIB- Alpine Space

Hernández M (2009) El paisaje como seña de identidad territorial: valorización social y factor de desarrollo, ¿utopía o realidad? Boletín de la Asociación de Geógrafos Españoles 49:169–183

Grimalt M, Blázquez M (1989) El mapa de marjades de la Serra de Tramuntana de Mallorca. Treballs de Geografía 42:43–47

Lasanta T (1989) Evolución reciente de la agricultura de montaña: El Pirineo aragonés. Geoforma Ediciones

Lasanta T, Arnáez J, Ortigosa L, Oserín M, Ruíz Flaño P (2009) Espacio agrícola y agricultura en Cameros a mediados del siglo XX. In: Lasanta T, Arnáez J (eds) Gestión, usos del suelo y paisaje en Cameros (Sistema Ibérico, La Rioja). Universidad de La Rioja e Instituto de Estudios Riojanos, pp 83–108

Llánes P, Herrera R, Gómez M, Muñoz A, Acosta J, Uchupi E, Smith D (2009) Geological evolution of the volcanic island La Gomera, Canary Islands, from analysis of its geomorphology. Mar Geol 264:123–139

Maiso E, Lasanta T (1990) El espacio agrario en el valle del Linares: características y utilización reciente. Berceo 118:53–62

Masson L (1994) Contribución al conocimiento de los andenes. Debate Agrario: Análisis y alternativas 19:1–27

Perera J (2015) La toponimia de La Gomera. Un estudio sobre los nombres del lugar, las voces indígenas y los nombres de las plantas, animales y hongos de La Gomera. Edición AIDER La Gomera

Reynés A (ed) (2007) Marjades i prevenció de riscs naturals. Publication in the framework of the European project TERRISC INTERREG III B SUDOE. Recuperació de paisatges de marjades i prevenció de riscs naturals. Consell de Mallorca, Palma de Mallorca

Rodríguez-Aizpeolea J (1992) Distribució espacial i evolució da l'agricultura a la vall d'Ebo. In: III Congrès d'Estudis de la Marina Alta. Institut de Cultura "Juan Gil-Albert"—IECMA, pp 545–553

Rodríguez W (2012) Agricultura y medio ambiente en Canarias. Mis reflexiones I (2001–2012). Fundación Canaria-Centro de Estudios Canarios Siglo XXI

Romero LE (2015) Consecuencias geomorfológicas del abandono agrícola en la cuenca del Guiniguada (Gran Canaria, Islas Canarias): Aplicación a la conservación del patrimonio de bancales. Unpublished doctoral thesis. Departamento de Geografía, Universidad de Las Palmas de Gran Canaria

Romero LE, González A, Ramón A (2016) Towards a new valuation of cultural terraced landscapes: the heritage of terraces in the Canary Islands (Spain). Ann Ser Hist Sociol 26(3): 499–512

Sabaté F (2012) Recuerdos del futuro: la experiencia vernácula y la sostenibilidad. In: Afonso J (ed) Agricultura en Canarias. Conciliando tradición y ciencia. Actas VII Semana Científica Telesforo Bravo. Puerto de la Cruz: Instituto de Estudios Hispánicos de Canarias, pp 105–143

Santana A, Villalba E (2008) Atlas Anroart de Canarias. Anroart Ediciones

Silva R (2009) Agricultura, paisaje y patrimonio territorial. Los paisajes de la agricultura vistos como patrimonio. Boletín de la A.G.E. 49:309–334

World Heritage Convention (1992) Operational guidelines for the implementation of the world heritage convention

Chapter 8
Terraced Landscapes in Perù: Terraces and Social Water Management

Lianet Camara and Mourik Bueno de Mesquita

Abstract Challenged by the area's imposing verticality, the societies inhabiting the Andes Mountains made constant effort to profitably exploit the slopes and develop the territory for supporting agriculture, which gave rise to characteristic terraced landscapes. Many studies on this terracing date back to the 1980s and 1990s, years when terrace use and recovery were given great importance. In 2010, institutions began renewing interest in the loss of productive land, current climate changes, and terrace abandonment, seeking to reduce adverse effects created on the terraced slopes, on rural economies, and on the population's socio-culture. This article intends to examine the geographical knowledge of Peruvian terracing, considering the Rural Agricultural Productive Development Program (AgroRural), located in Lima, Peru, and discussing the relevant results of AgroRural's inventory and characterization of terraces in areas with a high terracing density. The article will also highlight some issues, debated in recent years, which have raised the rural population's concerns over their territory and economy—specifically climate change and the abandonment and degradation of terracing. Water management, which is an integral part of the terraced structure, will be discussed as well. The second International Terraces Meeting, held in Cusco in 2014, animated the debate among scholars, institutions, and the rural population concerning problems affecting terraced areas and how to apply traditional knowledge to counteract them.

Lianet Camara is due to the drafting of Sects. 8.1, 8.2, 8.3 and Mourik Bueno de Mesquita of Sect. 8.4 and to the joint work of Sect. 8.5.

L. Camara (✉)
University of Verona, Verona, Italy
e-mail: camara.lianet@gmail.com

M. B. de Mesquita (✉)
Centro Bartolomé de Las Casas, Cusco, Peru
e-mail: mourik@casadelcorregidor.pe

M. Varotto et al. (eds.), *World Terraced Landscapes: History, Environment, Quality of Life*, Environmental History 9,
https://doi.org/10.1007/978-3-319-96815-5_8

8.1 Geographical Survey About Terraced Landscapes in Peru

Peru has vast terraced areas, and, since the 1980s, an attempt has been made to identify and develop an inventory of the terraced system. In 1982, Masson (1994) estimated one million hectares of terraces. That figure is based on the natural resource inventories and evaluations developed by ONERN (Oficina Nacional de Evaluación de Recursos Naturales; National Office for the Evaluation of Natural Resources) between 1968 and 1982 on the western side of the Andes. ONERN's goal was knowledge of the conservation status of terraced surfaces, but it should be emphasized that ONERN published neither the survey nor an official cartography of functional and abandoned terraces.

In 1996, INRENA (Instituto Nacional de Recursos Naturales; National Institute of Natural Resources) determined the existence of 256,945 ha of pre-Hispanic terraces in eight Peruvian regions. Until recently, no other national inventory had been conducted.

However, in 2010, AgroRural led the initiative. AgroRural (the Rural Agricultural Productive Development Program) is part of the Ministry of Agriculture, and it initiated a major project to inventory and characterize the agricultural terraces of 95 municipalities in 11 Peruvian regions. The municipalities were selected via several indicators: the presence of at least 50–200 terraced hectares, which justifies structural intervention for water access; the importance of the municipality itself and the role it plays within the province; the farmers' willingness to work on terraces; communication mechanisms between the municipality and markets; and the presence of municipal, rather than private, land. This latter indicator was considered because community work is more beneficial to the population than individual labor, which requires a larger workforce.

Operating along the Andes Mountains, the AgroRural project aims to gather a base of territorial knowledge, develop innovative technologies, and implement recovery interventions—promoting a large-scale territorial transformation. Fundamentally, the objective is coordinating social and institutional strategies to develop a recovery program for specific areas (micro-basins). These micro-basins can be the subjects of a basic pilot program—studying them, developing investment programs and public policies to recover abandoned terraces, and harnessing the terraces' potential to counteract the effects climate change has on agricultural production.

The AgroRural inventory has attempted to document areas with the highest concentration of terraces regionally. The program's objective is not to detect and describe all terraced areas, but, rather, to document maximum production capacity and the possibility of socio-economic investment returns.

As shown in Table 8.1, AgroRural's systematization and inventory of agricultural terraces has cataloged 340,719.11 terraced hectares, of which 259,319.04 are in use and 81,400.06 abandoned. AgroRural also estimates another 150,000 ha of

Table 8.1 Inventory of terraced areas in Peru

Region	State of terraces		Total (ha)
	In use (ha)	Abandoned (ha)	
Cusco	43,273.09	16,029.81	59,302.90
Lima	35,066.91	20,927.97	55,994.89
Ayacucho	36,655.33	9723.89	46,379.22
Apurimac	30,652.48	13,475.02	44,127.50
Arequipa	35,276.36	7680.05	42,956.41
Puno	20,705.62	2828.65	23,534.27
Huancavelica	17,634.44	4244.55	21,878.99
Tacna	11,174.67	2709.33	13,884.00
Moquegua	11,247.23	1739.77	12,987.01
Amazonas	11,121.77	539.68	11,661.44
Junin	6511.13	1501.34	8012.48
Total	259,319.04	81,400.06	340,719.11

Source Eguren and Marapi (2014)

terraces that were not included in the program. This report concludes that the regions with the most terraced hectares are Cusco, Lima, and Ayacucho.

Considering the municipalities involved in the program and fieldwork in different communities of four regions, the altimetry of densely terraced areas can be identified.[1] Most of the inter-Andean valleys are between 2500 and 4000 m above sea level. The terraced areas corresponding to the lowest altitudes (2300–3500 m) are the best preserved and used for agricultural production, as can be seen in Fig. 8.1. The terraces of Andamarca in the Ayacucho region are located at an altitude ranging from 3100 to 3400 m. Terraces higher than 3500 m are usually abandoned and, in a few cases, are covered by weeds and small shrubs.

The terraces at 2500–3500 m have a temperate climate, support a wide variety of crops, and have a vegetative cover of greater intensity than the higher terraces. In addition, irrigation facilities (canals, springs, lagoons, wetlands, etc.) are well-conserved, and rainfall is more regular, ranging from 500 to 1000 mm per year.[2]

The clivometric profile has very accentuated values, ranging between 15 and 75%. Among the terraces with slopes up to 50%, cultivable surfaces can have an area of up to 30 m, and wall heights vary from 1 to 4 m. Slopes above 70% show a decrease in the average surface of the embankments up to a maximum of 2.0 m, and wall heights vary from 2.5 to 10 m.

Terrace crops are characterized by a high degree of biological diversity and vary according to the altitude. Many species and varieties are cultivated: grains and

[1]Because AgroRural has not published its full study, the data analyzed here are the result of the author's field experience reconnaissance.

[2]SENAMHI (2009)

Fig. 8.1 Well-preserved terracing in Andamarca. *Photo* L. Camara

cereals, legumes, tubers, vegetables and spices, fruits, fruit trees, native trees, etc. (Tovar 1995; Tapia and Fries 2007). An important aspect of agronomic knowledge is crop management—optimizing sowing and harvesting, developing specific methods of conservation and soil fertilization, choosing cultivable varieties, and understanding the available technologies.

The mechanisms regulating soil fertility in the Andes are modulated by many variables—climatic, altimetric, pedological, topographic, crop intensity, etc. Farmers in the area practice crop rotation and cultivate different types of plants in the same terrace to encourage a symbiotic relationship between the crops—feeding, not just the harvest, but also the land. Such mechanisms imitate nature by diversifying plant species. In most agricultural systems, both practices combine to ensure biodiversity.

Most farmers grow 12–15 plots regularly and others on rotation. Crop distribution is extremely heterogenous. Some fields have only one crop, others more, sometimes in alternating rows, often cultivated in small plots of 20–25 m^2 with 30–40 different cultivars. Some may even have 100 different cultivars, thanks to a socially regulated exchange. This heterogeneity involves continuous experimentation, not only to avoid risks, but, above all, to create new varieties; this favors hybridization and crossbreeding between different crops. The local varieties, in fact, are most suited for the climate and the soil, and they express their best potential in the territory where they have acclimated over the centuries. For this reason, they are more resistant and require less external interventions. They are, therefore, more sustainable, both from an environmental and an economic point of view. In this sense, biodiversity is confirmed a unique and precious heritage: genetic but also cultural, social, and economic.

The following map (Fig. 8.2), elaborated with AgroRural's data, shows the regional distribution of terracing in the Lima region. Mostly, Lima has a high percentage of abandoned areas respect to inventory, most of them are also uncultivated lands or difficult to define. One of the causes is the emigration trends. For example, in 23 of 33 municipalities in the province of Yauyos the population does

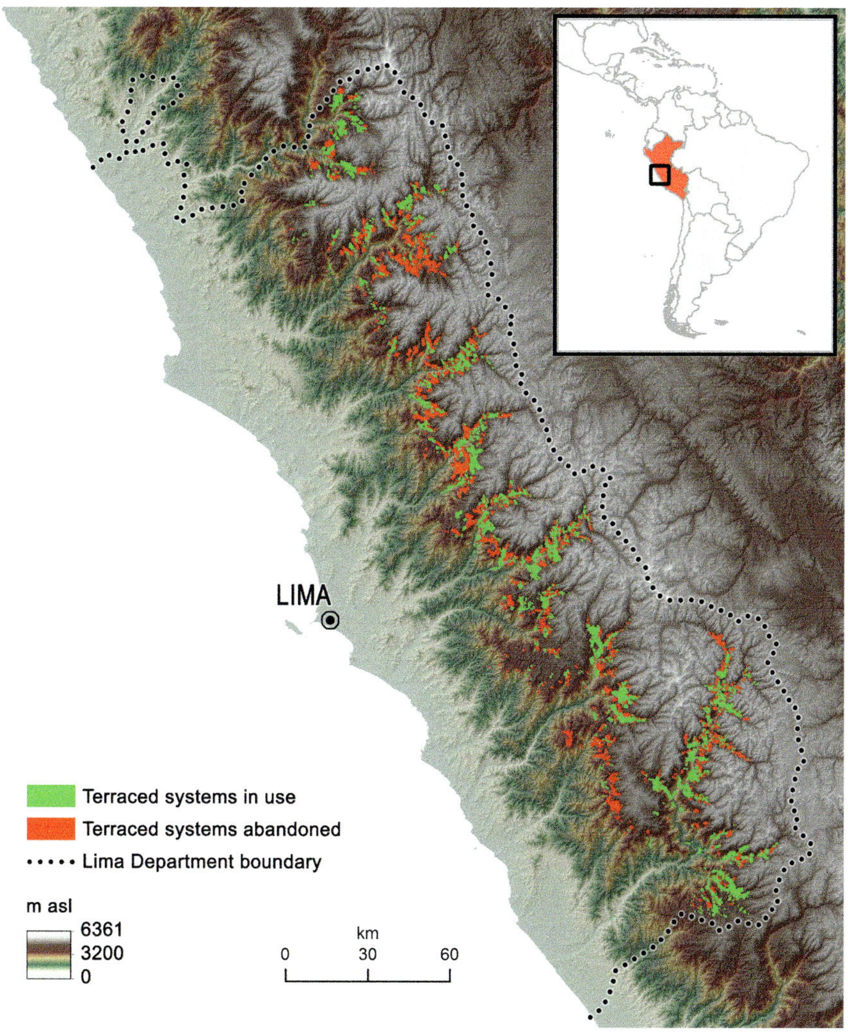

Fig. 8.2 Terraced areas in Lima region (elaborated by Francesco Ferrarese and Lianet Camara on data provided by AgroRural)

not exceed 1000 inhabitants, the population decrease puts the survival of the municipalities at risk. The expansion of a modern market on the coast with technically diversified crops and the mining sector are some of the factors of the demographic imbalance that bears the area. This growing uprooting of the rural population is causing a crisis in mountain agriculture and on vast terraced areas. The scarce agricultural production only satisfies the subsistence levels, the population therefore decides to emigrate and chooses the way of no return (Camara 2015).

8.2 Methodology

To identify and inventory terraced areas, AgroRural employed a methodology using satellite imagery to differentiate functioning from abandoned terraces via the following indicators: the vegetation color (green), the ground color (reflects the persistence of work), the quantity of shrubby vegetation on the walls, and the presence of irrigation infrastructures. Abandoned platforms evidence erosion problems, and the shrubs become straw-yellow in color. The soil is covered with permanent vegetation—dry, herbaceous plants and sometimes native trees. However, high-resolution images are not available in this satellite interface, and there is limited access to medium- and low-resolution images (Landsat TM and CBERS-2B images, with a resolution of 30 m on average). These conditions do not allow adequate visual interpretation to accurately identify the terracing in the images. However, this limitation was overcome through the generation of a "global mosaic" using only a panchromatic band of the high-resolution camera (HRC) sensor. This camera is one of three cameras on board the CBERS-2B satellite, producing images of a wide, 27-km band with a spatial resolution of 2.7 m. This material offers good resolution and very detailed images, excellent for observing and classifying surface terracing units. In this way, visual interpretation of the areas formerly seen in poor resolution has been completed. The images used belong to the INPE-BRAZIL (Instituto Nacional de Pesquisas Espaciais; National Institute of Space Research).[3]

Although satellite and cartographic information allows researchers to view and correlate variables that represent conservation, use, or abandonment, the results do not show if seemingly abandoned terraces have truly been abandoned or if they have simply been in a state of rest for a long time. Farmers leave the land to fallow according to different criteria, but it can be assumed that, if the land has been left uncultivated for 20 years, it is likely due to lack of workers. For this reason, it is difficult to establish precisely how many areas are truly abandoned.

Field work in communities of the Cusco, Apurimac, Ayacucho, and Lima regions in 2012, and the results of investigations by the non-governmental organization (NGO) Cusichaca (Kendall 2007) the Autoridad del Agua (Water Authority) (ANA 2010)

[3]The information referring to AgroRural's methodology was provided to the author during the research phase. However, this information was not published.

have observed that abandonment rates are much higher than the ratio developed by AgroRural. The AgroRural numbers substantially underestimate the real consistency of terraces due to the difficulty of recognizing those at rest and those covered with vegetation, a problem that could be partially solved by direct detection in the field (Camara 2015).

The possibilities offered by satellite and GIS analysis tools are powerful for linking different variables to understand and catalog the terraced heritage (full or partial conservation status, agricultural use, distribution of terracing, etc.). However, to assess the terraces' qualitative state, researchers must gain the collaboration and trust of farmers, who can define and identify structures in use and those in a critical situation of abandonment. The way this process relates to other geographical variables has not been sufficiently treated from a qualitative point of view. For cataloging, the agricultural production area's current conservation state (in use or abandoned) must be determined, and, on the basis of this evaluation, agronomic actions undertaken—inserting added value products into the market, calculating production, etc.

8.3 Threats to Terraced Landscapes in Peru

As mentioned in the previous paragraphs, Peru has a long history of terrace construction. The agriculture of the *sierra* is mountain agriculture, and it is important for the population living there. The agriculture practiced in this area is the absolute protagonist of food quality. It uses traditional techniques and production methods to sustainably exploit and integrate natural resources, such as land, water, pastures, and various types of forage crops. Animal products are also transformed with particular qualitative characteristics because they are suitable for agricultural productions of high nutritional and variegated quality. In these areas, agriculture is often the only opportunity for self-support and, to a lesser extent, for the development of incipient tourist-recreational and artisanal activities (Camara 2015).

Agriculture in the *sierra* has higher costs and lower profitability. Farming in the mountains means clashing with the limits of land use and shorter vegetative cycles, resulting in higher labor costs due to the slopes that require more complicated use of machinery and higher transport and logistics costs. This is also a consequence of splitting plots into small, agricultural units. The *sierra* economy is one of small plots, with 85% of farmers having less than 10 ha of land.[4] The wide dispersion of these plots limits production efficiency and increases production costs.

On the *sierra*, there is little entrepreneurship, and small family farms are always at high risk of divestment, especially since these structural "disadvantages" are added to pressures from the consequences human activities have on the natural environment, as mountain areas are particularly vulnerable to climate change and atmospheric phenomena. These territories are increasingly marginalized by

[4]INEI (2009)

socio-demographic changes, characterized by rural exodus events and the consequent reduction in agricultural employees. This accelerates the abandonment and degradation of terraced areas, at different levels of intensity.

The following paragraphs will focus on three areas of major importance to both farmers and public policies. Analyzing and identifying the critical aspects of terraced areas—including productive, economic, and social processes within the small communities investigated—reflects the realities happening in larger areas and at the national level.

(a) Climate change

Climate change is one of the most serious threats to sustainable agriculture in the Andes. Climate variables—such as temperature, precipitation, atmospheric pressure, solar radiation, and wind speed—structure and influence ecosystem development. In mountain climates, the increase in altitude is accompanied by a decrease in air temperature and humidity, as well as an increase in wind speed, turbulence, and solar radiation. These climatic factors—related to the altitude, vegetation cover, and topography—produce complex meteorological conditions that are expressed in a series of microclimates, some with only short distances between them. Climate variables are irregular in the Andes, even in years considered "normal". This has generated various climate scenarios with their own characteristics.

The Andes are particularly vulnerable to increasing temperatures and changes in rainfall seasons, all of which destroy habitats, affect plants' growth phases, and alter the times for sowing and harvesting—increasing the region's vulnerability and exposing agricultural production to risk. The trends and forecasts of the IPCC (Intergovernmental Panel on Climate Change) (2007) show that extreme climatic events are adversely affecting the flow and frequency of agricultural production and altering the geographical distribution of animal and plant species. The El Niño phenomenon also causes extreme climatic events on the inter-Andean valleys and in the highlands. The consequences impact the most vulnerable and economically weaker population groups, whose very survival is at risk because they live, above all, by agriculture (IICA 2016).

The increase in temperature and humidity has led to a greater spread of infectious wildlife diseases and has altered the geographical distribution of animal and plant species. Over the past 50 years, pastures have been able to stretch almost 300 m higher, and the potato crop has reached an altitude world record of over 4500 m. Farmers have expanded into higher altitudes in an attempt to tackle climate change, but the mountains' biodiversity and ecosystems are being compromised. Farmers are growing increasingly concerned as grazing areas are threatening to compete with high altitude farming systems (De Haan 2009; Camara 2015), and the areas for animal species adapted to the coldest climates are reducing. This displacement of production areas is altering the food cultivation patterns, the agricultural calendar (sowing, harvesting, etc.), and human settlements. Farmers in some areas also declare that climate events have become more unpredictable and severe. The rains occur in short cycles, start later than usual, and are more dense but short-lived, which does not allow the soil to adequately absorb moisture but instead erodes the surface. Climate change, therefore, risks making terrace agriculture impossible in the inter-Andean valleys, and even more marginal plateaus, as has already happened in several communities. These impacts to the Andes Mountain ecosystems are very worrying because of the consequences they have on people who depend on mountain resources. Josse et al. (2009 in Herzog et al. 2012) have estimated that about 40 million people find support in the Andean ecosystems.

Through a civilization process encompassing thousands of years, the Andean peoples have developed subsistence strategies to circumvent risks (Earls 2008)—converting the limits of a harsh landscape, with high variability and climatic instability, into opportunities. This process has included domesticating hundreds of species adapted to the diverse climate, terrain, and altitude. For rural Andean people, the morphological limits and the dispersion of risk are not new concepts. Most of the *sierra* farmers have always relied on these strategies, sowing "here and there, up there and down" a wide range of genetic products and different varieties of the same crop (Camara 2015). This dispersion of agricultural land and crops in various ecological zones and with different maturation times is a strategy developed to minimize losses caused by climatic phenomena. Irregularities in precipitation, drought, hailstorm intensity, or frost do not occur in a uniform manner or affect all soil in the same way, but rather to varying degrees and in dispersed places. Therefore, some plants resist more and are damaged less.

Diversification is maintained and enriched by farmers not resorting to a fixed plan but instead relying on a program organized locally to dissipate one area's specific problems. This logic is based on meeting local needs and addressing climate uncertainties as well as labor, capital, land, and other economic factors. This ecological-economic reasoning is evident in the way localities use and manage lands based on social and cultural networks of reciprocity and responsibility within the home and the community as a whole.

(b) The abandonment of the terraces

The spread of terracing throughout the Andean area of Peru is very substantial, but terraces have been considered marginal for intensive agronomic activity in various historical periods and were consequently exposed to degradation and abandonment processes.

Based on observations following a survey of the territory[5], it seems possible to trace the complex degradation phenomenon to two large categories of abandonment causes: natural and anthropic. The anthropic factors, which today spur terrace conservation, have recently changed sides, as they can be traced to the European raid and the ensuing socio-economic, political, and cultural transformation that hit the mountain areas and imposed upon them a system of resource exploitation. Additionally, archeological (Kendall 2005), anthropological (Treacy 1994), and geographic studies (Donkin 1979; Denevan 1988) identify abandonment of agricultural terraces with

[5]Field work was carried out in 2012 in various rural communities: the Chicha-Soras valleys, Andamarca, Laraos, Zurite, and Ollantaytambo.

changes taking place in rural societies and organization systems. Such studies have determined the decrease in available labor forces, lower crop profitability in the absence of adequate transformation companies, and (in answer to these problems) community strategies for valorizing their local products.

Agronomic studies show abandonment in response to modern technical procedures and alterations in traditional agricultural practices (Masson 1986; Tapia 1996). Mechanizing cultivation operations in mountain areas is difficult and strongly impacts the manpower able (and willing) to work. Land ownership is also gradually reducing and fragmenting, making it difficult to determine economies of scale for introducing mechanized systems, and complex social and economic cultural elements hinder organizing cooperation initiatives (Camara 2015).

Economic literature has extensively discussed the competitive advantages associated with reducing costs of agricultural activities in terraced areas. Gonzáles de Olarte and Trivelli (1999), who carried out economic research in three Peruvian regions, report abandoned terraces in areas where the level of development is inadequate. These areas are mostly characterized by high levels of poverty, lack of productive infrastructure, absence of public and private investments, and serious feeding and nutrition problems.

For Gonzáles de Olarte and Trivelli (1999), terraces do not produce positive economic externality because their very expensive construction and reconstruction is not a strategic way to stimulate rural development. The recovery of terraces is an unattractive investment unless traditional crops are exchanged for others the market demands. For example, compared to potatoes and beans, flowers would have greater export elasticity (Gonzáles de Olarte and Trivelli 1999).

Economically, abandoning terraces would result in the loss of positive effects in food production, biodiversity, the control and balance of slope stability, and hydrogeological risks. In a market economy, recovering pre-Hispanic agricultural terraces would increase the physical and natural capital of agricultural land and water (irrigation systems or rainwater exploitation), which would retain water, reduce erosion, and improve landscapes and microclimates. These changes, taken together, should improve farmers' productivity, augment their income and, consequently, increase their well-being. It would seem a promising formula, but, according to economic studies, its feasibility depends strongly on the national and regional economic context, the institutional framework of intervention, and the level of local involvement.

The current economic climate is dominated by growth and by a series of reform packages aimed at fiscal management, liberalizing trade and the market, and exporting so-called non-traditional products.[6] Because terrace recovery is considered costly, this scenario is attempting to redefine the recovery rules.

[6]From a technical point of view for international trade, a non-traditional product has an added value in its production process or is sufficiently important to transform its natural essence. In 1992 La Comisión Económica para América Latina y el Caribe (CEPAL) applied criteria and limitations to consider non-traditional products, these should not belong to the energy group, their export may be a new product, the product may appear in the market outside the traditional season, and the export of said product must have some importance for some sector of the country's economy.

Fig. 8.3 Abandoned terraces in the Chicha-Soras valley in Apurimac. *Photo* L. Camara

The growing marginalization of these territories is also the result of socio-demographic changes, characterized by rural exodus events and the consequent reduction in agricultural workers. Rural social instability, which, in the 1980s, was characterized by terrorist violence, has particularly hit the rural areas of the Central Andes, especially the Apurimac and Ayacucho regions.

The presence of livestock in the terraced fields observable in Fig. 8.3 must be considered, not only as a simple breeding activity, but also as a way to keep the terraces free from unwanted sylvan vegetation. However, in many cases, the overgrazing of horses and cattle has destroyed vast terraced areas. This phenomenon is observable, not only in the elevated areas of the valleys, but also in the lower areas where the terrace systems can be the object of potential recovery.

Another pressure terraced slopes and agriculture face is both direct and indirect competition for land and water. Many production areas and pastures have been used for mining concessions, whose water demands are growing rapidly, and the pollution from mines in the soil and water sources is devastating (Del Pozo and Paucarmayta 2015).

The most important non-traditional products are asparagus (fresh and canned), grapes, avocados, artichokes, mango, etc.

Terrace abandonment is a clear manifestation of the transformation of traditional production methods. The progressive change in Andean agriculture is leading to social, economic, and cultural consequences and to the loss of local resources, knowledge, and agricultural landscapes. This has made evident the complex relationships that allow the whole system to function effectively.

Nevertheless, terracing has aroused great interest and calls for recovery and restoration. Two distinctly different orientations predominate: the recovery of large-scale terraces by the Peruvian State has focused on areas of monumental interest, such as the terraces around Chinchero, Pisac, Ollantaytambo, and Machu Picchu—recoveries executed by the Instituto Nacional de Cultura Cusco (National Institute of Culture, Cusco), today the "Regional Directorate of the Ministry of Culture". The second orientation has to do with the rural development programs attempting to recover smaller terraced areas, for example DESCO SUR (Centro de Estudios y Promoción del Desarrollo Zona Sur), PRONAMACHCS (Programa Nacional de Manejo de Cuencas Hidrográficas y Conservación de Suelos), MARENASS (Proyecto de Manejo de los Recursos Naturales en la Sierra Sur); Cusichaca and others NGOs and associations recovered terraces in Arequipa, Ayacucho, Apurimac and Lima regions.

In general, this terrace recovery was almost always conditioned by an external intervention and several actors: NGOs, the State, research institutes, international cooperation agencies, private companies, development agencies, and through rural development programs. These development programs were intended to strengthen human, natural, material, financial, cultural, and social resources; combat rural poverty; improve living conditions; etc. They are implemented in the following thematic lines: environmental management and conservation of natural resources, sustainable agricultural production, orientation to agri-food for the market, management of rural tourism, rural education and technical productive training, food security, and support for strengthening institutions and local participation.

(c) Social management of water

Water management is more of a strength than a threat for the Andes peoples, who have a long tradition of managing water resources. Agriculture would not be possible without water, and various cultural, material, and religious factors were woven around these elements, consolidating a structure that would explain the importance of water in the life of Andean societies.

Terraced farming implies a great responsibility for water conservation in quantitative and qualitative terms. The careful management of water resources and the efficient use of rainwater for crops and pastures are criteria for good agricultural practices. Traditional Andean technologies are efficient in irrigation management, minimizing waste, and avoiding excessive leaching and salinization of the soil.

The hydrological cycle is constantly managed—from capturing to conserving rainwater, from infiltration processes to protecting places from which infiltrated water emerges, from domestic use to field irrigation, from ritual conception to social organization. This control is complex and integral. Each of these systems is unique and responds to the particular characteristics for which it was designed, but all the systems have a common root in the need to face lacking or excessive water.

Managing the natural spaces for collecting and concentrating water resources has proved to be effective for monitoring ecosystems and production activities. This approach considers the basin as the fundamental unit of territorial management and lays the foundations for a multi-sectoral production management and service provider.[7]

Andean companies' need to bring water from the source to areas of use has led to technological developments combining agronomic, water, and construction knowledge adapted to the territory's biological and topographical conditions—from the plateaus to the terraced slopes to the valley floor—as a single integrated system. The great capacity for social organization induced vast quantities of people to build and maintain large irrigation works in the inter-Andean valleys.

Currently, there is concern about the decrease in water in the valleys, which has led communities to find alternatives to rain-farming for agricultural production. Farmers say that, in the highlands, there are lagoons and springs, whose waters "do not seem to disappear." They wonder if their ancestors faced problems related to water availability and conservation. The existence of these lagoons and springs has pushed some NGOs and government programs to re-use the technologies already present in the territory and encourage farmers to recover them. Some of these sources are as follows.

- The *bofedales*, *ojonales* or "wetlands" or "*peatas*" are the most important grassland ecosystems in the arid and semi-arid Peruvian highlands above 3800 m. They allow infiltration and water retention in the water table and maintain dense vegetation in the surrounding areas.
- The *qochas* are lakes or ponds artificially excavated or built in a natural depression. They are an ancient agricultural technique, based on the use of a dense network of channels connected to each other by a lagoon. They provide a water and soil management system used alternately for crops and grazing.
- The lagoons, tanks, or water tanks exploit a natural depression of the ground or natural ponds, or they are artificially built with the purpose of capturing and storing rainwater, allowing, through a slow infiltration, the recharge of aquifers and the natural irrigation of prairies.

  The *amunas*[8] constitute a series of pre-Hispanic strategies for water, soil, and crop management based on the climatic-geographical challenge of the Andes. This technology works like a sponge, allowing rainwater to infiltrate inside the

[7]The river basin approach was used by the Ministry of Agriculture in 1974, but the program's actions began in 1980 with the creation of the PRONAMACHS program (Programa Nacional de Manejo de Cuencas Hidrográficas y Conservación de Suelos) as a response to dialectic territorial planning policies, plans, and strategies for development.

[8]An initiative of the GSAAC (Gestión Social del Agua y el Ambiente en Cuencas) was to find mechanisms to recover and exploit ancient ancestral water use practices, commissioning a study about how the *amunas* system functions, with the objective to promote alternative actions to increase the water supply for consumption and for production activities in Andean communities.

cracked volcanic rocks that are found in the highland areas or in the mountain slopes (Alencastre et al. 2006). It is an artificial water supply, a recharge system for aquifers, or a way to introduce water into the subsoil, retaining and discharging it naturally. The *amunas* or *millpu* are considered "water sowing" by the rural communities, and they are carried out through community action.

- The *ojos de agua* or *pukios* (springs), streams, small streams, lagoons, and lakes are sources that the Andean communities have used since ancient times to irrigate their terraced, flat, and hilly fields. The springs are the natural sources where underground waters emerge from areas with fractures or faults. This water is received and, in some cases, is directed to tanks specially built to conserve and protect captured water and facilitate its distribution downstream through the so-called *acequias amunadoras* or channels.
- The *acequias* are structures that allow the transfer and distribution of water from the source of collection (rivers, streams, wetlands, etc.) to tanks or lagoons. These types of channels are not covered with any impermeable material, but are rather cuts on the ground that allow moistening of the soil and maintain a vegetative cover through the infiltration of water. These channels cross entire slopes.

Andean companies are experienced in traditional organization for water management and irrigation, which has followed similar schemes for hundreds of years. Irrigation control is often divided into a set of community institutions.[9] Around traditional irrigation are interwoven various expressions of religious and cultural beliefs that are the basis of the whole social structure and explain water's importance in the lives of farmers.

Camara (2015) found that, in almost all communities studied, water distribution is managed by a *Junta de Regantes* ("irrigation committee"), whose function is to distribute water to all members of the community based on membership, quantity, areas, land preparation, and other criteria determined in assembly with direct collective deliberations (Tillmann and Bueno de Mesquita 2015).

Irrigation systems in rural communities are systems of "collective action". A family alone, or individuals, could neither (re)construct nor maintain an infrastructure necessary for the management, defense, and protection of complex irrigation structures. Within these irrigation systems, families and community members maintain the agricultural production system and participate in making collective decisions. Gerbrandy and Hoogendam (1998) believe that it is not appropriate to talk about collective rights in the internal management of an irrigation system, since the usufruct is at the individual level. Individual usufruct is not an absolute right. Instead, it starts from a context of collective decisions about the

---

[9]The studies of Guillet (1992), in the community of Lari, and of Valderrama and Escalante (1988), in the community of Yanque, both in the Colca valley, describe a complex and articulated agricultural calendar system based on a broad range of crops, with water distributed to individual fields. In both Lari and Yanque, a sort of micro-irrigation planning seems to operate under the control of the community authority.

irrigation system's management. This means that every member must use water in an appropriate and responsible way. In this context, the rights in water management belong to individuals or families within collective decisions. Customary law is dynamic and brings an adequate level of integral river basin management.

In irrigation systems, water plays an important socio-cultural role. It is an essential element linked to family and social cohesion, expressed in many rituals and festivities carried out around the water. This socio-cultural function is directly related to resource productivity.

In the cosmogony of the Andean peoples, the spiritual dimension is linked to all human activities, and water is one of the most important elements. Water is a being who belongs to no one but belongs to everyone, and its "being" is an expression of flexibility and adaptability. The various sources of water (the sea, the lakes, the rivers, the lagoons, etc.) are sacred spaces and places of creation and fertility. "Water belongs to the land that bathes" is a principle of the Andean vision. Gushing from the sources, the water irrigates the distinct ecosystems it crosses, integrating them into the same spatial unity. The religious and organizational characteristics of Andean societies, based on collective work for agricultural activities and irrigation, have formed a "water culture," which still exists today.

For these reasons, re-construction, maintenance, cleaning the canals, and most of the rural activities are celebrated through community festivals, at the beginning and end of each part of the agricultural cycle (sowing, harvesting, plowing, etc.).

The cleaning jobs are aimed at removing debris from canal beds and cutting weeds and dried leaves that have fallen into the canals and reservoirs. This work is done in a festive context, and, in many communities studied, the "water festivals" are fertile propitiatory ceremonies linked to the agricultural cycle in which the traditions guaranteeing water distribution are maintained and ratify the water rights and irrigation obligations. The festivities, rites, and legends reaffirm the bonds of friendship and exchange between different communities, making it possible to settle the reciprocity system among the members.

8.4 The Terrace Landscape Debate

The second International Conference on Terraced Landscapes, held in Peru in May 2014, was an opportunity to convey initiatives and projects related to terraced landscapes, which were no longer receiving the attention from institutions and NGOs that they had received in the 1980s and 1990s. Some studies in terraced valleys had continued working to recover and increase productivity. In the conference, farmers, researchers, professionals, government officials, and students from all over the world gathered to carry on an intercultural dialogue about the current state of terracing, investigating how to use and keep the terraces, production linked to agrobiodiversity, and the culture of the Andean peoples.

A space was created for the participation of farmers from different communities, who described their own valleys (see Fig. 8.4), exposing their problems concerning

Fig. 8.4 Discussion forum between farmers and technicians about terraces landscapes. *Photo* C. Ramos, promoter of farmer forum

the terraced system, discussing biodiversity, exchanging information, and seeking practical solutions. The work has analyzed the main critical issues for the future of agricultural terraces, including the following (which have been discussed in detail above).

(a) The governance of water, of the earth, and of the terraced landscapes and their rehabilitation. Territory, territorial planning, and river basin management.
(b) Policies and sustainable agriculture for small-scale economies in terraced areas with long-term public benefits.
(c) Intercultural dialogue, knowledge, and information on agricultural development on terraces, terrace rehabilitation, water harvesting, soil conservation, and adaptation to climate change.
(d) Viewing the terraced landscape and its environmental and cultural wealth as a viable alternative for better conditions and livelihoods for rural populations. Food quality, heritage, and the ability to choose community futures based in different worldviews.

Motivated by this second conference and several preliminary meetings related to Peruvian terrace farmers and the guidance of women, the "*Ayllu Pata Pata* platform" was organized. Its vision of current bottlenecks was discussed in 2016 and presented in a series of proposals and future policies for the terraced landscape. Farmers have also identified critical points in their perceptions:

i. Lack of formal and informal education on terraced landscapes and their different characteristics and benefits for communities.
ii. Supermarkets substituting local markets, the latter of which have a greater diversity of crops.
iii. Weakening community organization due to migratory processes and individualization.
iv. Limited attention and support from national, regional, and local governments, which give priority to the neoliberal economic models of mining that have seriously damaged territorial and social structures.
v. Inadequate water distribution and health risks caused by lack of drinking water and the socio-economic conditions of families already facing extreme climatic conditions.
vi. The need to recognize the "living heritage" of terraced landscapes.

After analyzing these critical issues, some thematic questions emerge. Is there a debate, concerning terraced landscapes, that involves Peru? What problems underlie the situations and phenomena mentioned above? Who is interested in the future of terraced agriculture?

8.5 Conclusions

Terracing systems are widely used by rural communities in the inter-Andean valleys and present themselves in slightly different ways throughout the territory. These systems are the main agents modeling the Peruvian territory, a landscape of extraordinary beauty which is often exposed to degradation and abandonment due to structural, social, and economic consequences.

The results provided by AgroRural are still valid qualitatively and quantitatively. However, the study's poor dissemination does not allow scholars, or the famers who collaborated with the program, access to important information.

The speed and breadth of the environmental consequences generated by both climate change, terrace abandonment, or the loss of water collection and distribution, have medium- and long-term impacts on the production system's sustainability on the terraced slopes.

To restore these areas, common strategies are needed between different actors (communities, public and private bodies, etc.) who desire not only preserving terraced structures but also making these places favorable in terms of production and services to improve the quality of life of the millions of people who live there. It is important to disseminate and implement strategies and policies to enhance these landscapes, through awareness, redevelopment and improvement, and dissemination of their culture and the heritage to which communities, and the entire nation, are heirs.

References

Alencastre A, Apaza DI, Arroyo R (2006) La Amunas de Huarochiri: Recarga de Acuíferos en los Andes. GSAAC, IICA, Lima

ANA (2010) Inventario de la Cordillera Blanca. ANA, Lima

Camara L (2015) Il recupero del "saper fare" andino per la produzione agricola. Doctoral thesis, University of Verona

De Haan S (2009) Potato diversity at height: multiple dimensions of farmer-driven in-situ conservation in the Andes. PhD thesis, Wageningen University, Netherlands

Del Pozo C, Paucarmayta V (2015) Cómo impacta la minería en la producción agropecuaria del Perú. In: Economia y Sociedad N° 87 Mineria y Desarrollo. CIES

Denevan W (1988) Measurement of abandoned terracing from air photos: Colca valley, Peru. In: Yearbook. Conference of Latin Americanist geographers, vol 14. University of Texas Press, pp 20–30

Donkin RA (1979) Agricultural terracing in the aboriginal new world. Wenner-Gren Foundation for Anthropological Research, University of Arizona Press, Arizona

Earls J (2008) El Conocimiento Andino es Clave para Enfrentar el Cambio Climático. In: La Revista Agraria N°92, CEPES, Lima

Eguren F, Marapi R (2014) Mesa Redonda: Perú, país de andenes. Rescatando el pasado para enfrentar desafíos del futuro. In: La Revista Agraria N°160. CEPES, Lima

Gerbrandy G, Hoogendam P (1998) Agua y Acequias: Los Derechos al Agua y la Gestión Campesina de Riego en los Andes Bolivianos. Ed. Plural, PEIRAV, Cochabamba

Gonzáles de Olarte E, Trivelli C (1999) Andenes y desarrollo sustentable. IEP-Consorcio para el Desarrollo sostenible de la Eco región Andina Condesan, Lima

Guillet D (1992) Covering ground: communal water management and the state in the Peruvian highlands. The University of Michigan Press, Ann Arbor

Herzog S, Martínez R, Jørgensen PM, Tiessen H (2012) Cambio Climático y Biodiversidad en los Andes Tropicales. IAI, SCOPE, MacArthur Foundation, São Paolo-Paris

IICA (2016) El fenómeno de "El Niño" en la agricultura de las Américas. Boletín Técnico. IICA

INEI (2009) Perú: Perfil del Productor Agropecuario 2008. Lima

INRENA (1996) Informe Técnico del Estudio de Inventario y Evaluación de Andenes, Ministerio de Agricultura. INRENA, Lima

IPCC (2007) Final Report: Cimate Change, UNEP, Geneve

Josse C, Cuesta F, Navarro G, Barrena V, Cabrera E, Chacón-Moreno E, Ferreira W, Peralvo J, Saito J, Tovar A (2009) Ecosistemas de los Andes del norte y centro. Bolivia, Colombia, Ecuador, Perú y Venezuela. Secretaría General de la Comunidad Andina, Programa Regional ECOBONA-Intercooperation, CONDESAN, Programa BioAndes, EcoCiencia, NatureServe, LTAUNALM, IavH, ICAE-ULA, CDC-UNALM, and RUMBOL SRL, Lima

Kendall A (2005) Applied archaeology: revitalizing indigenous agricultural technology within an Andean community. In: Sillar B, Fforde C (eds) Conservation, identity and ownership in indigenous archaeology. Public Archaeology. James & James, London

Kendall A (2007) Final report. Cusichaca Trust

Masson L (1986) Rehabilitación de Andenes en la Comunidad de San Pedro de Casta, Lima. In: De la Torre C, Burga M (eds) Andenes y Camellones en el Perú Andino: Historia, Presente y Futuro. CONCYTEC, Lima, pp 207–216

Masson L (1994) Contribución al conocimiento de los andenes. In: El Debate Agrario: Análisis y alternativas. CEPES, Lima

SENAMHI (2009) Escenarios de Cambio Climático en la Cuenca de los Ríos Mantaroy Urubamba para el Año 2100. Proyecto Regional Andino de Adaptación, PRAA, Lima

Tapia M (1996) Ecodesarrollo en los Andes Altos. Fundación Friedrich Ebert, Lima

Tapia M, Fries AM (2007) Guía de Campo de los Cultivos Andinos. FAO, Lima

Tillmann T, Bueno de Mesquita M (eds) (2015) II Congreso Internacional de Terrazas: Encuentro de culturas y saberes de terrazas del mundo. JICA, CBC, CONDESAN, Cusco

Tovar O (1995) El Perú como Centro de Domesticación de Plantas: Centro de Datos para la Conservación. Universidad Nacional Agraria La Molina, Lima

Treacy J (1994) Las Chacras de Coporaque: Andenería y Riego en el Valle del Colca. IEP, Lima, pp 213–226

Valderrama R, Escalante C (1988) Del Tata Mallku a la Mama Pacha: riego, sociedad y ritos en los Andes peruanos. DESCO, Lima

Chapter 9
Australian Dry Stone Terraces: An Historical and Contemporary Interpretation

Raelene Marshall

Abstract The topic of this paper will attempt to describe Australian terraced landscapes styles and their historical evolution through the prism of the range of practical, survival, cultural, social and aesthetic genres constructed from the 1800s early settlement era through to present-day contemporary designs. The story stretches from the Gold Rush times, through a creative surge after World War II, to the contemporary terraces built in the last decade or so at the Mount Annan Botanical Gardens in New South Wales. Of particular interest here is the historical background that in the 1850s and 60s Gold Rush period Swiss Italians from Ticino and Swiss immigrants from the southern part of Graubünden settled in the area around Daylesford, Victoria. Even today, their influence is ever present in the township of Hepburn Springs through the names of its residents, the names of its Mineral Springs (Locarno) and its buildings.

9.1 Introduction

It could be claimed that dry stone structures and terraces across the world are a creative endeavour made in nature, of nature. Albeit in the early days constructed as a practical agrarian craft, rather than for more esoteric reasons their makers chose the landscape as the setting to display their craftsmanship. Regardless of their end purpose, made by desire, design or default the landscape had become their museum or gallery.

Long before we put public art on street corners, Australia's indigenous and immigrant artisans were influencing and creating aesthetic and practical dry stone structures that would change the face of the landscape for centuries to come. Although the years have seen many changes, today's modern-day walls, terraces and sculptures made by equally dedicated craftspeople and artisans add to that genre and carry on the essence of this ancient craft in new and innovative ways.

R. Marshall (✉)
Dry Stone Walls Association of Australia, Melbourne, Australia
e-mail: raelenemar@optusnet.com.au

M. Varotto et al. (eds.), *World Terraced Landscapes: History, Environment, Quality of Life*, Environmental History 9,
https://doi.org/10.1007/978-3-319-96815-5_9

Thousands of years prior to the arrival of British and European immigrants to Australia in the 1800s, the Gunditjmara indigenous peoples of south west Victoria (Vic)[1] and the Ngemba from Brewarrina in New South Wales (NSW) in particular were well known for their sophisticated construction of dry stone fish traps and weirs. Designed to catch fish, eels and other aquatic animals, even today theirs and other traps elsewhere in the country are still visible along rivers and coasts.[2]

With European settlement in the early 1800s, came the need to survive in harsh and unfamiliar terrains. Distances were great, and people were forced to make do with what was at hand. In some places, the land was barren and infertile; other areas were flat and stony, and in the mountainous and undulating areas around the Victorian goldfields of Daylesford and Walhalla Italians, Swiss Italians and some British needed to tame slopes and or create dedicated flat land to build housing and grow crops.

However, in other more recent developmental periods of Australia's landscape history, the beauty and functionality of terracing was more creative and designed to enrich the artistic, visual and cultural amenity of domestic, public and private gardens.

9.2 Australia

The history of dry stone terracing in Australia has had quite a different trajectory to that of its European counterparts. Unlike the need in much of Europe to establish flat areas of land for survival needs in mountainous areas, in Australia's very early settlement days there was already an abundance of flat land available for cultivation. Australia is the lowest, flattest, and oldest continental landmass on Earth and has had a relatively stable geological history. To that end, squatter immigrants who arrived in the 1800s tended to settle on land that was relatively easy to work unless there was a pressing need to construct built-for-purpose terraces such as to establish vineyards or flat land for housing and crops in places such as the Victorian goldfields.

For about 50,000 years, before the first British and European settlement in the late eighteenth century, Australia was inhabited by Indigenous cultures who spoke languages classifiable into roughly 250 groups and are the oldest living cultural history in the world. After the European discovery of the continent by Dutch explorers in 1606, Australia's eastern half was claimed by Great Britain in 1770 and initially settled through penal transportation to the colony of New South Wales

[1]"National Heritage Places—Budj Bim National Heritage Landscape", *Australian Government Department of Environment and Energy*, https://www.environment.gov.au/heritage/places/national/budj-bim (accessed February 16, 2017).

[2]"National Heritage Places—Brewarrina Aboriginal Fish Traps (Baiame's Ngunnhu)", *Australian Government Department of Environment and Energy*, http://www.environment.gov.au/heritage/places/national/brewarrina (accessed February 16, 2017).

from January 26, 1788. The population grew steadily in subsequent decades, and by the 1850s, most of the continent had been explored and an additional five self-governing crown colonies had been established.[3]

Even today, due in some part to a scarcity water, much of the land mass is still uncultivated so although there are relatively few terraced landscapes, for those that do exist their historical and more recent content and context and is both complex and fascinating.

The craftsmanship of many freestanding boundary walls built in early settlement days can comfortably equal the style and quality of our British and European counterparts. However, given the tyranny of distance, the scarcity of amenities, the raw uncultivated landscapes and the extremely small populations it is probably unrealistic and almost unfair to try to compare the far less prolific and more rudimentary terraces with the beautifully crafted terraced walls of our British and European counterparts.

9.3 Survival Creativity and Dry Stones

In the mid to late 1800s, the need for work and the deprivations caused by war, poverty, crop failures and natural disasters motivated British and European migrants to leave their homeland to search for new survival opportunities in Australia. Some came in search of a new life or adventure, others in search of gold or land on which to grow crops or graze livestock and some brought with them traditional walling skills from their native villages and farming regions.

On arrival, they tended to gather in communities of their own nationalities where many used their homeland skills to recreate 'homes' in the challenging rural and bush landscapes and remote mountainous mining towns. In those early days, the need for shelter was ever present. Their story was about survival, existence and creating a sense of place for themselves, their family and their future.

In areas of dry stone constructions where the horizons are vast, the land is stony, the eucalypts are majestic, the bush is unrelenting and native grasses grow nearly as tall as men, those early walls and terraces blend so seamlessly with the landscape it is hard to believe that they once weren't there.

It is probably fair to say that in those early days the thought of dry stone craftsmanship as an art form would have been be the farthest thing from the minds of the Italians and Swiss Italians and others who settled and built terraces on the gold fields in Victoria. Bearing in mind that the early emerging communities often consisted of as few as one thousand or so people, they would have needed to be able turn their hands to a wide range of infrastructural and creative skills. Needless to say, their expertise would have been wide and varied.

[3]"The First Fleet", *Project Gutenberg Australia*, http://gutenberg.net.au/first-fleet.html (accessed February 12, 2017).

As time has passed, the imprint and influence they made in changing the face of our agrarian and cultural landscape should never be underestimated. As far as dry stone terracing is concerned and in terms of immigrant survival, probably the two locations, both in Victoria were most important; namely the Daylesford and Yandoit areas in the central goldfields and the areas surrounding the Long Tunnel goldmine around Walhalla in Gippsland.

9.4 Early 1800s Settlement

9.4.1 Hepburn Central Goldfields Victoria

Located within a wider area known as Victoria's Central Highlands is the other area known for its early settlement gold rush days and with that vernacular stone housing and dry stone terracing. The Hepburn Shire landscape is a diverse topography of forests, mountains and plains derived from volcanic and marine sedimentary rocks. The area's character could be described as a rolling but irregular topography crossed by ridge-like rises. Therefore, the terraces here are far less prolific than those around Walhalla and were built for the most part to tame areas of undulating terrain rather than create flat land on steep mountainsides.

In 1851, gold was discovered at Jim Crow (Daylesford district today) and the area soon became known for its extensive alluvial deposits. The varied geology provided the area with a rich collection of goldfields and mineral springs. It also provided two prime sources of construction stone for the early settlers to use, bluestone and freestone (Marshall 2000). The Jim Crow goldfields were also known for their mix of cultures. Among these were Chinese, a significant Anglo Saxon community and large communities of Northern Italians who came from areas of Piedmont, Lombardy, Veneto, Emilia-Romagna, the Duchies of Modena and Parma and the neighbouring Italian-speaking canton of Ticino in Switzerland.

Because of the diversity of cultures, the development of the area also provided opportunities to exchange traditional skills among which was craft of dry stone walling that leaves a legacy that shaped the making of a cultural landscape. The search for gold meant the land was turned 'upside-down'. Paddocks were cleared to grow crops and graze stock, and the stone was ideal for marking boundaries, for constructing practical low terraces and for 'solid housing' to keep out the cold.

By far, the most significant group of terraces in this area are those made as part of the construction of the Hepburn Swimming Pool located on Spring Creek within the Hepburn Mineral Springs Reserve. Whilst by European standards, they are neither finely crafted nor noteworthy as beautiful works of art they do, however, hold an important status as a major contributor to the heritage values and cultural and social development of the area. So important was this pool that after a period of dilapidation, volunteers using old photographs to replicate missing features such as dry stone terraced seating managed to refurbish it in 1993.

Fig. 9.1 Terraced Hepburn swimming pool (restored by volunteers in 1993) in Central Goldfields, Victoria. *Photo* R. Marshall

Built in the early 1930s by local Swiss, Italian and other interested citizens to protect the mineral springs from ruination due to mining, the pool sits in the natural basin formed in Spring Creek and is financed by prominent Swiss Italian settlers the pool of the 'pre-Olympic' class that predates the modern Olympic Pool and such is its importance that as an early Olympic size competition pool it hosted Victorian swimming championships in the 1930s. Today, it is included on the Victorian Heritage Register[4] and was named Victoria's Favourite Built Place in 2004 as part of the Victorian Government's involvement in the International Year of the Built Environment (Fig. 9.1).

9.4.2 Gold Mines Walhalla West Gippsland Victoria

Elsewhere in Victoria, 150 km east of Melbourne is the sixth largest gold-producing region was known to have over 400 known reef and alluvial gold deposits in the area. Walhalla country is a tribal area of the indigenous Braiakalung

[4]http://vhd.heritagecouncil.vic.gov.au/places/5639.

clan that is part of the Kurnai group. Located in a steep sided, heavily timbered valley at the foot of the Baw Baw Ranges in west Gippsland, geologically the main strata rocks are of the Silurian-Lower Devonian period. In Upper Devonian times intrusions occurred in this stratum giving rise to the gold carrying quartz reefs including the famous Cohen's Reef. Historically, Cohen's Reef was the most productive gold deposit in the Walhalla-Woods Point Goldfield.

Gold was discovered in the early 1860s, and a township initially known as 'Stringers Creek' was established. The 'gold' word soon spread, and by 1870, the township named Walhalla named after the mining company and a Norse myth was laid out. Progress of the area continued to grow so that by 1871 there were 1484 inhabitants, numerous stores, butchers, bakers, hotels, banks, a post office, a school, a Mechanics' institute hall, two breweries and a newspaper.

The buildings were crowded along the narrow floor of the valley and perched precariously on the hillsides, and unlike elsewhere in Victoria and other parts of Australia where dry stone walls are mostly constructed on flat or undulating terrain, the Walhalla area is known for a variety of European style terraced stone walls built to create flat land for housing and support the much-needed roads and waterways (Marshall 2000: 1).

Those brave early settlers came from Cornwall in England, Valtellina in Italy and the Poschiavo Valley in Switzerland. As mines were established along Cohen's Reef, the population of Walhalla continued to grow and people including Irish, Cornish, Chinese and other European immigrants began to move out to establish homes further up the sides of the valley.

Most were employed by the quartz crushing mines such as the Long Tunnel and Long Tunnel Extended Mines, and as luck would have it, among them was a good mix of skills that enabled them to survive. The tyranny of distance and isolation became the mother of creativity as they were forced by circumstance to use what was at hand. Italian woodcutters cut wood and supplied wood burning charcoal to the lucrative Long Tunnel gold mine; the Irish and Cornish were skilled miners, and the Chinese and Italians became food producers (Aldersea and Hood 2003).

To retract the stone, they had to blast the hillsides with gelignite. Fortunately, many had come from not dissimilar terrains where the dry stone terracing craft used for survival purposes was quite familiar. So, despite the challenges, they were soon able to conjure up memories of home to utilise and reinvent already familiar and practical skills in new and creative ways.

Today, the road to Walhalla is still well off the beaten track with one road in and out. So it is hard to imagine just how isolated they must have felt in those unrelenting mountains and extremely challenging bush with only their wood, stone and other traditional skills needed to create the buildings, terraces, chimneys and house walls indispensable for day today survival.

Today, Walhalla caters principally for the tourist market. The township and surrounding area is protected as a conservation area under the Baw Baw Shire Council's Special use Zone Planning Scheme. It is also listed on the Register of the National Estate and as an Historic Park. The valley is peaceful, and the hills are again covered with trees. Several buildings are heritage-listed, the post office, band

Fig. 9.2 Rudimentary terracing to create flat land for housing in Walhalla, Gippsland, Victoria. *Photo* R. Marshall

rotunda, fire station, shops and houses are restored. The train line has been rebuilt and the dry stone terraced flat land areas are recognised for the significant and important contribution they made to the establishment of the area (Fig. 9.2).

9.4.3 Goonawarra Vineyard Terraces Sunbury Victoria[5]

The Sunbury area is particularly important as the nucleus of pastoral settlement by prominent Port Phillip founders. After exploratory trips to Port Phillip in 1835, pioneering figures including Aitken, Evans, and Jackson returned to take up land and make permanent settlements near Sunbury. John Aitken's settlement in the Sunbury-Gisborne area at Mt. Aitken has been described as Victoria's "first inland occupation of sheep country" (Dixon 1981: 26–27; Boys 1935: 30; Milbourne 1978: 26).

The Sunbury settlement also gains immense significance through its association with George Evans and William Jackson who were members of the small *Enterprize* expedition which settled Melbourne. They were the only two founding settlers who are known to have had an extended stay at Port Phillip.[6] In addition to

[5]The following is an edited extract from *The City of Hume Heritage Study of the Former Shire of Bulla* (1998) by David Moloney and Vicki Johnstone, reproduced with permission of the author and copyright holder David Moloney, February 2017.

[6]National Trust of Victoria, National Trust File No. 3020.

these Port Phillip pioneers, the study area was soon settled by others who accrued massive pastoral fortunes, such as WJT Clarke and James Malcolm.

Constructed c. 1863–69 the former vineyard terracing and associated structure is located beside Jacksons Creek, north of Goonawarra in Sunbury Victoria. Of state, historical and architectural significance, the dry stone terrace walls and gravel roadway of the former *Goonawarra* vineyard are the only known vineyard terraces to have been constructed in Victoria in the nineteenth century. Classified for their association with the politician James Goodall Francis a former Premier of Victoria who commissioned their construction as evidence of the extravagant sums of money, he was prepared to invest in his *Goonawarra* country estate.

Apparently built as much for their aesthetic as utilitarian benefit and fashionable Continental inspiration, the terraces are an integral part of the adjacent *Goonawarra* winery and homestead which was the largest of the wineries established during Sunbury's vineyard boom which followed the 1862 Land Act's Novel Industry provisions.

In the early 1860s, intensive new agricultural pursuits, such as the cultivation of vines, were provided with government support. Section 47 of the 1862 Land Act—the 'novel industries' clause—enabled leases, under very favourable conditions, to be issued for areas of Crown land less than 30 acres in size. The rent was fixed separately for each individual lease and, after a period of 5 years, the lessee had the right to purchase the leasehold land at £1 per acre.

The homestead and winery were erected on land which was leased JG Francis under this clause. Francis applied for a full 30 acres of land in April 1863, stating that he intended to grow vines and tobacco. A five-year lease of the 30 acres of allotment D in the Parish of Bulla, right on the eastern edge of Sunbury, was granted to Francis on 1 July 1863. The conditions of the lease required that 10 acres of vines for the manufacture of wine and 5 acres of tobacco were to be planted over that five years at a rate of not less than 3 acres per annum.

At the time of his application, Francis was a prominent member of the Victorian Legislative Assembly, and of Melbourne's business community, and he set up the Goonawarra vineyard as both a hobby business and a country retreat. In 1870, the Balliere's Victorian Gazetteer described it as his summer residence. Francis was later to become the Premier of Victoria.

It appears that some sort of dwelling, if not part of the original homestead, was erected soon after he was granted his lease. The entry for this property in the Bulla District Road Board rate book, made in November 1863, describes it as 'house and enclosure for vineyard'. It also seems from the high valuation beside this entry that Francis had accumulated considerably more land than the original 30 acres.

The terraces were an innovative measure taken in an apparently unfulfilled hope of producing superior wines. As well as their extraordinary cost, contemporary commentators often noted the conspicuous position and picturesque effect of the terraces. Whilst some are substantial and sturdy, they are mostly small and at the time of the Heritage Study undertaken in 1998 by Historians David Moloney and Vicki Johnston were found to be in a poor and fragile condition. The associated brick building, presumably a store room, has an unusual barrel vaulted roof, and an unusually angled chimney or vent at one end. The road cut into the hillside above the terraces is also original and contributes to an understanding of the place (*The Australasian*, 8 July 1882: 56).

Set on a steep part of an escarpment of Jacksons Creek the fabric consists of the remnants of the terraces and the stone walls that supported them. At least, one is quite high and in substantially better condition however most of the walls are small, badly eroding and vandalised. The area is also substantially overgrown.

There are no records of tobacco having ever been planted on Francis' property, but it is recorded that about 12 acres of vines were planted in 1864, located on "a hot and dry but deep and thoroughly drained hill of volcanic debris" (Shire of Bulla 1864). This was probably the north–west facing hillside to the south of the present homestead.

On 2 November 1866, the architect John Flanagan advertised for tenders for "rubble and cut work (labour only) of a wine store at Sunbury". This was for the large bluestone winery, which was completed in early 1867 at a cost of £2500. In April 1867, the winery was described in a newspaper article as "the great feature of the place".

Newspaper reports that relate to the vineyard and the terracing can be viewed in Appendix 9.1. They note the elaborate nature and the extraordinary expense of the terrace work. They also note the picturesque nature of the vineyard and that the terraces were "ornamental as well as useful. A cursory inspection of the surroundings of the vineyard is sufficient to show that utility is not the only object aimed at in its management".

9.4.4 Ha-Ha Wall the Mansion at Werribee Park Melbourne's West Victoria

Whilst many early settlers came to seek their fortunes, there were some few among them who came to improve the family fortunes using money they brought from their wealthy farming families back home. It is not surprising that once they had established themselves they then began to try to recreate the larger house and garden styles they had left behind.

Home to one of the relatively rare ha-ha retaining walls in Australia, the large property at Werribee Park was developed by Thomas Chirnside a leading Victorian Scottish pastoralist who arrived and settled in 1839 and owned large areas of land across the western basalt plains. By 1870, he and his brother Andrew both well-educated and astute businessmen had acquired 250,000 acres of freehold land in Victoria, including the property at Werribee. The Mansion completed in 1877 was modelled on the ideal grand English country estate.

Werribee Park is set on approximately 1000 acres of land 35 km southwest of Melbourne. The area includes the Werribee Mansion (1873) with formal garden, grotto, mansion gates and gate lodge, freestanding laundry, the Chirnside homestead (1865) with ha-ha wall, ration store (built by 1861), blacksmith shop, men's quarters, stables, implement shed, Bellenger's house, ford, woolshed. It is believed that the concept of the ha-ha wall is of French origin attributed to the reaction of Louis, Grand Dauphin to encountering such a feature in the gardens of the Château de Meudon circa 1700.

Used in landscape design to prevent access to a garden by grazing livestock and or to deter vehicular access, the recessed design element creates a vertical barrier, which preserves an uninterrupted view of the landscape beyond, when on approach, the vertical drop suddenly becomes visible.

When the need for public open space nearer to the cities developed, the State Government Instrumentality Parks Victoria became public custodians of many properties that were once working farms or large pastoral estates. Among them were farms that once boasted a range of functional boundary and dividing dry stone walls. One was the Mansion at Werribee Park (Marshall 2000: 10).

Today, the unique ha-ha wall that surrounds the smaller early Werribee Park homestead is maintained under the jurisdiction of Parks Victoria. The property, now part of the outer Melbourne urban sprawl, is of aesthetic significance as one of the few properties in the State to achieve the ideal of an English country house. The retention in general of the original layout, boundaries, early plantings and many garden features is fundamental to the significance of the mansion.

The property was listed in April 1987 on the Victorian State Heritage Register.[7] The bluestone Chirnside homestead (1865), with associated ha-ha wall (1867), was built primarily as a permanent base for Thomas's nephew Robert, who managed the sheep station from 1859 until 1866 and was then a lessee until 1873. The garden surrounding the homestead features an impressive *Ficus macrophylla*, and framing the central view are two *Lagunaria Patersonia*, near the ha-ha wall. Today, the unique ha-ha wall that surrounds the smaller early Werribee Park homestead is the only dry stone construction that remains on the once heavily walled estate.[8]

9.5 Garden Terraces Post-World War II

The early settlement gold rush period had brought with it much wealth, and from the 1860s onwards, significant changes were taking place. The visual landscape was altering dramatically. Fences and stone walls had begun to clearly define rural boundaries as large pastoral estates were subdivided. Churches and schools were built. Times were good. People were settling down. Homesteads were becoming grander. Pastoralists were protecting their stock with fences, but the abundance of stone meant that dry stone walls still remained economically competitive.

---

[7][VHR] H1613; Heritage Overlay H02.

[8][VHR] H1613, http://vhd.heritagecouncil.vic.gov.au/places/1207/download-report.

The cultural and social landscape was also undergoing important change. In the decades after the 1850s gold rush, wallers were in demand and in and around the Victorian goldfields areas, groups of people from many nationalities began to build communities and put down roots.

However, the good times were to about to come to an end. A depression in the 1890s brought with it the beginnings of a steady decline in lifestyle and income. Food was scarce, and in many dry stone wall areas, people began to pull the walls apart to reach rabbits that took refuge among the stones.

By the time, the second Great Depression of the 1930s arrived those walls that had fallen into earlier disrepair become fertile breeding grounds for "underground mutton". Dry stone walls changed from "containing" to "container", and the walls became a steady source of food and income to poor families. Fewer wallers were working at the trade. Old wallers were dying. Times were tough. World War II was imminent.

The period after World War II brought about even more change. People were learning new trades and "making do" with recycled materials. The lack of skilled wallers and the economic climate meant that post- and wire fencing had become a competitive alternative. So by the 1950s, dry stone walls were no longer an important form of fencing for agricultural purposes. Yet even as early as the 1920s, their slow decline and fall in favour for farming use began to be paralleled by an increase in popularity for use in public and private garden settings.

In the period after the Second World War, the craft of dry stone walling began to experience a revival through the influence and work of a small group of innovative landscape designers. This was a time period when people began to re-connect with things of beauty, a time of creativity, of being at one with nature and her materials (Proudfoot 1989).

In their gardens, people began to value recycled materials to create a sense of place. Nature and the natural form suddenly became important. They had talk of the influence of early 1900s landscape designers such as Edna Walling from the UK and Paul Sorensen from Copenhagen and their physical and spiritual affinity with stone (Ratcliffe 1990; Dunn 2002; Hamilton-Smith 2002).

Their talent lay in blending terraced stonework with native plantings to create gardens of exquisite beauty. Each of these designers valued and respected the form, shape and placement of local stone. They understood and appreciated the inherent qualities of dry stone terracing for their aesthetic value as well as their practical application (Marshall 2000: 9).

Arguably Australia's best-known and most influential garden designer Edna Walling began her career in Melbourne Victoria 1916. Since commencing her career, Walling played a dominant role in the creation of Australian gardens as we know them today. Her love of the natural countryside and bush where she experienced a spiritual and intellectual awakening inspired the designs that interpreted the fragile relationship between garden and landscape (Dixon and Churchill 1998).

The landscape gardens for which she is renowned employed expansive bodies of water, magnificent vistas and were replete with collections of plants from the northern hemisphere. Her gardens carried clear messages of success and power,

Fig. 9.3 Post World War II people wanted to create things of beauty. Edna Walling garden near Lilydale Victoria. *Photo* R. Marshall

of good taste and refinement. Her vision was to achieve a unity between house and garden evidently influenced by Italian and Spanish gardens in her use of pergolas, walls, steps and paths.

Simplicity in all elements of design so that nothing detracted from the integrity of the setting was a hallmark of her work. Designs that respected scale and the great bones created by terraced stonework were integral to her design outcome (Fig. 9.3).

Buildings were simple, made of rock and timber, with wooden shingles. Each of her beautifully rendered garden plans demonstrates her use of formal axes and structure. Of sweeping dry stone terraces and retaining walls, flights of steps and well-considered bodies of water and her expert treatment of sloping land with terraces bound by dry stone walls of local stone (Churchill and Dixon 1988).

By far, the most unique example of terracing as both a practical solution to taming a landscape and an art form in its own right can be found in the Everglades a five-hectare property on the very edge of the spectacular Jamison valley in the rugged Blue Mountains 100 km from Sydney in New South Wales.[9]

The site of a small weatherboard cottage on a rocky sandstone hillside bought as a weekender by Henri Van Velde a wealthy Belgian businessman, Van Velde

[9]"Everglades Gardens: How An Historic Garden Developed", http://www.acsgarden.com/articles/garden-travel/everglades-gardens-how-a-historic-garden-developed.aspx (accessed January 12, 2017).

employed designer Paul Sorenson to transform three hectares of the untamed bush into the formal gardens as we know them today. Built in the 1930s as a weekend retreat the steep extremely challenging sloped site with a one in two gradient in some places presented Sorensen with major challenges.

In his design, Sorensen decided to separate the views of the valley from the more formal garden areas thus creating separate garden rooms, each with their own charm and grandeur. For some, it would have seemed strange to purposefully obscure the view from the formal terraces; but for Sorenson, the garden had greater interest if it was made impossible to see the total layout from any single position.

Sorensen's horticultural background was perfectly attuned for the problems that confronted him. His early studies supervised in Denmark by the leading landscape architect at the time and his work in Berlin, Dresden, Frankfurt and Paris, but undoubtedly, it was the skills he acquired in the Swiss Alps that would be the perfect fit for the challenges that lay ahead.

The earth was poor in nutrients thin, sandy and full of ironstone fragments. To ensure easy access and suitability for planting, Sorensen employed the European dry stone tradition thereby creating large flat land areas by taming the landscape with a series of terraced walls.

Sorensen overcame the lack of fertile soil by digging by hand deep areas for planting. The excavated soil was enriched with humus and distributed to the garden beds. Ironstone was removed and graded according to quality, and colour and the best pieces were used to create the massive, dry stone walls which are a main feature of the garden.

The labour-intensive work was executed before the days of bulldozers and bob-cats. As in earlier times, levers pulleys and gantries were used to move boulders and shape the landscape. Swathes were cut into the hillsides, stakes and lines were marked out and men labouring in pairs trundled barrows along narrow planks and catwalks. Few were skilled tradesmen; the lack of heavy machinery and the physical labour required to create the walled areas and planting areas was quite daunting however Australia was still in the grip of the Great Depression so fortunately there were plenty of able-bodied men glad of the opportunity for work (Roland 1989). The property is now managed by the National Trust of New South Wales (Australia).

9.6 Garden Walls Contemporary Terraces

A short walk from the Sydney Opera House in New South Wales is the beautiful sandstone terraced retaining walls designed to display a wide range of Australian natives. Sandstone is the foundation of the city, and its warm and richly coloured stone abounds in natural outcrops and majestic heritage buildings. An ideal material for constructing dry stone walls.

The first European farm in Australia, the Royal Botanic Gardens Sydney is the oldest scientific institution in Australia. From the earliest days, it has played a major role in the acclimatisation of plants from other regions and the scientific study of native plants of New South Wales and the South Pacific. The terraces were completed in time to celebrate the Opera House's 30th Anniversary and to showcase Sydney to the world during the 2000 Olympics.

However, the most significant concentrated complex of contemporary dry stone terraces can be found at the Australian Mount Annan Botanic Gardens in New South Wales. The development commenced in 1985 and required a creative solution to turn a 170 year-old dairy farm on heavy clay soils into interesting garden displays and public recreational space.

Dry stone walls and terraces were part of the answer. Aesthetically appealing and long lasting, they gave character and identity to the overall design. Constructed over several years by fee-paying students who attended dry stone wall learning workshops, the cost neutral project was the brain-child of Geoff Duggan, who was until recently, a senior Landscape Planner with the Royal Botanic Gardens at both Sydney and Mount Annan (Marshall 2000: 13).

As part of the development Geoff and his colleagues were confronted with the dilemma of how to grow the diverse collection of Australian native plants in the very heavy clay soil. The terraces constructed with quarried sandstone provided by far the best opportunity to plant using different soil mixes and to facilitate directional drainage focussed on the garden beds.

9.7 Preservation and Regulations

Whilst dry stone terraces can be found in selected areas, by far the majority of walls in Australia are free standing. As a result of suggested recommendations made in the 1999 Federal Government funded *Touring Exhibition A Stone Upon A Stone*, the Dry Stone Walls Association of Australia (DSWAA) Inc. was founded in Ballarat, Victoria in 2002.

A not-for-profit organisation the DSWAA is an all-encompassing and diverse group of rural property owners, pastoralists and farmers, urban dwellers, environmentalists and other professionals with a special interest in the identification, documentation and conservation of dry stone walls to be found Australia-wide, as well as a number of practitioners engaged in the craft of dry stone walling and dry stone sculpture.

Broadly, its goals can be summarised, firstly as being: that National, State and Local governments, as well as the wider community, recognise the significance of dry stone structures built by indigenous peoples, European explorers, early settlers and modern craftspeople as valued artefacts of our national identity; secondly, that following recognition by governments, landholders and the wider community of the cultural significance of dry stone walls throughout the Australian landscape, statutory protection is considered as an appropriate means of protecting and preserving the heritage of dry stone walls in Australia.

As mentioned above, the DSWAA was founded in Victoria and this State remains the centre of the Association's activities although there are many members in other States. Victoria has the majority of Australia's dry stone walls; a reflection of the volcanic geology of the State's south–west and the history of European settlement that led to the construction of many hundreds of kilometres of dry stone walls in the mid- to late-nineteenth century.

It is therefore not surprising that it is in Victoria where statutory planning controls to protect dry stone walls are most comprehensive of any State.

In December 2008, a provision was written into all of the 78 municipal Planning Schemes in Victoria to give those municipalities the capacity to protect historic dry stone walls. This provision requires a permit to be obtained to demolish, remove or alter a dry stone wall constructed before 1940. However, to activate this provision, the relevant municipality has to have undertaken a survey or study to identify the specific dry stone walls that it wants this provision to apply to. Only a few municipalities have undertaken studies to identify walls that meet required criteria to be designated as worthy of preservation. It is hoped that this number will increase, and that States other than Victoria will introduce statutory measures into their planning schemes to provide the same capacity for protection as Victoria.

As well as the above, there is a limited number of instances in States other than Victoria where historic dry stone walls enjoy formal protection, but these are localised and the result of a recognition of the value of a small group of walls in a particular area rather than as the result of a comprehensive or municipality-wide research or survey.

In terms of ongoing preservation and maintenance, although there are relatively few wallers in Australia, those that do exist are both passionate and dedicated to maintaining the ongoing future of the craft. However, at present, there are no dedicated facilities to undertake any formal qualifications. Some wallers have travelled to the Dry Stone Walls Association of Great Britain[10] to gain their qualifications, others have gained theirs from Geoff Duggan[11] who is qualified to undertake the British test in Australia and some have learned the technique as part of other landscaping courses or experiences.

The DSWAA conducts occasional workshops and some maintenance during its field trip activities, and their website (http://dswaa.org.au/) maintains a Directory of Wallers. Some have formal qualifications whilst others do not. Therefore the Association advises potential clients that listing does not imply endorsement of wallers.

---

[10]"Craftsman Certification Scheme", *The Dry Stone Walls Association of Great Britain*, http://www.dswa.org.uk/craftsman-certification-scheme.asp (accessed February 6, 2017).

[11]"Dry Stone Walling Workshops", http://www.geogenic.com.au/dry-stone-walling-workshops.html (accessed March 6, 2017).

9.8 Art, Craft, Survival and Creativity

If we reflect on Maslow's theory on human motivation, can we argue that despite the scarcity of terraces in Australia, those that do survive, serve to constantly remind us of the tenacity of immigrants with the flexibility to re-interpret a harsh and unfamiliar geological landscape in new and creative ways? That, by building dry stone terraces their basic needs for survival and shelter were addressed and due to the need to develop new skills, establish communities and make new friends their need for creativity and companionship may have also been fulfilled.

Located in the Grizedale Forest Sculpture Park, a working forest in the Lakes District National Park in Cumbria in the UK is an excellent and creative example of how we might understand, interpret and challenge in a contemporary context the relationship between art, craft, history and modernity. The award-winning Ridding Wood Trail offers the visitor easy access and provides an exciting introduction to the visual arts. But it is much deeper in the forest, where the natural extension of the ancient craft of dry stone walling has been re-thought and reinterpreted for a modern-day audience.

Challenging the notion of cultural landscape as museum or gallery the forest is the perfect setting for Andy Goldsworthy's site-specific dry stone sculpture 'Taking a Wall for a Walk'. Sited among the scrub and bracken, the wall snakes in and out of the edge of the woodland. Because this serpentine-like dry stone section of wall, a modern-day piece of sculpture, joins so seamlessly with fragmented sections of an old agrarian wall, with it raises as many questions as it does answers.

Where does history end and modernity begin? Where does the dry stone craft end and the dry stone art begin? Who is the artist? Andy for his artistic vision and design of the work, or the professional wallers who built the wall with and under his guidance? Perhaps its both. However, it could be argued that in the end, the degree of cultural significance is best demonstrated by the extent to which a society's socio-political processes endeavour to influence changes in behaviour and devote resources and processes to ensure its conservation.

To that end, where better in Australia to showcase and promote the beauty and benefits of dry stone terracing than in the unique and historic 1788 first fleet landing area of Sydney Cove, a small bay on the southern shore of Sydney Harbour. In the nearby Royal Botanic Gardens the single most distinct landscape feature is the historic hand-hewn sandstone seawall. The visible complex of impressive dry stone terracing designed to display a rockery of Australian natives curves around Farm Cove from Mrs. Macquarie's Point to the Sydney Opera House. The well-trodden pathway delineates the garden from the harbour and exposes the craft of terracing to provide an aesthetic and focal point for locals and visitors alike (Fig. 9.4).

Fig. 9.4 Early terraces overlooking Sydney Harbour. *Photo* R. Marshall

Appendix 9.1

1. *The Leader, 20 April 1867, p. 9*

The vineyard of the Hon J.G. Francis, M.L.A., said the Leader, was one of the first to be set up in Sunbury locality under the "novel industry" clause of the 1862 Land Act. It was 'the most extensive concern in the hands of one proprietor'.

Francis' vineyard now exceeded his original selection and had extended onto land "since acquired at auction"; already 38 acres have been planted with vines and an additional 10 acres were to be planted during the coming season. The oldest vines were in their fourth season and covered about twelve acres located on a hot and dry hillside of deep volcanic debris; the natural drainage was said to suit the vines which were expected to produce wines of "first-class quality".

The sites selected by Francis were chosen not only for their natural suitability for vine-growing but also keeping in mind the particular requirements of different grape varieties:

> … the steep bank of six acres selected for part of the current season's operations will not be easily or inexpensively worked, but these disabilities are overlooked in favour of the thorough adaptability of the site for the production of superior wine. The loose black soil is so intermixed with stones large and small that it can be worked in no other way than be hand, and the estimated cost of preparing this land is £30 per acre. This, added to the cost of planting, tending for three years, and staking, makes vine growing a rather costly investment.

The already established vines were supported on trellises made of ironbark stakes and iron wires. The grape varieties grown were Chasselas, Riesling, Hermitage, Mataro, Verdeilho (sic), Espart and White Pineau. Some wine had already been made of the Riesling grapes. Mr. Baldini was the overseer of the vineyard.

2. *The Australasian, 2 January 1869, p. 24*

Francis now had 37 acres planted and 3 acres more to be planted to complete the vineyard as intended, said the Australasian. It reported that: A great part of this land has been rendered fit for planting at heavy expense, being almost precipitous, and full of great stones and rocks. However the soil is naturally good, and where the stones have been taken out, to be used as terrace walls or for filling up the hollows, the vines thrive admirably.

Mr. Baldini was the vineyard manager. The report noted that the 1866 claret had been tasted; the 1867 claret was not yet bottled; the chasselas of 1867 was light and pleasant to drink; the riesling of 1867 was still in wood; and the verdeilho of 1867 was full bodied and rich in flavour.

3. *The Australasian, 7 May 1870, p. 601*

"The steep banks beside Jackson's Creek are peculiarly fitted for the vine, the soil being rich and the stones numerous; but it has been costly work in some places, terracing, and leveling, and filling up holes".

The Illustrated Australian News, 10 June 1878

The Illustrated News reported that Goonawarra vineyard, formed in 1864, was nearly 37 acres in area. It was situated on slopes facing North West, West, and South West, with a frontage to part of Jackson's Creek. The formation had cost, in some parts of the vineyard, as much as £100. The ground was said to have originally been 'rocky spurs of the Mount Macedon range'. The vineyard was nearly in full bearing.

4. *The Australasian, 8 July 1882, p. 56*

The vineyard occupies a somewhat narrow strip of land extending along the creek for some distance, the vines extending from the water's edge well up towards the crown of the sloping banks. In places there is an alluvial flat of considerable width between the creek and the gently sloping hill, and in other places the banks rise bold and steep from the edge of the stream.

This portion of Jackson's Creek would be pleasing in a state of nature, but with the vines covering such a long stretch of its winding and undulating banks, its appearance is exceedingly beautiful, so that from the residence a charming view is obtained of vine-clad slopes and the waving line of water.

Along the upper boundary of the vineyard a road has been cut in the hillside and substantially formed with stone, while on a portion of the hill too steep for vines to grow, terraces are being formed for their cultivation. The terraces appear to be formed to complete the outline of the landscape. The road is ornamental as well as useful and a cursory inspection of the surroundings of the vineyard is sufficient to show that utility has not been the only object aimed at in its management.

5. *The Australasian 2 June 1883, p. 696*

The vineyard was 50 acres in extent, of which less than 40 acres were in full bearing, the balance of 10 acres having been added recently. The Australasian looked back on the unusual terraces:

> This vineyard, situated on the eastern side of the stream, is confined entirely to the hill slope and takes the shape of a crescent. The situation is a pleasant one indeed … At the northern end the decline is rather precipitous, and here a system of terracing, which must have entailed considerable expense, has been carried out. In order to prevent the storm waters from the hills washing away the soil, the bank of each terrace was faced with stone, and this elaborate work which is carried out for some distance, must have added considerably to the first cost of establishing the vineyard.

References

Aldersea J, Hood B (2003) Walhalla valley of gold. Walhalla Publishing, Trafalgar Victoria Australia
Beames RO, Whitehill JAE (1988) Some historic gardens of South Australia. National Trust of Australia (SA) and Botanic Gardens of Adelaide
Churchill J, Dixon T (1988) The vision of Edna Walling Hawthorn Victoria Australia. Bloomings Books, Hawthorn
Crittenden V (2002) Shepherd, Thomas (c. 1779–1835). In: Aitken R, Looker M (eds) The Oxford companion to Australian gardens. Oxford University Press-Australian Garden History Society
Dixon HP (1981) Early settlement of the Gisborne "Bush Inn" district: 1802 to the gold rushes. Typescript 26–27
Dixon T, Churchill J (1998) The vision of Edna Walling, garden plans 1920–1951. Bloomings Books, Hawthorn
Dunn P (2002) Mount Macedon. In: Aitken R, Looker M (eds) The Oxford companion to Australian gardens. Oxford University Press-Australian Garden History Society
Frazer SP (1987) Historic Tasmanian gardens. Mulini Press
Hamilton-Smith E (2002) Mount Buffalo. In: Aitken R, Looker M (eds) The Oxford companion to Australian gardens. Oxford University Press-Australian Garden History Society
Hughes J (2002) Macleay, Alexander. In: Aitken R, Looker M (eds) The Oxford companion to Australian gardens. Oxford University Press-Australian Garden History Society
Marshall R (2000) A stone upon a stone exhibition. http://dswaa.org.au/wp-content/uploads/2016/04/A-Stone-Upon-A-Stone.pdf. Accessed 20 June 2016
Moloney D, Johnstone V (1998) The City of Hume heritage study of the former Shire of Bulla
Morris C (2008) Lost gardens of Sydney. Historic Houses Trust of NSW
Mossop E (2006) Contemporary Australian landscape design. BT Latitude
Proudfoot H (1989) Gardens in bloom-Jocelyn Brown and her Sydney gardens of the '30s and '40s. Kangaroo Press
Ratcliffe R (1990) Paul Sorensen—Australia's master gardener. Kangaroo Press
Raworth B (2002) Butler, Walter. In: Aitken R, Looker M (eds) The Oxford companion to Australian gardens. Oxford University Press-Australian Garden History Society

Read S (2015) Early vineyards and viticulture in the Sydney basin. Paper to Australian Garden History Society, State Library of South Australia, 15 Oct 2015
Roland R (1989) Steeped in antiquity. Your Garden: Pacific Magazines, pp 50–53
Stockton B (1978) The gardens of Jocelyn Brown. Thesis for B. of Landscape Architecture, University of NSW School of Landscape Architecture
Tanner H, Begg J (1976) The great gardens of Australia. Macmillan

Chapter 10
Agricultural Terraces in Mexico

José Manuel Pérez Sánchez

Abstract The Mexican territory's geographical characteristics allow the development of diverse agricultural systems, which have been practiced since ancient times. The terrace system is one of these, and it is found in various challenging landscapes, such as hills, cliffs, and ravines. This work aims to show the diversity of agricultural terraces in Mexico and present a general description of this system's characteristics. Studies in this field identify several types of terraces: terraces with stone walls, *metepantles*, and *lama-bordo*. To understand the characteristics of these terraces, a documentary review of five Mexican regions was made: Tlaxcala, the basin of Mexico, the valley of Toluca, the Mixteca Alta, and the Mayan area. Traditional agricultural systems in Mexico are an important subject, studied by archeologists and anthropologists from the perspective of traditional agriculture, agroforestry systems, and traditional ecological knowledge. This paper's introduction emphasizes the role of agricultural terraces for the development of civilization in Mesoamerica, specifically in central Mexico, the Mixteca Alta, and the Mayan area. Traditional agriculture and the different agricultural systems practiced in Mexico are also addressed. The document continues with the subject of terraces, presenting the various names given to them in different studies and regions. Terraces in Mexico have always been important for societal development—from ancient cultures to the rural communities managing the systems today. Through terraces, models may be generated to help improve soil management, control erosional processes, and support food security for the population.

J. M. Pérez Sánchez (✉)
Faculty of Anthropology, Universidad Autónoma del Estado de México,
Mariano Matamoros sur s/n, Colonia Universidad, Toluca CP 50130, Mexico
e-mail: jmps9@hotmail.com

M. Varotto et al. (eds.), *World Terraced Landscapes: History, Environment, Quality of Life*, Environmental History 9,
https://doi.org/10.1007/978-3-319-96815-5_10

10.1 Introduction

In Mexico, agricultural terraces are a system of ancient origin that, along with other agricultural systems (*chinampas*), allowed the development of pre-Hispanic societies. Most studies on terraces highlight archeological and geographic–cultural analyses, leaving aside social organization, construction, and maintenance methods, as well as changes in the terraces themselves. However, some recent anthropological research is beginning to examine these important, left-behind considerations (Pérez 2014). The objective here is to examine these considerations by presenting a general overview of the terraces in five Mexican regions: Tlaxcala, the Toluca Valley, the basin of Mexico, Oaxaca, and the Mayan area.

In Mexico, the debate concerning intensive farming systems was crucial to the origin and development of Mesoamerican civilization. Three important, representative cases are the basin of Mexico, the Mayan area, and Oaxaca. In the early 1950s, Palerm (1972) became interested in the relationship between cultivation techniques and civilization development in the basin of Mexico. He studied and compared the systems of slash-and-burn, fallow, and irrigation as they relate to productivity, demography, and urbanism. After analyzing each agricultural system, Palerm concludes that the Mesoamerican civilization did not emerge from areas of slash-and-burn agriculture. Its origins might be in an area with fallow agriculture, which was slightly more settled and semi-permanent. However, Palerm ultimately claims that urban culture originated and developed in places with irrigated agriculture, and from there it spread. Both the environment and the technological levels of agriculture are elements that limited and conditioned the spread of civilization in Mesoamerica.

Palerm and Wolf (1972) describe the importance of the irrigation system in the old Acolhuacán Dominion (Northeastern Mexico City) and relate the agricultural terraces to the irrigation system in the *Somontano* and the Sierra. They document that, according to ancient remains, past irrigation systems were associated with terraces. The terraces contained a network of canals that carried water from springs to deposits for domestic purposes, and the inhabitants cultivated the terraces and built houses on them.

Pérez (2006) considers that the development of agriculture and complex societies occurred independently in different places in Mesoamerica. For example, in the Mixtec Region such societies developed around 1200 BC, when there was high economic potential due to elevated population densities, regional markets, and sociocultural development. In the Mixtec Region, the degree of agricultural and socioeconomic complexity was equal to that of the valley of Mexico and the Mayan area. In a complex environment such as the Mixtecan, both society and agriculture had to adapt and did so by constructing various terraced systems for both housing and agriculture. Pérez (2006), like Palerm, argues that, in the Mixteca Alta, *lama-bordo* terraces and their agricultural production were important for civilization development. This contributes to the discussion of the role of society, the State, and domestic units in landscape transformation and the creation of intensive agricultural terrace systems in the region.

The hydraulic hypothesis for the development of civilization in Mesoamerica's central highlands suggests that elsewhere (e.g., in the Mayan area) civilization originated through hydraulic agriculture—that irrigation systems similar to those in the central highlands were required for development. However, this supposition is not valid in the Mayan area because the environmental conditions of the humid tropics do not necessitate large irrigation works, or other agricultural irrigation systems, for complex societies to develop (Lobato 1988). In the Mayan area currently, the most extensive agricultural system is the slash-and-burn technique, which is purported to be the technique used by the ancient Mayan people.

Around the early 1970s, other agricultural systems in the Mayan area, such as terraces, drained fields, and orchards, were discovered, revealing that Mayan agriculture was more diverse than previously thought. In the mid-1970s, a more intensive study on the terraces began (Harrison and Turner II 1978). Turner II (1974) reports agricultural terraces between the states of Campeche and Quintana Roo on the banks of the Becan River. Interest in studying terraces reinforced the claim that the Mayan people practiced other intensive systems adapted to the humid tropic environment, not merely the slash-and-burn technique (Lobato 1988).

10.2 Traditional Agricultural Systems

The study of traditional agricultural systems began in Mexico with anthropological and archeological research (Palerm 1967; Palerm and Wolf 1972). Archeology has played a key role in documenting the origins of agriculture and the domestication of plants and animals. Traditional agriculture, by feeding human groups, allowed the emergence of irrigation, urbanism, and the State in ancient societies. Traditional agricultural systems are of ancient origin and are adapted to local environmental conditions such as relief, soil, altitude, rainfall, humidity, and climate (González 2016).

In Mexico, a wide variety of agricultural systems are known: slash-and-burn, fallow, intensive rainfed moisture and irrigation, temporary, *tlacolol*, rotation, *coamil*, *chinampas*, *camellones*, *marceño*, *tecallis*, *mawechis*, marsh agriculture, terraces, *banquetera*, *kool*, *milpa*, *huamil*, oasis, cacao plantations, agroforests, *kuajtakiloyan*, *calal*, *metepantles*, *te'lom*, and orchards (Aguilar et al. 2003; Palerm 2008; Moreno et al. 2013, 2016; González 2016). These agricultural systems result from a long process of adapting to the conditions and requirements of the geographical and socioeconomic environment, and they have resulted in a repertoire of cultivable plants with alimentary and commercial value, agricultural technology, and family organization for agricultural work. The importance placed on family units is visible in the way farmers manage different agricultural systems through crop diversity, production destinations, family work, and traditional ecological knowledge (Moctezuma et al. 2015).

10.3 The Agricultural Terraces

In Mexico, terraces were implemented by the ancient inhabitants and developed throughout the Mesoamerican area. Agricultural terraces are a technique for conserving water and preventing soil erosion (Donkin 1979; García Cook 1986; Rojas 2001; Pérez and Juan 2016; González 2016). Terraces are part of the intensive rainfed system and are associated with the moisture system, because they collect enough water to maintain the humidity of the farmland (Palerm 2008). This system was important for the development of the great pre-Hispanic cultures of Mexico, with high population densities and urban centers (Palerm and Wolf 1972; Pérez 2006).

Terraces dominate sloping mountainsides. They are constructed with support walls of stone or adobe, protect the soil from erosion, retain rainwater, and allow the accumulation of alluvial soil (Palerm 2008; Doolittle 2004). Also classified as earth embankments, the terraces can combine borders and canals constructed perpendicularly to the mountainside, whose vertical slopes are of stone, lime, *tepetate*, or blocks of wall (Trautmann 1981; Martínez et al. 1999). The terraces are artificial, staggered or vertical, constructions with walls of different materials, canals to capture water, and a surface to cultivate.

These, however, are not all the characteristics of terraces. They also increase water infiltration into the soil, diminish the volume of runoff downstream, evacuate excess water at “non-erosive velocities,” reduce sediments in the runoff, and refurbish lands for agriculture. The terraces have been adapted to different environments where climate, soil, topography, and stoniness play an important role in their construction and operation (Martínez et al. 1999).

10.4 Terrace Distribution

In 1979, R. A. Donkin first documented the distribution of terraced landscapes in the Americas, and particularly in Mexico, based on archeological data, historical sources, the characteristics of pre-Hispanic agriculture, and agricultural implements used by ancient populations. Donkin proposes two types of considerations in the study of terraces: continental–regional and local characteristics. The former includes territorial distribution, climate, and cultural associations; the latter includes the form and construction of the terraces, a social base and purpose, crops, and irrigation. The abandonment of terraces due to environmental and cultural issues must also be considered.

Donkin identifies three regions of agricultural terraces on the American continents: North America, Central America, and South America. In Mexico, he classifies six sub-regions (Fig. 10.1) according to the diversity of environments and the territory’s physiographic characteristics. The different terraced landscape zones in Mexico cover the north, center, and south (Table 10.1).

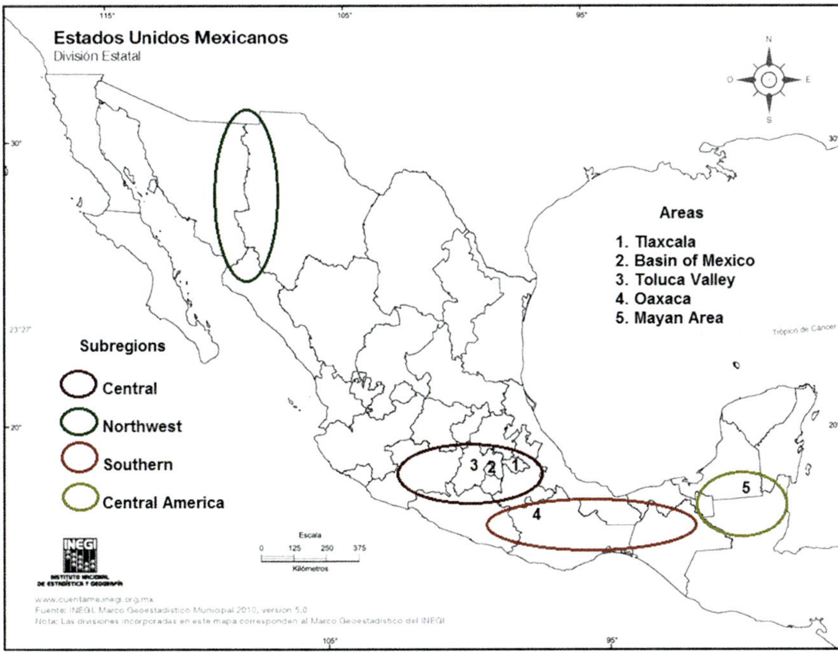

Fig. 10.1 Sub-regions and terraced areas in Mexico. *Source* Donkin (1979)

Although Donkin's proposal was made about 30 years ago, it is still relevant, and several more recent studies deal with current conditions of the terraces in various areas of Mexico. In Tlaxcala, the studies are of the regions of the volcano Matlalcueye and the Tlaxcala Block (Patrick 1977). In Oaxaca, research focuses on the Mixteca Alta region (Pérez 2006, 2015), and, in the valley of Toluca, on Calixtlahuaca (Smith 2006; Pérez and Juan 2016). Though covering different sites and regions, the studies share the common perspectives of cultural geography, agroecology, cultural ecology, anthropology, history, and archeology.

10.4.1 Terraces in Tlaxcala

Plains, hills, ravines, and the volcano Matlalcueye characterize the landscape of Tlaxcala (Fig. 10.2). In this region, the problem of soil erosion is not recent. In pre-Hispanic times, the inhabitants knew the problem, and they had, to some extent, controlled it by modifying the surface with terraces (Cook 1986). Archeological studies show that pre-Hispanic populations had already modified the slopes of hills by 1500 BC, as the settlers in the Xochitécatl area did by building terraces on the slopes to construct their houses and cultivate the soil (Cook 1986; Heine 1975). In the Tlaxcala Block region, the construction of the pre-Hispanic terraces allowed the

Table 10.1 Regionalization of agricultural terraces in Mexico

Region	Sub-region	Area
A. North America	I. Central Mexico	1. Basin of Mexico (east and south) 2. The valley of Teotihuacan 3. The valley of the Rio Tula 4. Basin of Toluca 5. South of the Piedmont of the Mesa Central: Tenancingo, Coatepec, Tepoztlan, Ozumba-Chimalhuacan, San Miguel Atlauta, Tochimilco-Atlixco
	II. East Central Mexico	6. The basin of Puebla-Tlaxcala 7. East of the Piedmont of the Mesa Central
	III. West Central Mexico	8. The Western Highlands and the basin of the Balsas-Tepalcatepec 9. The basin of Lake Patzcuaro 10. The valley of the Rio Chilchota
	IV. Northwest Mexico	11. Southwest Chihuahua and northern Sinaloa 12. Northwest Sonora and north Chihuahua
B. Central America	V. Southern Mexico	13. The Mixteca Alta and the basin of Tehuacan: northern, western, southern, and eastern Mixteca Alta 14. The Barranca Grande: basin of Tehuacan
		15. Oaxaca: the valley of Oaxaca-Tlacolula 16. The Highlands of Oaxaca
		17. Chiapas: Chamula-Saklamanton, Amatenango—Aguacatenango, Comitan—San Francisco, Ocosingo
	VI. Central America	18. The Mayan lowlands

Source Donkin (1979)

inhabitants to identify and use hard soils (*tepetate*) as a base to build both houses and canals to deflect water (González 2016).

In Tlaxcala, there are two types of farming systems: the terraces and *metepantles*.[1] The latter is a local term used to name a specific type of terraces commonly found on slopes, hills, and canyons. Instead of having a rock wall, *metepantles* have walls of earth with *agave* plants. In the Tlaxcala Block, terraces have two distinctive features: retaining walls and ditches. The walls are defensive and anti-erosive elements; they vary in size depending on the slope of the land and cultivated area (Patrick 1977; Bilbao 1979; Mountjoy 1985; Pérez 2014; González 2016). The ditches are parallelepiped-shaped, from 0.40 to 1 m deep, with a length of 1–35 or 40 m (Bilbao 1979). The ditches capture water, sediment, and organic matter, and they are constructed according to the land's slope and the type of soil (Bilbao 1979; Pérez 2014; González 2016).

[1]A term derived from the Náhuatl word *mepancle*: row of agaves that separate land. *Me-pantli* of *metl*: maguey [*agave*], and *pancli*, *pantle*, or *pancle*: board, row, wall. *Metepantle* is the name given to the set of things put in rows: plants, trees, or *agave* (Diccionario del Náhuatl 2007).

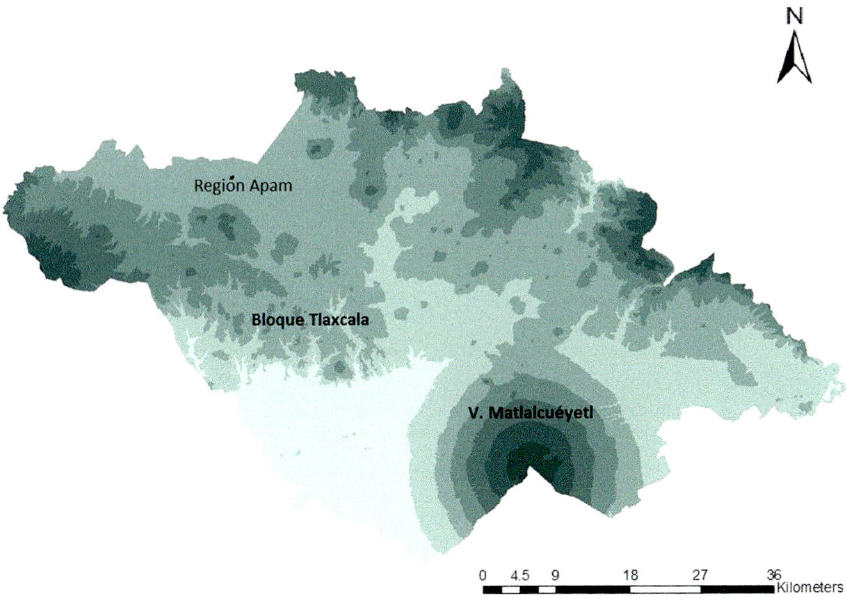

Fig. 10.2 Map of Tlaxcala, Mexico

In *metepantles*, ditches are subdivided into stages following the slope of the land. In the rainy season, the slope of the ditch lets water pass to the next level, capturing rainwater, mud, other plants, and insects, which, while decaying, form organic matter that the farmers deposit on the surface as fertilizer (Bilbao 1979; Pérez 2014; González 2016). The ditches are in three places: (1) along the roads, (2) along the edges of the farmland, and (3) at the base of the retaining walls in order to catch rainwater or divert excess water toward the canyons (Pérez 2014; González 2016).

Patrick (1977) describes the *metepantles* system around the Matlalcueye volcano, explaining its retaining walls and row of plants, the gradient and width of the surface, the size of the slope, and the characteristics of the maguey (*Agave sp.*) plant (its size and the space between plants). In the *metepantles* terracing technique, planting area, board retention, and the row of *agave* are fundamental for the development of agricultural practices.

In the municipality of Ixtacuixtla, Pérez (2014) describes the *metepantles* system in the town of La Caridad Cuaxonacayo—a system which is associated with water deposits and characterized by its cultivation surfaces, vegetation, and containing levees, canals, or ditches. The *metepantles* contain walls or embankments with different types of vegetation—agave, prickly pear (*Opuntia sp.*), or fruit trees (pear, apple, plum)—that support the wall (Fig. 10.3). Reservoirs collect rainwater, and a network of water canals provides moisture to crops in the dry season and diverts excess water from the *metepantle* when necessary.

Fig. 10.3 *Metepantle* with Agave, Ixtacuixtla, Tlaxcala. *Photo* J. M. Pérez Sanchez

Maintaining and reconstructing *metepantles* is the responsibility of farmers and their families. There is also collaboration between families for the harvest (Pérez 2014). However, because young people choose to work in urban environments like Tlaxcala, Puebla, or Mexico City, farmers often hire workers to plant and harvest corn (Fig. 10.3).

10.5 Terraces in the Basin of Mexico

In the early 1950s, Ángel Palerm and Eric Wolf conducted the first terrace studies in the basin of Mexico in the northern Acolhuacán (northwest Mexico City) region. They divided it into four zones: the Sierra, the *Somontano*, the arid zone, and the valley (Fig. 10.4). Palerm and Wolf (1972) identified, in the *Somontano* area, agricultural terraces in several villages where maize was cultivated for subsistence and flowers for sale in Mexico City. In San Miguel Tlaixpan and San Nicolas Tlaminca, people cultivated on adobe and stone irrigation terraces. Local administrations were responsible for distributing the terraces' irrigation water according to each terraces' size, and people built their houses on or near the terraces.

Palerm and Wolf (1972) report other sites with terraces in the hills of San Joaquín to the north and Tetzcutzingo to the south of Acolhuacán. In both hills,

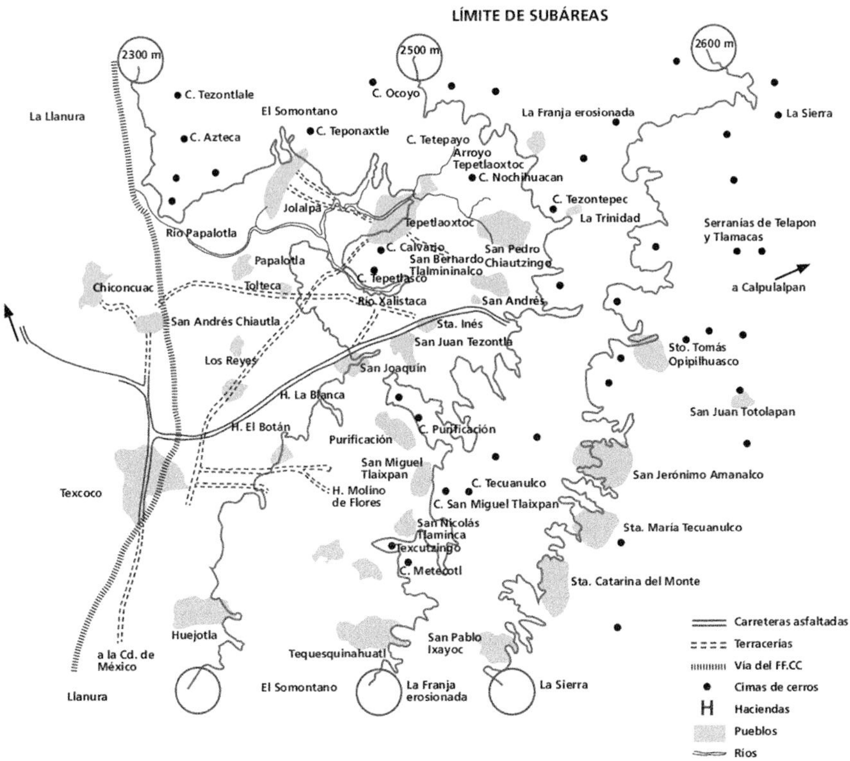

Fig. 10.4 Map of Acolhuacán, basin of Mexico. *Source* Sanromán (2013)

the terraced landscape is associated with water canals for irrigation. The hill of Tetzcutzingo is completely terraced, and there is archeological evidence of baths and terraced gardens where the king Netzahualcoyotl had his recreation area. Water came from the springs of the Sierra de Tlaloc through a network of canals specially made for this purpose.

Studies from the 1990s (Sokolosky 1995; Rodríguez 1995) describe the conditions of terraces in various villages of Acolhuacán. For example, in San Jerónimo Amanalco and San Juan Tezontla the terraces are located between the villages and are associated with irrigation systems through a network of canals and water deposits. The walls or edges, constructed with stones, retain the terraces. Other types of walls have maguey plants to hold the soil on slopes and prevent erosion, as in the *metepantles* of Tlaxcala. The main crops are maize, beans, barley, flowers, and fruit trees.

It was still common in the 1970s, in the town of Purificación Tepetitla, for houses to stand next to the approximately 3000 m^2 agricultural terraces, where crops of maize, fruit trees, flowers, and medicinal plants prevailed. In the following decades, this village's terraces were impacted by population increase, subdivision of land, and abandonment of agriculture (Ennis-McMillan 2001).

10.6 Terraces in the Toluca Valley

Mountain ranges and a volcano surround the Toluca Valley (Fig. 10.5). The terraces are located on slopes of hills and ravines. In the otomí region of Lerma and the Sierra de la Cruces, terraced landscapes predominate. In Ocoyoacac, there are terraces south of the archeological zone, which is in the north of the town center. In the Sierra Morelos, a conurbation of the city of Toluca, the terraces are located on the hill Tenismo of Calixtlahuaca, as well as on different slopes in Tlaxomulco (Fig. 10.6), San Marcos Yachiuacaltepec, and San Mateo Oxtotitlán. South of the city of Toluca, there are terraces on the hill of Tlacotepec and in some areas of Nevado de Toluca's northern slope (Smith 2006; Pérez and Juan 2013, 2016).

Studying the archeological zone of Calixtlahuaca, Frederick and Borejsza (2006) identify two terrace types on the slopes of the Tenismo hill, next to the archeological zone. In the middle and upper slopes, terraces dominate *metepantles*; the terraces have a stone wall, and the *metepantles* have a land wall with vegetation of maguey or fruit trees. Donkin reports that, in the terraces of Calixtlahuaca, ground erosion destroyed diverse areas of cultivation; this situation has not changed much at the present time, as reported by Frederick and Borejsza (2006) and Pérez and Juan (2016).

Some terraces are no longer cultivated and are used for grazing, while, in others, farmers still grow corn, beans (*Phaseolus sp.*), and pumpkin (*Cucurbita sp.*) for local consumption, as well as endemic fruit trees such as the wild cherry (*Prunus serotina*) and hawthorn (*Crataegus mexicana*) (Pérez and Juan 2016). The terraces

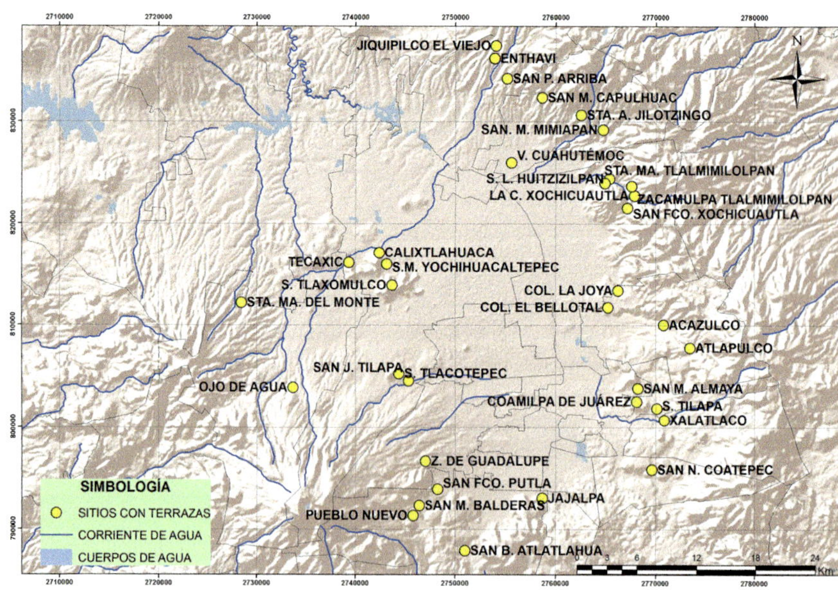

Fig. 10.5 Map of Toluca Valley

Fig. 10.6 Terraces of Tlaxomulco, Toluca. *Photo* J. M. Pérez Sanchez

of the high slope are in poor condition and have been abandoned; maguey is cultivated on some of them because it does not need permanent care, given that it resists low temperatures.

Pérez and Juan (2013) report that there are terraces in other areas of the Toluca Valley—for example, in the villages of Zacamulpa Tlalmimilolpan, San Isidro, La Concepción, and Xochicuautla in the Sierra de Las Cruces, whose altitude ranges from 2400 to 2800 m above sea level. In Ocoyoacac, on the southern slope of the archeological zone, there are terraces with rock walls and drainage canals, where the endemic vegetation consists of wild cherry (*capulín*), hawthorn (*tejocote*), and maguey (Fig. 10.7). South of the Toluca Valley, in Santa María Jajalpa, there are records of corn cultivation in *bancal*-type terraces, which are associated with ditches. Other sites with possible terraces are the hill of Tlacotepec, the northern slope of Nevado de Toluca, San Miguel Balderas, Santa María del Monte, and various hills in the State Park Sierra Morelos of Toluca. On the terraces of Santa María Jajalpa, family groups carry out work and maintenance, and, occasionally, they hire one or two workers for planting and harvesting (Medina 2010).

10.7 Terraces in Oaxaca

Oaxaca is one of the places with the highest presence of terraced landscapes, according to Donkin's research (1979) and recent studies, such as those by Pérez (2006, 2015, 2016) and Heredia et al. (2008). In the Mixteca Alta, Pérez Rodríguez describes terraces in Jazmín Hill (Nochistlán), while Donkin identifies terraces in Coixtlahuaca, Nochistlán, Huajuapan, Tamazulapan-Teotongo, Yolomecatl,

Fig. 10.7 Terraces of Ocoyoacac, Toluca Valley. *Photo* J. M. Pérez Sanchez

Huamelulpan, Achiutla, and Tilantongo-Monte Negro (Fig. 10.8). In Tamazulapan and Yolomecatl, Donkin found abandoned terraces with fractured walls and eroded surfaces, and, to the north of them, he found terraces, which were more conserved and functioning, irrigated by stone canals that flow into ravines. Other terraces have smaller stone canals measuring only a few meters long.

In the southern half of the Sierra de la Mixteca, terraces are predominant, from the North and East of Nochistlán to Coixtlahuaca. Northeast of Nochistlán; East of Yanuitlán; and West of Apaola, Santa Maria Apasco, and Coixtlahuaca, most of the terraces are cultivated, in contrast to the contour terraces which have been abandoned. In another region, the valley of Oaxaca-Tlacolula, stone terraces are in use, but some sloping terraces have not been cultivated and are abandoned.

From archeological studies of the Jazmín Hill landscape in the Mixteca Alta Central, Pérez (2006) identifies terraces of the *lama-bordo* type—a long series of terraces built in the drains and ravines that capture adjacent slope floors. These terraces spread from the hilltops to the bottoms of the valleys and can measure up to four kilometers in length. Despite the terraces being abandoned for a long time, they are still present and contribute to reduce soil erosion (Pérez 2015).

In some excavated sites on Jazmín Hill, terraces are associated with urban centers, and narrow terraces—one and two meters wide—have been found, which were modified over time and have served to "stabilize soils, create broader areas for living and working, serve as roads or as possible green areas […] to produce food,

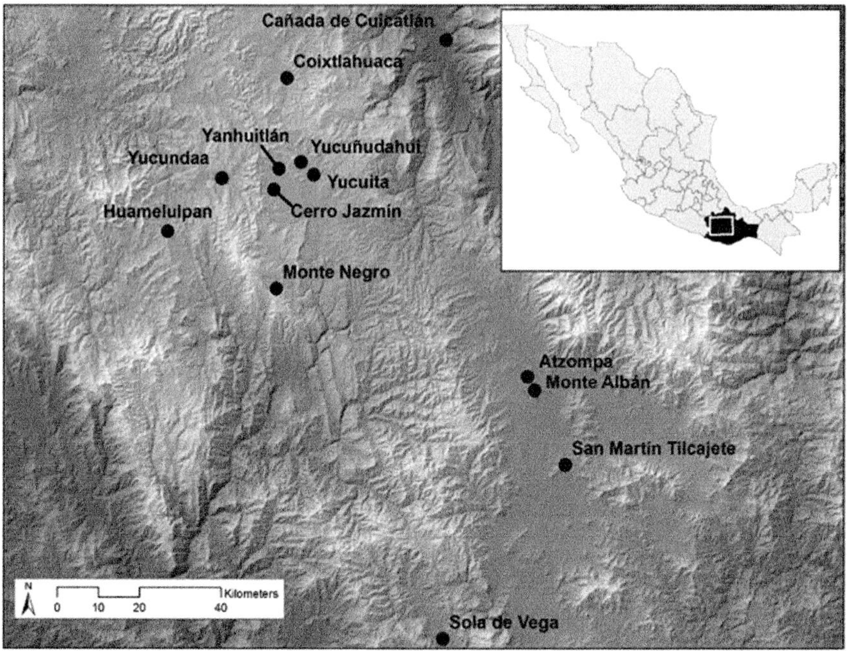

Fig. 10.8 Map of Mixteca Alta, Oaxaca. *Source* Pérez et al. (2017)

for [recreation] and [...] to create green barriers between neighbors" (Pérez 2015: 83). Currently, farmers using *lama-bordo* terraces cultivate maize (*cajete* and temporary; Zea mays), beans (*Phaseolus sp.*), pumpkin (*Cucurbita sp.*), and *chilacayota* (*Cucurbita sp.*) (Pérez 2006).

A recent study by Pérez (2016) shows the construction method for the terraces in Huamelulpan, Santa María del Rosario, and San Pedro Coxcaltepec:

> [...] process in which green and dry brush is cut and piled in mountain drainages in a cross-channel direction to retain soil and sediments washed down from the hills during the rainy season. Called *bordos*, these brush barriers are lined in the front with stones found nearby to create a permeable retention wall. The stones are carefully fitted like a jigsaw puzzle, but they are not cut or bound with mortar. The placement of stones is progressive and begins after the brush *bordos* have already begun to collect sediment. Low, vertical rows of stone are positioned in front of the brush *bordos* to create the base of a terrace wall. As the rains continue to transport more sediment, additional stone rows are placed slightly upslope and at a tilt, increasing terrace height. (Pérez 2016: 22)

This type of terrace construction protects the crops and prevents the wall from collapsing from the strong water currents that seep into the walls. Family groups (four to eight people) are in charge of terrace maintenance, and, when the work is too much for one group, the cooperation of two or more family groups (*gueza*) is required. Due to the migration processes in which young people participate, families here must occasionally, as in the Tlaxcala region and the Toluca Valley, hire workers for planting or weeding (Pérez 2016).

10.8 Terraces in the Mayan Area

Studies of terraces in the Mayan area (Fig. 10.9) are important and have contributed to an understanding of how agricultural systems helped develop Mayan civilization and society in relation to the environment. There is evidence to suggest that the

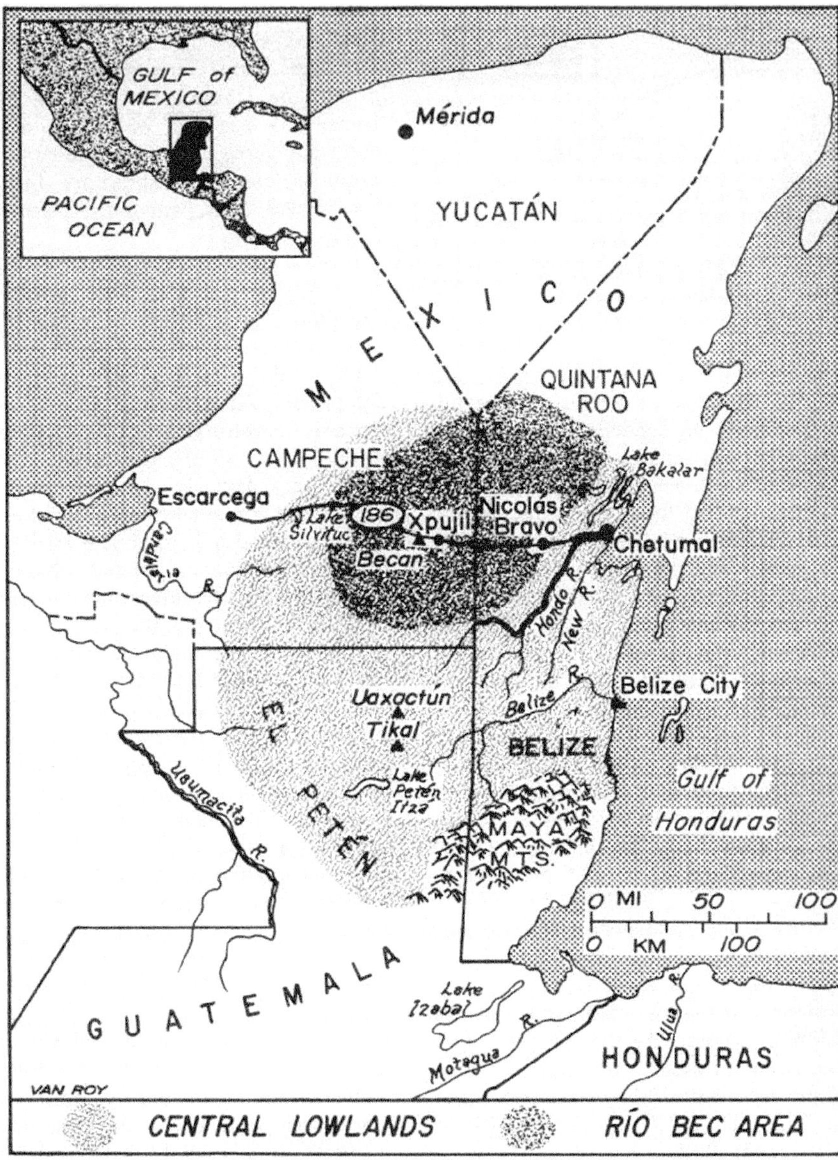

Fig. 10.9 Map of the Mayan area. *Source* Turner II (1974)

Mayans developed other intensive cropping systems, and not only the slash-and-burn system, as formerly believed. Turner II (1974) and Lobato (1988) report evidence of intensive terrace systems and elevated fields in the Río Bec region, between Quintana Roo and Campeche and the Usumacinta River basin in the state of Chiapas, Mexico.

Turner II (1974) describes the Río Bec region's terraced systems, mounds, and archeological ruins, which are located in limestone hills that range from 20 to 60 m high. These hills contain shallow, drained soils and are susceptible to erosion in the rainy season. Between Xpujil (Campeche) and Nicolás Bravo (Quintana Roo) terrace surfaces cover an area of ten square kilometers. Two types of terraces were identified there. The first consists of limestone walls, with a height of 0.80–1.40 m, filled with debris. The second type is made of smaller embankments formed by walls of unfilled limestone rocks or rows of vertically buried rocks; the walls are less than 0.40 cm high, and the width of the base fluctuates between 0.60 and 0.80 cm. Such terrace construction facilitates drainage of excess water via guide walls, infiltration outlets, and lateral inclined embankments. The small walls and the paths between the terraces allow control of water flow.

Another case in the Mayan area has been registered by Lobato (1988), in the Santo Domingo River valley, which is in the municipality of Ocosingo in Chiapas (region of the Usumacinta River). This valley has an altitude of 400 m above sea level. It has several hills between the sierras of Piedras de Bola and Güiral, and it has an average annual rainfall of 2566 mm. In this region, three terraced systems have been identified. The first consists of a set of seven bench terraces with stone walls facing east, covering an area of 930 m^2. The average slope of the terraces is 39%. The terrace walls are made of sedimentary and calcareous rock, and they are generally 3.0 m high by 2.50 m wide; the soil is clay-calcareous with a depth of 25 cm.

The second system consists of terraces with walls of small calcareous rocks, different from those of the first system, which are brought from a nearby stream and are, on average, 0.60 cm high. The third system is made of four terraces covering an area of 7200 m^2. The walls, which oscillate between 1.50 and 2.50 m high, are constructed with rocks of greater volume than those used in the other systems. Lobato (1988) considers that these terraced systems were carefully constructed under proper planning, and, at the time of their discovery, types one and three were in agricultural production under the slash-and-burn system where maize and beans were grown with good yields. Toward the 1980s, the farmers continued to use the terraces under the slash-and-burn system with corn and bean crops. The time, effort, and organization required to build and maintain the large-scale terraces were great, so it is inferred that the work on the terraces was intensive to produce high yields per unit area.

10.9 Conclusions

In Mexico, terraces are one of the traditional agricultural systems of Mesoamerican origin, and there are two predominant types—terraces with rock walls and *metepantles*. The terraces and *metepantles* are considered part of traditional agriculture and agroforestry and are associated with traditional crops, such as maize, pumpkin, green beans, and fruit trees, which provide food security for farm families. The terrace use and management continue to be carried out by peasant and indigenous families, who have accumulated a body of knowledge that allows them to develop this system and control soil, vegetation, and forms of cultivation.

Most studies of Mexican terraces have been archeological, discussing the origin and development of civilization and giving an account of the terracing system's historical development, the growth of complex societies, techniques of building walls and water canals, the relationship between terraces and irrigation through canals and reservoirs, and the relationship between terraces and former urban centers.

In Mexico, agricultural terraces are distributed in various regions: northern, western, central, southern, and eastern. Although there are records of terraces in other regions, it is necessary to carry out more ethnographic studies that describe the current state of conservation, use and construction techniques, social organization, crops and medicinal plants, flora and associated fauna, ecological knowledge concerning terrace development, and terrace contribution to environmental conservation and its relationship with the market. Different types of terraces have been identified, and the nomenclature used to refer to them varies in different regions—for example, *metepantles* and milpa (cornfield) in the Central Highlands, levees in Hidalgo, *lama-bordo* in the Mixteca Alta of Oaxaca, and terraces in Jalisco, Chiapas.

Agricultural terraces are the basis for modeling soil conservation efforts, such as those in Oaxaca under the leadership of *Milpa Intercalada con Árboles Frutales* (MIAF) (Cortés et al. 2012), which aimed to capture coal and boost the economy of farming families, alternating traditional crops with fruit trees, as well as increasing the amount of organic matter in the soil, controlling soil erosion, using rainwater more efficiently, and contributing to the food security of indigenous families. Another strategy that could be developed to boost agricultural terraces is agro-ecological tourism, which combines archeological sites with terraced crops.

References

Aguilar J, Illsley C, Marielle C (2003) Los sistemas agrícolas de maíz y sus procesos técnicos. In: Esteva G, Marielle C (eds) Sin maíz no hay país. Consejo Nacional para la Cultura y las Artes, pp 83–122

Bilbao J (1979) Sistemas y prácticas agrícolas en una comunidad de Tlaxcala. In: Ponencia, 43° Congreso Internacional de Americanistas, Vancouver, Canadá

Cook A (1986) El control de la erosión en Tlaxcala: un problema secular. Erdkunde 40:251–261
Cortés FJ, Turrent A, Hernández E, Nicolás N, Torres J, Zambada A, Díaz P (2012) Milpa Intercalada con Árboles Frutales (MIAF). Colegio de Posgraduados, Secretaría de Agricultura, Ganadería, Desarrollo rural, Pesca y Alimentación
Diccionario del náhuatl en el español de México (2007) UNAM
Donkin RA (1979) Agricultural terracing in the aboriginal new world. The Werner-Gren for Anthropological Research, The University of Arizona
Doolittle W (2004) Canales de riego en el México prehistórico. La secuencia cultural del cambio tecnológico. Primera edición en español. Universidad Autónoma Chapingo
Ennis-McMillan MC (2001) La Purificación Tepetitla. Agua potable y cambio social en el somontano. Universidad Iberoamericana, CNA-CIESAS
Frederick C, Borejsza A (2006) Reconocimiento geoarqueológico de Calixtlahuaca. In: Smith M (ed) Proyecto: Calixtlahuaca. Organización de un centro urbano posclásico. Informe Técnico parcial. Instituto Nacional de Antropología e Historia, Arizona State University, pp 33–38
García Cook Á (1986) El control de la erosión en Tlaxcala: un problema secular. Erkunde 40:251–262
González JA (2016) Sistemas agrícolas en orografías complejas: las terrazas de Tlaxcala. In: Moreno AI, Casas A, Toledo V, Vallejo M (eds) Etnoagroforestería en México. Universidad Nacional Autónoma de México, pp 109–145
Harrison PD, Turner II BL (1978) Pre-hispanic Maya agriculture. University of New Mexico
Heine K (1975) Sobre la disposición y antigüedad de las terrazas de la ladera poniente del cerro Xochitécatl, Tlaxcala, México. Comunicaciones 11
Heredia EVY, Kowalewski SA, Pérez RV (2008) Cerro Jazmín: La morfología de un centro urbano en la Mixteca Alta. In: Guadalupe Mastache A, Cobean RH, García Cook A, Hirt KG (eds) El Urbanismo en Mesoamérica. Instituto Nacional de Antropología e Historia and Pennsylvania State University, pp 423–446
Lobato R (1988) Terrazas prehispánicas en la región del río Usumacinta y su importancia en la agricultura maya. Estud Cult Maya XVII:19–58
Martínez M, Rubio E, Palacios C (1999) Terrazas. SAGARPA, Colegio de Posgraduados
Medina AI (2010) El sistema agrícola tradicional: el cultivo del maíz en Santa María Jajalpa. Tenango del Valle Estado de México. Tesis de Maestría en Antropología Social, Facultad de Antropología, Universidad Autónoma del Estado de México
Moctezuma S, Pérez JM, Rivera G (2015) Aportes alimenticios de los agroecosistemas tradicionales en el México rural. In: Padilla S (ed) La crisis alimentaria y la salud en México. Castellanos editores, UAEM, pp 83–100
Moreno CAI, Toledo VM, Casas A (2013) Los sistemas agroforestales tradicionales de México: Una aproximación biocultural. Bot Sci 90(4):375–398
Moreno AI, Casas A, Toledo V, Vallejo M (comp.) (2016) Etnoagroforestería en México. Universidad Nacional Autónoma de México
Mountjoy D (1985) Adaptation and change in a local agroecosystem of Tlaxcala, Mexico. Senior Thesis. University of California
Palerm A (1967) Agricultural systems and food patterns. In: Nash M (ed) Handbook of middle American Indians, anthropology social. University of Texas, pp 26–52
Palerm A (1972) Agricultura y civilización en Mesoamérica. SepSetentas
Palerm A (2008) Sistemas agrícolas en Mesoamérica contemporánea. Guía y lecturas para una primera práctica de campo, Segunda edición. Universidad Autónoma de Querétaro, pp 243–281
Palerm A, Wolf E (1972) Agricultura y sociedad en Mesoamérica. SepSetentas
Patrick L (1977) A cultural geography of the use of seasonally dry, sloping terrain: the "Metepantli" crop terraces of Central Mexico. PhD University of Pittsburgh
Pérez RV (2006) Sociedades complejas y paisajes agrícolas: un estudio regional de asentamientos y terrazas en la Mixteca Alta, Oaxaca, México. In: Grau Mira I (ed) La aplicación de los SIG en la arqueología del paisaje. Universidad de Alicante, España, pp 247–254

Pérez SJM (2014) Agricultura de terrazas en Tlaxcala. La Caridad Cuaxonacayo. Gobierno del Estado de Tlaxcala

Pérez RV (2015) Investigaciones resientes sobre el urbanismo temprano en Cerro Jazmín, Mixteca Alta, Oaxaca. Cuadernos del Sur 21(40):62–91

Pérez RV (2016) Terrace agriculture in the Mixteca Alta Region, Oaxaca, Mexico: ethnographic and archeological insights on terrace construction and labor organization. Cult Agric Food Environ 38(1):18–27

Pérez SJM, Juan PJI (2013) Caracterización y análisis de los sistemas de terrazas agrícolas en el valle de Toluca, México. Agric Soc Desarrollo 10(4):397–418

Pérez SJM, Juan PJI (2016) Agricultura de terrazas en el cerro Tenismo, Toluca. Terra Nueva Etapa 31(51):163–184

Pérez RV et al (2017) Chronological trends in the use of valley of Oaxaca ceramics and ceramic styles at Cerro Jazmín, Mixteca Alta, Oaxaca. J Arch Sci Rep 12:580–590

Rodríguez RAR (1995) San Juan Tezontla. Lucha por el agua. Colección Tepetlaoxtoc. 6. Universidad Iberoamericana

Rojas RT (2001) La tecnología agrícola. In: Manzanilla L, López LL (eds) Historia antigua de México. Aspectos fundamentales de la tradición cultural mesoamericana, vol IV. CONACULTA, INAH, UNAM, Porrúa, pp 13–68

Sanromán, RJC (2013) "Ayuda" y "respeto" en San Pedro Chiautzingo, agencia e interdependencia en las relaciones de sus pobladores. Tesis de maestría en antropología social, Universidad Iberoamericana, México.

Smith M (2006) Proyecto: Calixtlahuaca. Organización de un centro urbano posclásico. Informe técnico parcial. Temporada 2006. Instituto Nacional de Antropología e Historia, Arizona State University

Sokolosky J (1995) San Jerónimo Amanalco. Un pueblo en transición. Colección Tepetlaoxtoc. 5. Universidad Iberoamericana

Trautmann W (1981) Las transformaciones en el paisaje cultural de Tlaxcala durante la época colonial. Franz Steiner Verlag GMBH, Weisbaden

Turner BL II (1974) Prehistoric intensive agriculture in the Mayan lowlands. Examinations of relic terraces and raised fields indicates that the Río Bec Maya were sophisticated cultivators. Science 185:118–124

Part II
Towards a Multifunctional Vision of Terraced Landscapes

Chapter 11
Mapping Agricultural Terraces in Italy. Methodologies Applied in the MAPTER Project

Francesco Ferrarese, Salvatore Eugenio Pappalardo, Alberto Cosner, Stefano Brugnaro, Kaodi Alum, Angelica Dal Pozzo and Massimo De Marchi

Abstract Agricultural terraced systems and their geographical distribution are not often represented in maps, even if they play an important role in hydro-geological hazard: Moreover, detailed cartography of these features could improve understanding and valuing their environmental, social, and economic complexity. The general aim of the MAPTER project is to map and quantify the extension the Italian terraced systems. To do that, it was necessary to apply different methodologies according to different data sources locally available: (i) traditional cartographic and photo analysis, (ii) orthophoto and high-resolution satellite images, (iii) DTM LiDAR analysis, (iv) Web Map Services and Geobrowser analysis, (v) participatory mapping and Voluntary Geography Activities, (vi) the use of unmanned aerial vehicles (UAVs). At the time of this paper, the LiDAR DTM allows us to highly improve the traditional methodologies of cartographic and aerial photos analyses, in terms of precision, costs, and time-consuming: Many terraced systems in Italy appear to be underestimated by the previous, traditional, methodologies in comparison with the results obtained by LiDAR technology and LiDAR data elaboration. This work tested different methodologies and approaches to survey study areas in the Italian context, representing a starting point to coordinate researches among academic institutions, public administrations, voluntary geographers, and citizens science, in order to implement the geographical knowledge for terraced systems and landscapes, key heritage of Italian territories.

F. Ferrarese (✉) · K. Alum · A. Dal Pozzo
Department of Historical, Geographic Sciences and the Ancient World, University of Padova, Padua, Italy
e-mail: francesco.ferrarese@unipd.it

S. E. Pappalardo · S. Brugnaro · M. De Marchi
Department of Civil, Environmental and Architectural Engineering, University of Padova, Padua, Italy

A. Cosner
Archaeologist Freelance Professional, Trento, Italy

M. Varotto et al. (eds.), *World Terraced Landscapes: History, Environment, Quality of Life*, Environmental History 9,
https://doi.org/10.1007/978-3-319-96815-5_11

11.1 Introduction

The spatial dimension of terraced landscapes currently represents a challenge to perform a synoptic observation of such peculiar agricultural ecosystems. The terraced landforms are paramount not only in terms of cartographic representation but also in understanding and valuing their environmental, social, and economic complexity that range from change of agricultural practices to geo-hydrological hazards. Hence, locating and mapping agricultural terraced landforms are gaining even more attention both in the academic and in the policy-making spheres, especially for agricultural terraces that are still not represented in official maps or for those which are disappeared from the physical and cultural landscape. Within such framework, the University of Padova launched the MAPTER project, by providing an overview of the terraced systems on the national territory. In Italy, terraced systems are widely diffused as documented in the first geographical survey at national scale which was developed by ISPRA (2013) and Bonardi and Varotto (2016). The MAPTER project was developed in preparation of the Third International Meeting "Terraced Landscapes: choosing the future," held in Italy (Venice-Padua, October 6–15, 2016). Due to different data sources, MAPTER project adopted a multi-scale approach comparing different methodologies to investigate characteristics about the morphology of the terrace systems. The aims of the MAPTER project are (i) identifying and mapping agricultural terraces; (ii) extracting terraced features such as terrace risers, especially dry-stone walls; and (iii) quantifying the extension of the agricultural, or abandoned, terrace system areas. During the project, different methods were tested, according to data source availability: LiDAR Digital Terrain Model (DTM), high-resolution satellite images, participatory mapping and Voluntary Geography spatial data, and the use of unmanned aerial vehicle (UAV) . Finally, terraced systems were analyzed in order to calculate some geographical parameters such as intensity and extension indexes and to perform a specific classification according to altitude, slope, lithology, and agricultural use (see Varotto et al. in this book).

11.2 Materials and Methods

To map and to survey terraced systems, different methodologies were tested and implemented in the last 20 years, according to geospatial technologies evolution and data availability.

11.2.1 Traditional Cartographic and Photo Analysis

The first and the long-standing methodology is based on traditional cartography analyses, often supported by aerial photos study and field survey. Aerial photos usually are made with an overlap that allows a three-dimensional view—by mean of stereoscope—that could enhance the understanding of the terrace risers. To note that the geometric distortion in aerial photos often leads to relief displacements (Lillesand et al. 2015). These spatial errors required a huge amount of field surveys in order to perform data validation and ground truth to complete the terrace system mapping. This method is strongly related to the geographic acquisition processes which include the analogical use of both aerial photos and field data survey, manually transferred on static paper maps and, sometime, georeferenced and processed in geographic information system (GIS) environment by the use of a digitizer.

11.2.2 Orthophoto

In the early 2000s, high-resolution orthophotos at 1.0 or 0.5 m were available and they could just improve the survey in order to have more accuracy both in geometry and radiometry of data. At the time of this paper, this geospatial information presents the advantage of ortho-rectification and high geometric resolution (sometimes under 0.2 m cell size). Therefore, mapping and extracting terraced systems by visual analysis, or supervised/unsupervised classification, are not affected by position errors. Moreover, orthophotos often make more detectable dry-stone walls from different types of terrace risers.

11.2.3 WMS and Geobrowser

Important geospatial data are at present also provided by Web Map Services (WMS) and Geobrowsers, which currently represent important resources to acquire and process high resolution, georeferenced, and ortho-rectified satellite images, which are mostly freely available to any Web users. These Web-based services are very useful whereby no other kind of spatial data are available or in case of terraced systems located in remote areas which would require field survey in situ. WMS are usually provided by institutional Geoportals from Regions, Ministry of Environment, and Geographical Military Institute (*Istituto Geografico Militare*, IGM) as well as no profit organizations; Geobrowsers are usable by Google Earth™, Bing Map™, QGIS and ESRI™ platforms. Geobrowser and WMS service are particularly useful for participatory and voluntary geography that has usually less skill of a GIS analyst.

11.2.4 LiDAR Survey

In the early 2000s, a new remote sensing technology based on laser light (Light Detection and Ranging: LiDAR) boosts a worldwide revolution in high-resolution geomorphometry analysis and landform survey (Jasiewicz et al. 2015). LiDAR DTM is a powerful set of elevation data that could represent dry-stone walls and terraced systems with a high level of accuracy at local scale. Scientific literature documented several morphometric studies about methodologies to identify, extract, and measure agricultural terraced landforms (Ninfo 2008; Passalacqua et al. 2010; Ore and Bruins 2012; Sas et al. 2012; Sofia et al. 2014; Tarolli et al. 2015). In addition to the cited literature about terrace features extraction, it is worth to also consider Hengl and Reuter (2009), Jasiewicz et al. (2015) since their works offer a wide and complete overview of morphometric analysis and feature extraction proper of DEMs and DTMs. The greatest strength of LiDAR data is related to terrain morphology detection under the canopy, enabling mapping surveys in wider areas and allowing further analyses to verify presence and geometry of agricultural terraces previously mapped only by traditional methodology. LiDAR detection has also the capability to obtain different surface models: Digital Surface Model (DSM, the earth surface with tree canopies, building, aerial wires, etc.) and Digital Terrain Model (DTM: the bare soil surface). LiDAR data require specific software for their management: first to pre-elaborating raw data (as Terrascan™) and then for data elaboration (as GIS software). In GIS environment is extremely performing the tools of surface processing—that allow the calculation of parameters such slope, aspect, curvature—and the tools of spatial analysis, for the calculation of drainage channels and direction, or for smoothing and filtering noise from DTM surface.

11.2.5 UAV Survey

Another more recent methodology to map and extract terrace landforms is the use of optical sensors on unmanned aerial vehicles (UAVs). The main advantages in using UAV are based (i) on the very high spatial resolution of aerial photos (0.05–0.3 m) due to the low altitude of the survey flight that give a strong accuracy of details and (ii) the temporal resolution which allows monitoring the biophysical, the hydraulic, and the land use status. On the contrary, if the very high spatial resolution is one of the strengths of this mapping methodology, it could also represent its same weakness, due to time- and resource-consuming, obstruction of vegetation and costs to survey wide areas. UAV data require specific software (as Pix4D™) for the elaboration of orthophotos and DTM and DSM. Then, digital models and orthophotos could be analyzed in GIS environment.

11.3 Result and Discussion

In some cases, especially remote zones which are not covered by high-resolution spatial data such as satellite imageries or LiDAR data, the use of traditional cartographic analysis, i.e., the long-lasting methodology described, is often required. In the maps of Italian Military Geographical Institute (IGMI) and in the maps of *Carta Tecnica Regionale* (CTR, Regional Technical map), the presence of cartographic symbols which represent supporting structures like dry-stone walls is often useful to directly identify terraced features and to directly extract these features by selection (in case of digital topographic map) or to indirectly map them by on-screen digitizing. The nominal scale of the maps is contingent upon both the accuracy of individuation and extraction of the features, and the spatial accuracy of the survey employed in the map construction. As an instance, the Italian Regional Technical Map (CTR) is usually an appropriate scale (1:10,000 and 1:5000) which make possible to represent dry-stone walls, for terrace steps major or equal to 10 m, among different supporting structures; however, regional technical maps are recent, edited after 1980, and their information derived from spring or summer aerial survey of the same time: Therefore, in case of abandonment of agricultural activities, growth of vegetation could hide the presence of terrace landforms. Hence, the use of this methodology has required the implementation of field and/or dedicated aerial surveys such oblique and zenithal photos in the winter season to reduce the canopy hiding effect. In Italy, many previous studies tested or applied this methodology when LiDAR or UAV data were not yet available (Varotto 2004; Tres and Zatta 2006; Rizzo et al. 2007; Varotto and Ferrarese 2008; Agnoletti et al. 2015; Modica et al. 2015). In Sicily Region, a whole mapping of agricultural terraced systems was performed at the regional scale, combining traditional cartographic analyses with use of aerial photos, field survey and on-ground photos as well. In this case, ground photos were often useful due to the arid climate and the seldom vegetation that let the terrain morphology to be seen from remote (Barbera et al. 2010). A survey on LiDAR DTM of the Filicudi Island (Fig. 11.1: 1) suggests that the total extension of terraced systems in Sicily could be underestimating (Terranova 2016). A digital map survey of the Campania Region was performed both by data collected in a topological geodatabase in GIS environment and by traditional cartographic analysis (Ronza 2006; Gravagnuolo and Di Martino 2015); however, a LiDAR DTM analysis suggests a wider extension of the terraced systems with respect to the previous surveys (Brugnaro 2016).

Often, the use of remote sense imagery lets to detect few or none features in case of high and dense canopies or wooded lands, but it is worth noting that dense vegetation could be related to the season of the aerial survey and to the environmental and climate context of the area. Forest cover generally makes optically barely visible—sometimes totally invisible (Rühl et al. 2005)—from aerial imagery terrain landforms such as agricultural terraces: So, mapping these geometries is arduous. Thus, historical maps and aerial photos produced before 1960—the beginning of abandonment of terraced agriculture—are important documents which

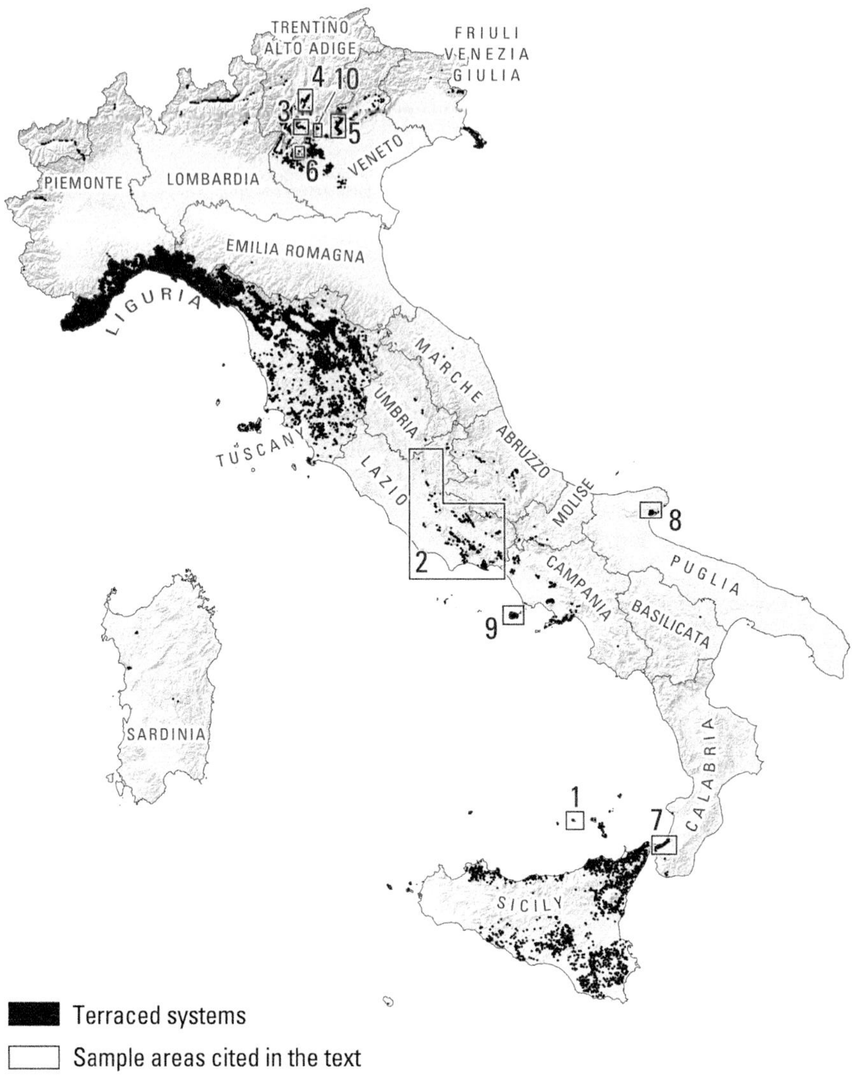

Fig. 11.1 Final result of the MAPTER project. In black terraced systems. Numbers of zones are cited in the text: (1) Filicudi Island; (2) Lazio region; (3) Terragnolo Valley; (4) Cembra Valley; (5) Brenta Valley; (6) Fumane area; (7) Costa Viola; (8) Gargano area; (9) Ischia Island; (10) Rio Freddo Valley

should be considered for mapping and surveying abandoned agricultural terraces hidden under the forest canopy. Indeed, photos and orthophotos analyses are usually combined with traditional cartographic analysis and field survey (Tres and Zatta 2006; Varotto and Ferrarese 2008). The first map of terraced systems obtained by using orthophotos, at a regional scale, was about the Liguria Region, in which the analysis has been validated with the aid of cartographic and field survey (Brancucci and Masetti 2008; Brancucci and Paliaga 2006; Brandolini et al. 2008; Brandolini 2010).

Within the MAPTER project, WMS and Geobrowsers were paramount in mapping terraced systems, especially in arid or Mediterranean semi-arid environments, like the mid-southern regions (36–42°N), where dense vegetation canopies are quite rare and so the morphology is easily visible (See Varotto et al. in this book). At the same time, it is worth noting that terraced landforms' detection need high-resolution images, in addition to a landscape not covered by a dense canopy. Moreover, hardware needs broad bandwidth to load high-resolution imagery of WMS. Currently, an interesting geotool is represented by the Street View Service, implemented into the Geobrowser Google Earth™: By means of on-ground 360° georeferenced photos, this application may be used as a powerful tool to perform ground truth on remotely sensed data previously analyzed, identifying terraced systems on the ground. But a strong limit of this survey is that terrestrial imagery is provided only from roads, sometimes far from terrace systems. In some study areas, during the activity of MAPTER project, we combined visual analysis on satellite images with Street View ground photos to survey terraced systems: According to GIS analysis, this methodology showed good results in the Lazio Region (Figs. 11.1: 2 and 11.2), by 5673 ha of agricultural terraces mapped (the fifth Italian Region for terraced systems on percentage of regional area); on the contrary, the same combined methodology performed poor results in Marche and Veneto Regions, due to the presence of wide canopy areas and for the high value of abandonment of terraced areas.

The MAPTER project (see Varotto et al. in this book), thanks to the availability of LiDAR DTMs for many terraced areas, was carried out by performing features extraction by the analysis of the profile curvature, which is the amount of concave or convex profile along the maximum slope (Minar and Evans 2008; Hengl and Reuter 2009). We also performed a cluster analysis of topographic derivatives, such as slope, profile curvature, high-pass filters. Moreover, we tested and compared results in three different ways: (i) by DTM derivatives for automatic features extraction; (ii) DTM and its derivatives as a background for on-screen digitizing; (iii) by traditional cartographic analysis combined with display analysis of orthophotos. The DTM derivatives analysis may need some pre-elaboration processes—such as a gentle smoothing of the surface or a striping removal—compared to the display analysis on orthophoto and satellite imagery. The profile curvature parameters strongly detect the concave high value of the top of terrace risers, so the reclassification of this parameter lets us identify the top of risers and then convert it into a vector format (shapefile) more suitable for length measure. The first application of this method was made over two different areas, of about 20 km^2, in Terragnolo and Cembra valleys (Autonomous Province of Trento) (Fig. 11.1: 3 and 4 respectively). To notice that Cembra Valley LiDAR data were more suitable for an accurate landform analysis due to the non-abandonment status of agricultural terraces, characterized by no high canopy cover on the terraces. On the contrary, in the Val Terragnolo area many agricultural terraces have been abandoned and the

Fig. 11.2 Google Earth™ and Street View™ show a clearly terraced system in the Vico nel Lazio Municipality (Frosinone Province, Lazio Region). The hot-summer Mediterranean climate and the high resolution of the imagery make the remote survey a quite easy and satisfactory tool

area is now completely re-vegetated by woodland with dense canopy: Here, LiDAR DTM survey allows us to quantify the extension and geometry of terraces but it is hard to detect terraces edges (walls) without extracting together a lot of spurious features. These tests were made by applying surface tools of QGIS 2.8 GIS software.

In the study area of the Brenta Valley (Veneto Region) (Fig. 11.1: 5), we were able to compare the results from a previous survey—which has been made by using traditional cartographic analysis on orthophotos plus field surveys—with the results carried out from LiDAR DTM feature extraction using both ESRI ArcGIS10.2™ and IdrisiGIS™ software. It is interesting to note that the use of the traditional methodology (topographic maps, aerial photos, and field survey) led to the identification of 320 ha of terraced surfaces in comparison with the 317 ha obtained by LiDAR DTM analysis; but, with regard to the dry-stone walls length, we have counted, respectively, 223 against 253 linear kilometers drawn or extracted by GIS analysis (Fig. 11.3): This data comparison highlighted, through the bigger feature density, the capability of LiDAR DTM for more accurate terrace wall identification.

Of course, the width of terraced steps is a fundamental parameter for the identification of these landforms. Narrow terraces, of 1 or few meters, could not be

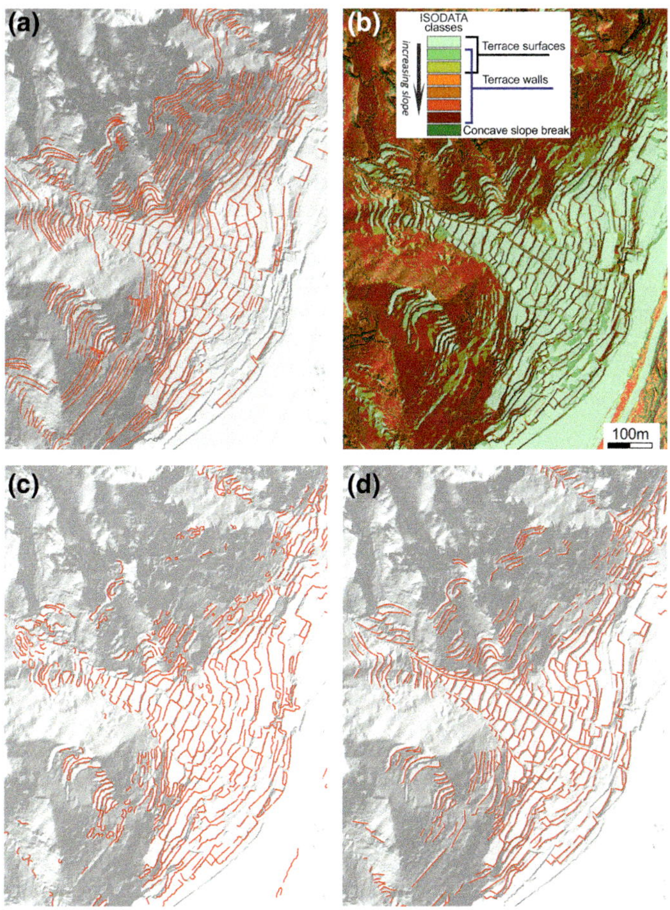

Fig. 11.3 San Gaetano Village in Brenta Valley (Vicenza Province, Veneto Region) lies in a terraced fan. The traditional survey shows an extensive terraced landscape, while the LiDAR survey locally performed a more accurate result. In **a** the first survey by aerial photos, maps, and field work: Here, it is to notice the missing of drainage channel walls, not recognized both in the aerial photos and in the map. In **b** a result of ISODATA routine in IDRISIgis^{TM} to extract eight cluster classes from five DTM derivatives. In **c** wall features extraction and selection from cluster grouping shown in **b**. In **d** the same data of **c**, refined and completed by means of display analysis on slope

detected also by an airborne LiDAR survey (read further in the text). However, aerial photos available for this area were made in spring or summer seasons, with yet a dense canopy cover. In the Brenta Valley area, there are only dry-stone walls, so the landforms extracted are only of this type. On the contrary, in a sample area of 3070 ha in the Fumane Municipality (Province of Verona, Veneto Region) (Fig. 11.1: 6), the traditional methodology, by means of cartography, aerial photo, and field survey, led to the quantification of 86.4 ha of terraced surfaces and 62.9 linear kilometers of dry-stone walls. The LiDAR DTM analysis of this same area was performed both by automatic extraction from profile curvature and from on-screen digitizing, using a background of topographic parameters such slope, hillshading a profile curvature, led to the identification of 1130 and 1104 ha, respectively, with 821 and 578 km of dry-stone walls or different terrace risers. This difference is due to the different target: only dry-stone walls in the first, traditional, survey and all terraced landforms in the LiDAR DTM analysis. The difference, instead, from automatic extraction and on-screen digitizing on LiDAR DTM is due, in this study case, to the capability of automatic extraction to detect a series of narrow and dense terraces (3 m wide), made for recent vineyard cultivation, not detected from the eye of the on-screen digitizing operator.

The automatic LiDAR DTM features extraction was performed also in two famous and wide Italian terraced systems: Costa Viola (Calabria Region, Tyrrhenian seaside) and Gargano Promontory (Puglia Region, Adriatic Sea) (Fig. 11.1: 7 and 8 respectively). The LiDAR DTM data were provided by the *Ministero dell'Ambiente e della Tutela del Territorio e delle Acque, Geoportale Nazionale*. It is a 1 or 2 m cell size DTM (2 m for coastal zones) that has high evidence of terraces. In these two areas, the automatic extraction by profile curvature (ESRI ArcGIS10.2^{TM}) and cluster analysis of topographic variables (IdrisiGIS^{TM}) produced a similar result, so we focused on the profile curvature reclassification, which is quicker and easier. Moreover, a striping effect in the LiDAR DTM was successfully mitigated applying a directional filter over the polylines carried out from profile curvature reclassification (Table 11.1).

The same method of reclassifying curvature profile was applied to a wide area of the Costiera Amalfitana (Salerno Province, Campania Region) and to the Ischia island (Napoli Province, Campania Region) (Figs. 11.1: 9 and 11.4). In the former case, the narrow terraced surface was locally insufficiently detected from the 2 m cell size DTM. At this moment, we have no other available LiDAR data (such raw data) to try a different interpolation for a new and finer DTM (Brugnaro 2016).

Finally, in the MAPTER project we were able to apply a UAV method. UAV survey of a terraced system was developed in a learning environment context, where the students were involved in geospatial data collection (ground control points and GPS survey) and photogrammetric pre-processing on the field.[1] Then,

[1]A fieldwork survey was organized in March 2016, with a classroom of the Second Level Professional Master in GIScience and Unmanned Aerial Systems for the integrated management of territory and natural resources, University of Padua.

Table 11.1 Results obtained by the on-screen digitizing and by the automatic feature extraction from profile curvature of the DTM in five polygons of Costa Viola terraced landscape (Calabria Region) and in five polygons of the Gargano landscape (Puglia Region)

	Surface	Length of dry-stone walls or terrain slopes					
		On-screen digitizing	Automatic DTM extraction	Automatic extraction with directional filter	On-screen digitizing	Automatic DTM extraction	Automatic extraction with directional filter
	(ha)	(m)	(m)	(m)	%	%	%
Costa Viola 1	4.96	6795	64,304	5077	100	94.6	74.7
Costa Viola 15	8.17	56,313	96,572	6952	100	1715	1235
Costa Viola 43	842	100,539	121,293	878,640	100	1206	874
Costa Viola 107	1103	38,252	54,714	496,980	100	1430	1299
Costa Viola 152	242	28,079	29,593	263,310	100	1054	938
Gargano 1	126	6046	10,024	6085	100	1658	1006
Gargano 6	811	35,825	52,032	43,624	100	1452	1218
Gargano 10	394	25,519	29,072	26,231	100	1139	1028
Gargano 11	460	44,469	58,795	44,815	100	1322	1008
Gargano 18	050	4701	6620	6099	100	1408	1297

These 10 polygons were randomly chosen from a set of 7 polygons for the Costa Viola, and a set of 20 polygons for the Gargano. Every polygon was on-screen digitized and represents a continue terrace system, divided from neighbor systems by morphology landforms (steep valley flanks, scarps). At the on-screen digitizing was given the best result, so that the percentage of the automatic extraction from DTM and the percentage of the same automatic extraction plus the following linear filter are shown in the last three fields. The removing of spurious features due to striping noise often gives a better result. The removing was applied using a directional filter in the set of extracted features

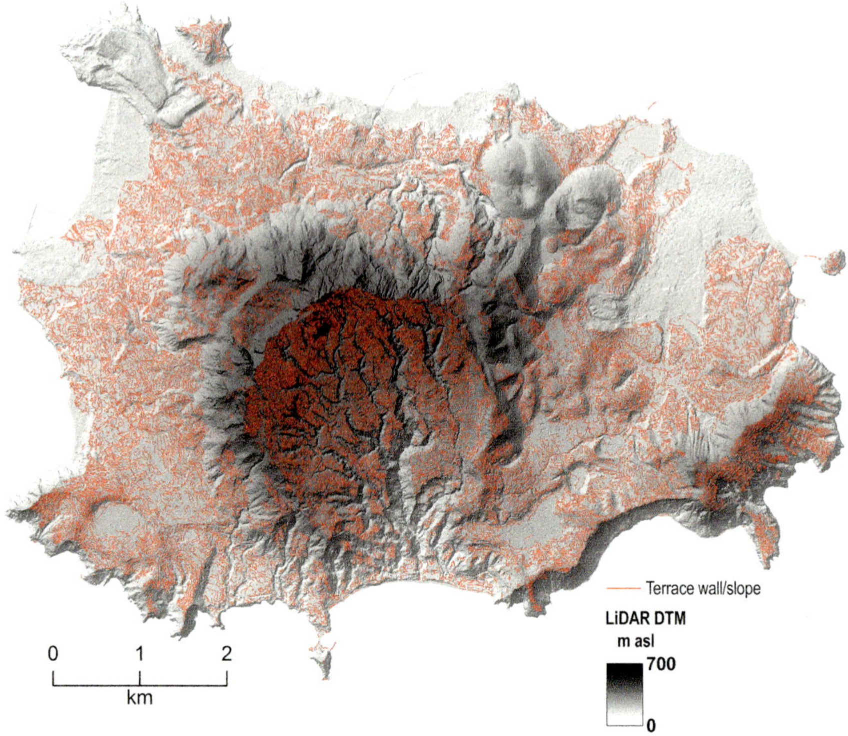

Fig. 11.4 Ischia Island in front of the Sorrento and Amalfi coasts (Campania Region). The extraction of walls/slope of terraces, by means of the profile curvature reclassification, led to the count of about 1100 km of linear features

a photogrammetric survey was performed in an 8 ha area of the Veneto Region Prealps (Rio Freddo Valley, Vicenza Province, Veneto region) (Fig. 11.1: 10). In this study case, a historical terraced agricultural system was set up in a steep slope of the valley; however, due to the sudden abandonment after the drastic flood of 1966, they were totally re-vegetated by dense deciduous woodlands which made invisible both terraced features and morphology in any recent (post 1970's) aerial or satellite image. As a result, agricultural terraces were still unmapped in any cartographic representation. A UAV survey was planned and performed in late winter, in order to have less vegetation as possible; therefore, a very high-resolution DTM at 0.05 m cell size was modeled, from stereoscopic aerial photos. Terrain analysis has clearly shown a complex terraced landforms system, generally not involved in drastic degradation processes, which is structured along the southern flank of the valley (Fig. 11.5). This UAV survey methodology highlights the possibility to map, with high accuracy and relatively at low-cost, areas of abandoned agricultural terraces for many purposes such as environmental, agricultural, and hydrogeological monitoring.

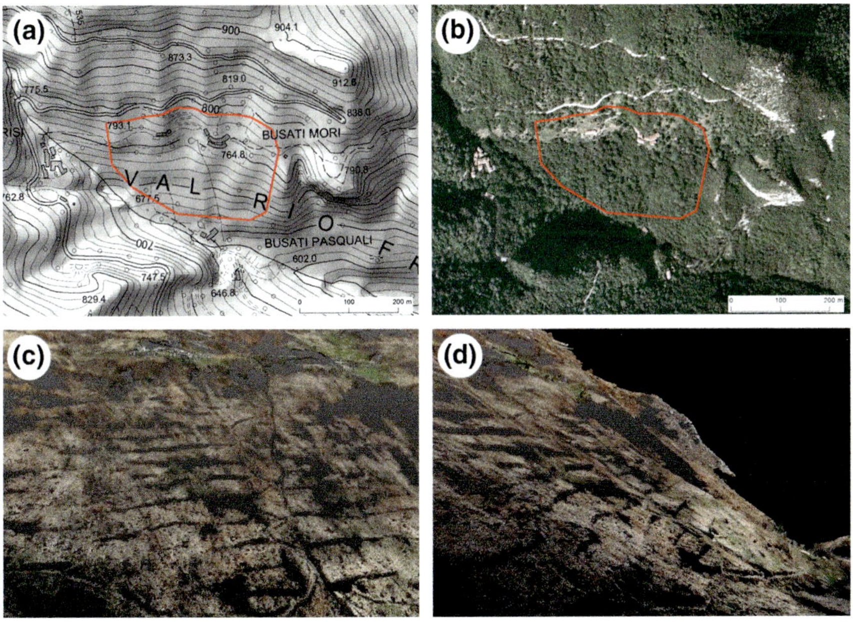

Fig. 11.5 UAV survey and mapping in Rio Freddo Valley, (Prealps of Vicenza Province) in the area bounded in red. In **a**, it is to note the 1:10,000 map edited in 1997, and so in the orthophotos of the year 2007 (**b**) a "cartographic desertification" of the abandoned, but still persisting, anthropic terraces. In **c** and **d,** a 3D drape of drone photos over the drone photogrammetric DTM. These terraces were cultivated with beans, potatoes, and corn until about 1960

11.4 Conclusion

The MAPTER project tested different methodologies and approaches to map and to extract different typologies of terraced systems widely distributed along the Italian territory. Terraced systems are extremely heterogeneous in relation to their physical conservation status, morphology, natural vegetation. They are shaped by the climate (arid or semi-arid in south and center of Italy, versus a continental climate in north Italy) and/or by the morphology of flank slope (inclination, bedrock, soil). By collecting and processing all the geospatial data we estimated, in Italy, a total surface of agricultural terraced systems of about 169,127 ha, and a total length of dry-stone walls of about 180,000 linear kilometers.

Mapping survey at a regional scale with the presence of vegetation canopy, LiDAR technology seems to give the most performance in terms of terrace risers features extraction; however, management of such big geospatial data requires suitable hardware/software resources and it may be time-consuming. Moreover, LiDAR data may be affected by technical problems, such as striping, during to the flight acquisition phase or to the elaboration techniques during the DTM production. According to recent DTM LiDAR analysis, two of the three regions with wider terraced systems (42,636 ha in Liguria and 22,730 ha in Toscana) appear to be underestimated by the traditional survey. Sicily Region, according to the MAPTER comparative analyses, seems to have the widest extension showing 63,554 ha of terraced systems (Barbera et al. 2010), even if an integrated analysis, at very large scale, combining LiDAR data with high-resolution satellite images showed a different extension in the study case of the Filicudi Island in the Aeolian Archipelagos of Sicily (Terranova 2016).

Considering that LiDAR data are becoming widely accessible, in the next years we could continue and maybe complete the LiDAR features extraction of terrace risers. The coastal landscape, for example, where there are some of the widest terraced systems, it is completely surveyed by LiDAR topography, but DTMs were interpolated for 2 m cell size, which are not very suitable for very narrow terrace steps (less than 5 m). Moreover, for some areas the use of UAV survey could provide high-resolution DTMs and pictures, useful data to monitor the state of terrace risers and the risk of landslide. At the same time, the development of lightweight LiDAR suitable for UAV systems can improve the availability of data in densely vegetated landscapes.

Geographic information supplied by Web Map Services and Geobrowsers has opportunities to be improved by participatory approaches, voluntary geography crowdsourcing of georeferenced information. The UAV survey of Rio Freddo Valley was planned after information collected by hikers of the valley. Involvement of hikers, trekkers, and local people shows a possible huge development in mapping terraced systems.

However, mapping terraced system showed some limitations in geospatial data management due to the high degree of complexity and variety of the Italian terraced systems. Hence, an exhaustive geographical survey, integrating different methodologies and techniques, should be implemented to carry out a more detailed map of terraced systems in Italy.

Acknowledgements The MAPTER project was realized thanks to the contribution of many universities, research institutes, public institutions, organizations, and individuals. We wish to thanks G. Brancucci (University of Genova), L. Bonardi, R. Madoi (University of Milan), G. Mauro, A. Giadrossi (University of Trieste), F. Lucchesi, C. A. Garzonio (University of Florence), E. Bonari, D. Rizzo (Scuola Superiore Sant'Anna—Pisa), A. Riggio, S. Modugno (University of Cassino and Southern Lazio), L. Fusco Girard, A. Gravagnuolo, M. Ronza (University of Naples Federico II), S. Di Fazio, G. Modica, S. Praticò (University of Reggio Calabria), G. Barbera, S. Cullotta (University of Palermo), G. Sistu, A. Pirinu (University of Cagliari), G. Tecilla, A. Cosner, G. Altieri (Autonomous Province of Trento), F. Alberti, U. Trivelloni (Veneto Region), M. Baldo, P. Giostrella, D. Giordan, D. Godone (CNR-Irpi, Turin), S. Costetti, D. Genovese, F. Laganà, E. Micati, M. Micheletti, T. Saggiorato (Italian Alpine Club).

Moreover, we are grateful to the "Ministero dell'Ambiente e della Tutela del Territorio e delle Acque, Geoportale Nazionale" for the availability of LiDAR data of many areas of the Italian coasts and islands and to the Department of "Regione Veneto, Ambiente e Territorio" for the regional LiDAR data.

References

Agnoletti M, Conti L, Frezza L, Monti M, Santoro A (2015) Features analysis of dry stone walls of Tuscany (Italy). Sustainability 7:13887–13903

Barbera G, Cullotta S, Rossi-Doria I, Rühl J, Rossi-Doria B (2010) I paesaggi a terrazze in Sicilia: metodologie per l'analisi, la tutela e la valorizzazione. In: Collana studi e ricerche dell'Arpa Sicilia, vol 7. ARPA Sicilia, Agenzia Regionale Protezione Ambiente

Bonardi L, Varotto M (2016) Paesaggi terrazzati d'Italia. Eredità storiche e nuove prospettive. Franco Angeli

Brancucci G, Masetti M (2008) I sistemi terrazzati: un patrimonio, un rischio. In: Scaramellini G, Varotto M (eds) Paesaggi terrazzati dell'arco alpino. Atlante. Marsilio, Venezia, pp 46–54

Brancucci G, Paliaga G (2006) The hazard assessment in a terraced landscape: the Liguria (Italy) study case in Interreg III Alpter project. In: ECI conference on geohazards, Lillehammer, Norway

Brandolini P (2010) I versanti delle Cinque Terre: un patrimonio ambientale e paesaggistico a rischio. Sopra il livello del mare. La Rivista dell'Ente Italiano della Montagna. Numero speciale: il paesaggio costruito. Salvaguardia e valorizzazione dei terrazzamenti artificiali 36:54–56

Brandolini P, Faccini F, Pescetto C (2008) I paesaggi terrazzati d'Italia. I Terrazzamenti della Liguria: un bene culturale e del paesaggio a rischio. L'Universo 88(2):206–221

Brugnaro S (2016) Estrazione mediante approccio Semi Automatico da dati LiDAR, per la quantificazione dei terrazzamenti agricoli: il caso di Amalfi. Unpublished thesis of the 'II level Master in GIScience and Unmanned Aerial Vehicle', University of Padova

Gravagnuolo A, Di Martino F (2015) GIS tools for mapping and classification of agricultural areas at risk of abandonment in terraced cultural landscapes. In: Sessa S, Di Martino F, Cardone B (eds) GIS DAY 2015. Il GIS per il governo e la gestione del territorio. Aracne Editrice, Roma

Hengl T, Reuter HI (eds) (2009) Geomorphometry: concepts, software, applications. In: Developments in soil science, vol 33. Elsevier

ISPRA (Istituto Superiore Per la Ricerca Ambientale) (2013) Linee guida per la valutazione del dissesto ecologico e la sua mitigazione attraverso misure e interventi in campo agricolo e forestale. Manuali e linee guida, 85. Ministero dell'Ambiente e della tutela del territorio e del mare, Ministero delle politiche agricole, alimentari e forestali

Jasiewicz J, Zwoliński Z, Mitasova H, Hengl T (eds) (2015) Geomorphometry for geosciences. Adam Mickiewicz University in Poznań—Institute of Geoecology and Geoinformation, International Society for Geomorphometry, Poznań

Lillesand T, Kiefer RW, Chipman J (2015) Remote sensing and image interpretation, 7th edn. Wiley

Minar J, Evans IS (2008) Elementary forms for land surface segmentation: the theoretical basis of terrain analysis and geomorphological mapping. Geomorphology 95:236–259

Modica G, Merlino A, Solano F, Mercurio R (2015) An index for the assessment of degraded mediterranean forest ecosystems. For Syst 24:2171–9845

Ninfo A (2008) Valutazione della pericolosità in aree terrazzate nel Canale di Brenta: un approccio su base LiDAR. In: Fontanari E, Patassini D (eds) Paesaggi terrazzati dell'arco alpino. Esperienze di Progetto. Marsilio, Venezia, pp 28–30

Ore G, Bruins HJ (2012) Design features of ancient agriculture terrace walls in the Negev Desert: human-made geodiversity. Land Degrad Dev 23:409–418

Passalacqua P, Tarolli P, Foufoula-Georgiou E (2010) Testing space-scale methodologies for automatic geomorphic feature extraction from lidar in a complex mountainous landscape. Water Resour Res 46:W11535

Rizzo D, Galli M, Sabbatini T, Bonari E (2007) Terraced landscapes characterization. Developing a methodology to map and analyze the agricultural management impacts (Monte Pisano, Italy). Rev Int Géomat 17:431–447

Ronza M (2006) Versanti terrazzati tra tufi e calcari: valenze ambientali, destinazioni agronomiche, scale di osservazione. Geotema 29:83–99. Patròn Editore, Bologna

Rühl J, Pasta S, La Mantia T (2005) Metodologia per lo studio delle successioni secondarie in ex-coltivi terrazzati: il caso di studio di Pantelleria (Canale di Sicilia). Forest@ 2(4):388–398

Sas RJ, Yu JP, Pau CYY, Styles KA (2012) Detection of old agricultural terraces in steep, vegetated terrain using airborne LiDAR: case study from Hong Kong. In: Proceedings of the 33rd Asian conference of remote sensing, vol I, Pattaya, Thailand, 26–30 Nov 2012, pp 408–417

Sofia G, Marinello F, Tarolli P (2014) A new landscape metric for the identification of terraced sites: the Slope Local Length of Auto-Correlation (SLLAC). ISPRS J Photogram Remote Sens 96:123–133

Tarolli P, Sofia G, Calligaro S, Prosdocimi M, Preti F, Dalla Fontana G (2015) Vineyards in terraced landscapes: new opportunities from LiDAR data. Land Degrad Dev 26:92–102

Terranova L (2016) Estrazione e valutazione dei terrazzamenti agricoli nel contesto eoliano: il caso di Filicudi. Unpublished thesis of the 'II level Master in GISscience and Unmanned Aerial Vehicle', University of Padova

Tres M, Zatta E (2006) Paesaggi terrazzati nel Canale di Brenta. Unpublished dissertation thesis in Landscape Planning, IUAV—University of Architecture of Venice

Varotto M (2004) Geografie dell'abbandono. Valstagna e la fine della civiltà del tabacco. In: Perco D, Varotto M (eds) Uomini e paesaggi del Canale di Brenta. Cierre Edizioni, Sommacampagna (VR), pp 213–226

Varotto M, Ferrarese F (2008) Mapping and geographical classification of terraced landscapes: problems and proposals. In: Scaramellini G, Varotto M (eds) Terraced landscapes of the Alps. Atlas. Marsilio, Venezia, pp 38–45

Chapter 12
Terraced Landscapes: Land Abandonment, Soil Degradation, and Suitable Management

Paolo Tarolli, Davide Rizzo and Gerardo Brancucci

Abstract Agricultural terraces are among the most visible and extensive human signatures on different landscapes of the world. Terraces are generally built to retain soil and water, to reduce erosion, and to support irrigation. They reduce slope gradient and length, and facilitate the infiltration of water in areas with a moderate to low soil permeability by controlling the overland flow velocity. Thereby, they create positive effects on agricultural activities. Since ancient times, agricultural terraces have been built in different topographic conditions (e.g. coastal area, hilly, and steep slope mountain landscapes) and used for the cultivation of various crops (e.g. vineyards, fruit and olive groves, cereals, tea). Their management however arises relevant critical issues. Historical terraces are often of the bench type with stone walls and require adequate maintenance. Poorly designed and controlled terraces can lead to slope failures, often due to walls collapsing, increasing potential soil water erosion. Also, terraced areas are often served by agricultural roads that can profoundly influence surface hydrologic processes and erosion. Land abandonment and ageing of the local population, which affected several regions of the world during the last decades, are among the main reasons for the poor maintenance of terraced landscapes. As a consequence, a progressive increase of land degradation and loss of soil functions (e.g. food production, environmental interaction such as water storage, filtering and transformation, biological habitat, physical and cultural heritage) is observed. The purpose of this chapter is to highlight the main critical issues of terraced landscapes, providing a few case studies, and a possible solution for proper long-term management.

P. Tarolli (✉)
Department of Land, Environment, Agriculture and Forestry, University of Padova, Legnaro (PD), Italy
e-mail: paolo.tarolli@unipd.it

D. Rizzo
UP 2018.C102 INTERACT, UniLaSalle, Beauvais, France

G. Brancucci
Department of Architecture and Design, Polytechnic School, University of Genova, Genoa, Italy

M. Varotto et al. (eds.), *World Terraced Landscapes: History, Environment, Quality of Life*, Environmental History 9,
https://doi.org/10.1007/978-3-319-96815-5_12

12.1 Introduction

The terracing of slopes is universally used to obtain cultivable land in steep areas since ancient times, today being one of the major morphological transformations implemented by humans in the world (Tarolli et al. 2014). The cultivable flat land is obtained by reducing the overall slope angle with the construction of soil reservoirs usually supported downstream by a dry-stone wall (terracing) or, in other cases, contained by clods (embankments). Building techniques are presented in detail in many texts (e.g. Brancucci et al. 2000; Rizzo 2008). The reduction of the slope has a positive feedback on the landscape regarding the reduction of the erosive power of the surface water flow and, consequently, of hydrogeological-related problems (Tarolli et al. 2014). Moreover, the decrease of the slope improves the pedogenetic processes. A necessary precondition to preserve such benefits is to guarantee the maintenance with the presence of farming. A terrace system can "resist" to erosive agents only thanks to the constant support of its components that include dry-stone walls and painstaking water regulation structures. Since the First World War, and definitely after the end of the Second World War, the development and promotion of the industrial activities accelerated the progressive abandonment of agriculture, especially in these highly demanding and poorly mechanised farming systems. The abandonment of terraced areas results in a new interference with the geomorphic system. As the lack of maintenance of a human-altered landscape persists the more, the geomorphic system gets the control causing erosion processes and land degradation. The final result is the increase of hazards with diffuse problems of instability and huge solid transport in the rivers (Brancucci and Paliaga 2006). In this chapter, we address the Liguria Region (Northern Italy) as a paramount example of the complex balance between the pristine morphology and the anthropic modifications. On these premises, we then highlight some potential hazards to propose, finally, key recommendations for designing suitable management strategies.

12.2 Case Study

12.2.1 Liguria: A Natural Laboratory for the Analysis of a Terraced Landscape

Liguria is a Northern Italian region that covers about 5400 km^2, distributed in a thin strip of land ranging from East to West for about 250 km, bounded on the South by the Ligurian Sea, to the North by the Alpine and Apennine chains. Altogether, Liguria presents a very complex geology (Fig. 12.1).

The region is crossed by valleys, inherited from the local tectonic events, with East–West and North–South direction across the Tyrrhenian side and Po Valley. The climate is Mediterranean, characterised by heavy rainfall events occurring

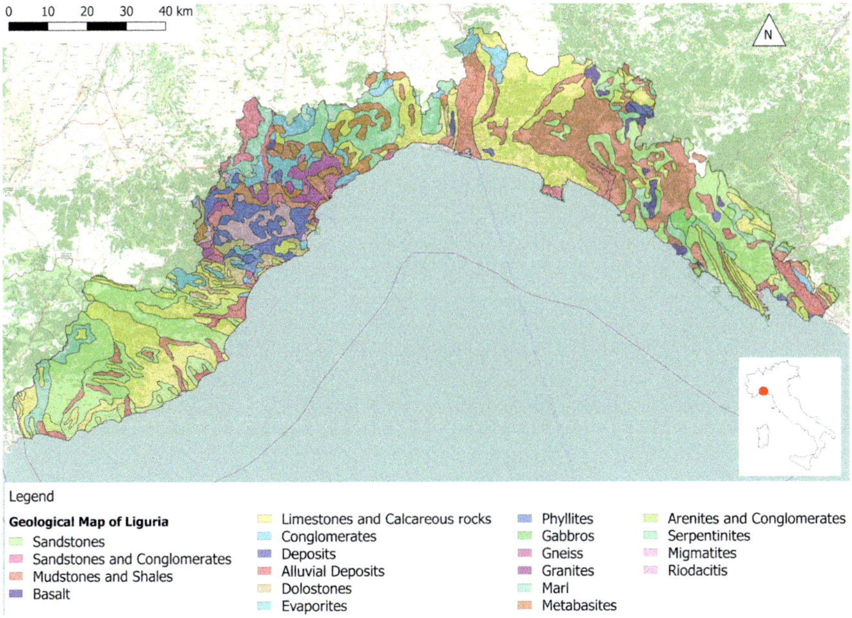

Fig. 12.1 Simplified geological map of Liguria (Brancucci and Ghersi 2018). Base map from http://www.openstreetmap.org/copyright

mainly in late summer and during the autumn. The rainfall events are becoming more intense in recent years (cf. Vallebona et al. 2015). High steep slope and rainfall are the primary agents of the erosion processes. Such processes do not allow to generate top soils suitable for farming practice. In the last centuries, terracing agricultural practice has reshaped entirely the Liguria landscape. For several years, the presence of humans, who tenaciously maintained efficiently such land, has prevented any hydrogeological criticalities. However, the changed socio-economic conditions of the last century determined a progressive land abandonment that appears difficult to revert. People migrated from the mountains to the coastal areas where industrial and tourist activities could guarantee economic prosperity. Such land abandonment was related also to an increase in soil erosion processes and landslide events (Fig. 12.2).

In several situations, the lack of proper vegetation management (such as chestnut coppices) exacerbated such instability. The complex morphological context makes farming hard, by increasing the costs of production and consequently the prices of products. The consequence is that only a few traders are willing to invest in these areas. Terraces are perceived as part of the "identity" of the Ligurian landscape; in such area the Cinque Terre National Park is the most "representative image". It is unlikely to recover the whole system, but it is imperative to break apart the problem into its various components to be able to provide adequate support to land planners and let them being constantly updated and updateable.

Fig. 12.2 View of the coastline of the Cinque Terre National Park: erosive forms of abandoned terraces near Volastra (SP) © Google, DigitalGlobe

12.2.2 Land Abandonment and Landslides Occurrences

The lack of maintenance of terraced systems results in progressive land degradation (Brandolini et al. 2018; Tarolli et al. 2014). Surface water infiltrates in subsoil due to altered permeability mainly caused by agricultural improper practices (Brancucci and Masetti 2008). Consequently, the continuous soil saturation can destabilise the dry-stone wall (Fig. 12.3).

Fig. 12.3 Different kinds of terrace disorders, from bulging (left) to series of collapses (right). *Source* Brancucci et al. (2000) adapted in Rizzo (2009)

Fig. 12.4 Landsliding and terrace failures (white rounded arrows) within vineyards triggered by the intense rainfall event of 25 October 2011 (Liguria, Italy) (Agnoletti et al. 2012; after Tarolli et al. 2014)

This coincides with an increase in the occurrence of small landslides in the steps between terraces, which can evolve downslope as debris flows (Tarolli et al. 2014) (Fig. 12.4).

A similar mechanism occurred on 24 November 2000 in Ceriana (province of Imperia) where intense precipitation affected an entire terraced area triggering several mass movements and causing two fatalities. The mitigation of such phenomena demands tremendous efforts especially where the land tenure is fragmented. The ownership of terraced fields is frequently divided into small plots owned by multiple stakeholders in which the property is difficult to document. This situation leads to an obvious and additional difficulty of the system management.

12.2.3 Ancient Terraces: A "Soil Reservoir" to Be Monitored

Figure 12.5 shows an ancient terrace system located in the inland of Liguria Region (Italy). This terrace has been built at the end of the eighteenth century for cultivating mainly potatoes and grapes. However, the very steep slope did not provide a suitable place. The structure is quite complex; it is characterised by a dry-stone wall of 7 m in height, complemented by an underground hydraulic system to drain water

Fig. 12.5 Ancient terrace system located in the inland of Liguria Region (Italy). The white arrow highlights the deformation in the dry-stone wall. In the two boxes are showed the details of the underground drainage system. *Photo* P. Tarolli

(box in Fig. 12.5). Nowadays the terraced field is not cultivated, except for grass cover and a single row of grapevine. The dry-stone wall presents a deformation (white arrow Fig. 12.5) that could represent a critical issue during intense rainfall events.

There is a significant amount of soil in such as steep hillslope; thus, if the underground drainage system does not work properly or if the deformation of the wall would cause its failure, the entire system will collapse and evolve downslope as a debris flow. It is clear that in a future perspective such ancient terrace system needs to be monitored carefully.

12.3 Terraced Landscape Management

12.3.1 Monitoring

Proper design, planning, and maintenance of terrace systems represent necessary steps to avoid land degradation. All these actions start from monitoring of the terraces and an identification of the failure mechanisms sources and consequences.

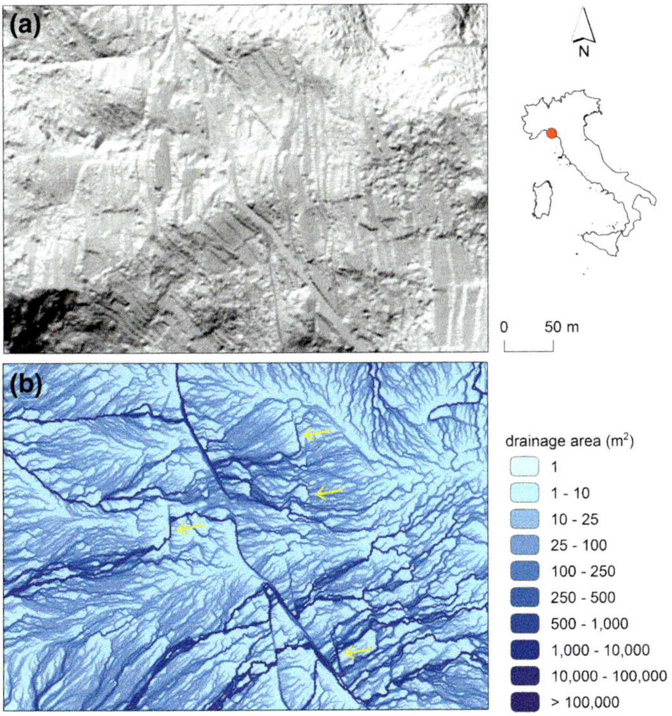

Fig. 12.6 High-resolution topography of a terrace system in Liguria (Italy). The figure shows: **a** a shaded relief map obtained by a 1 m LiDAR-derived DTM; **b** drainage area calculated according to the d-infinite algorithm (Tarboton 1997). Yellow arrows indicate the water surface flow direction deviations due to the presence of terraces. (LiDAR data were provided by the Italian Ministry of the Environment and Protection of Land and Sea, within the framework of the Piano Straordinario di Telerilevamento Ambientale, PST-A)

In the last decade, a range of new remote sensing techniques led to a dramatic increase in terrain information, providing new opportunities for a better understanding of Earth surface geomorphic signatures (Tarolli 2014). Many recent studies proved the reliability of LiDAR (Light Detection and Ranging) technology, both aerial and terrestrial, in many disciplines concerned with Earth surface representation and modelling (Booth et al. 2009; Lin et al. 2013; Sofia et al. 2014a,b; Lo Re et al. 2018). In terraced landscapes, high-resolution LiDAR data can be useful. They allow to readily recognise the topographic signatures of terraces (Fig. 12.6), including those in areas covered by vegetation. The capability of LiDAR technology to derive a high-resolution ($\sim$1 m) DTM (Digital Terrain Model) from bare ground data, and by filtering vegetation from raw data, underlines the effectiveness of this methodology in mapping abandoned and vegetated terraces (Tarolli et al. 2014; Sofia et al. 2016).

DTM surface derivatives such as landform curvature, if obtained with small grid cell sizes, can be useful to automatically identify the location and the shape of terraces through statistics (e.g. Sofia et al. 2014a). Terraces can be considered as ridges on the side of the hill being in a much sharper shape than natural terrain features. Consequently, they represent outliers of surface curvature. It is possible to detect such outliers in a box plot automatically and mapping them as terrace features. This approach can be used for a first and quick assessment of the location of terraces, particularly those abandoned that might require restoration efforts. However, high-resolution topography can be advantageous for other applications. It allows a better recognition of surface water flow paths (Fig. 12.6). Terrace failures are generally related to wall bowing due to subsurface water pressure (Preti et al. 2018). To identify the topographic influence on such pressure or the surface water flow direction, it is necessary to model the presence of the terraces and the surface morphology, considering the importance they have in influencing hydrological (surface flow paths) and geotechnical processes at the slope scale. Using 1 m LiDAR-derived DTM, it is possible to calculate flow direction and drainage upslope area and identify areas where the water flow is deviated or concentrated in specific paths (Fig. 12.6). This process, if not properly managed with a suitable drainage system, can trigger soil erosion that may evolve in gully or, in worst cases, in landslides (Tarolli et al. 2015). However, this approach (purely topographically based), while providing a first useful overview of the problem, needs to be improved with more accurate and physically based analyses because wall failures cannot be related only to surface flow direction changes. In such case, the use of a terrestrial laser scanner, supported by geotechnical instruments, should monitor a portion of a dry-stone wall providing more deep insights into failure mechanisms (Tarolli et al. 2014). A 3D model of a dry-stone wall, obtained with a centimetre grid cell resolution, could allow simulation of the behaviour of the wall in response to back load with high detail and without many artefacts or approximations. Also, the analysis of a direct shear test can offer an estimation of the Mohr-Coulomb failure envelope parameters (friction angle and cohesion) to be considered for modelling. Reference portions of dry-stone walls can be monitored, measuring the lateral earth pressure at the interface between dry-stone wall and its backfill, and the backfill volumetric water content (both in saturated and unsaturated states) and groundwater level. A permanent monitoring system can be implemented in such critical areas, where failures can involve houses or roads, with (a) pressure cells to measure the stress acting on the wall surfaces, (b) piezometers to measure the neutral strains, and (c) rain gauge for rainfall analysis.

12.3.2 Landscape Agronomic Approach

Once the geomorphological features of the terraced systems have been correctly mapped, it is necessary to revitalise their management, which was mainly ensured by farmers and farming activities. Terraced landscapes have been long neglected by agronomists due to their hard optimisable production. We claim here that the landscape perspective could help agronomy to address new management strategies of terraced systems, primarily to pursue soil conservation goals. This requires shifting the focus from the terrace field level to embrace the whole landscape for the management of surface water and soil in cultivated areas. Although the agronomist can be defined as the "overseer of the land", accordingly to the Greek etymology of the word (Buisson 2013), in recent history agronomists have frequently replaced the "land" by the cultivated field. This led agronomy to narrow the focus on the improvement of field performances finally addressing only issues related to crop yield, crop growth, and soil–crop relations (Cañas-Guerrero et al. 2013). However, agronomy intrinsically embraces multiple levels of the land, spanning from the cultivated field to its wider context. To this end, the cultivated field can play a critical role for interdisciplinary approaches to land management issues because it is a piece of land that is managed according to its socio-ecological context (Deffontaines 1991; Duru et al. 2015). In general terms, an agricultural field is bounded by natural or anthropic features (e.g. edges, channels and rivers, roads), by changes in cover or management (e.g. different crops, cultivation, and abandonment) as well as by land tenure (e.g. Deffontaines 1991; Inan et al. 2010; Levin and Nainggolan 2016). For terraced fields, we can add clearly visible attributes of risers, taluses and the intensive network of ditches required for the surface water and soil management on the hillside. Beyond the field level, multiple perspectives coexist because different stakeholders' land property may overlap (Rizzo et al. 2013). Accordingly, the description, understanding, and management of terraces require inherently to be addressed from a system perspective to relate the spatial configuration of agricultural practices and the natural resource management. Significantly, Cavazza (1996) pointed out two complementary perspectives within the single discipline: farm agronomy and land agronomy. On the one hand, agronomy should tackle the farmers' decision-making process in a clear organised spatial context defined by the farmers. On the other hand, agronomy should also deal with the context of farming activities and the interaction with the various stakeholders operating in a given area. The landscape agronomy approach (Benoît et al. 2012) recently proposed a transdisciplinary conceptual framework pursuing the integration of knowledge and perspectives from different scientific disciplines and societal stakeholders (Lardon et al. 2012; Moonen et al. 2016). An important pillar in this framework comes from the Italian Constitution (1948 art. 9) that reads: "The Republic promotes the development of culture and of scientific and technical research. It safeguards natural landscape and the historical and artistic heritage of the Nation". So it joins culture, research, and protection of the landscape. "Landscape" can be further defined, according to the European Convention, as "[…] an area, as perceived by people, whose character is the result of the action and interaction of natural and/or human factors" (COE 2000 art. 1a). This convention also inspired the Terraced Landscape Manifesto that calls for the protection and long-term maintenance of terraces (cf. http://www.terracedlandscapes2016.it/en/outcomes/). The landscape agronomy approach provides a timely answer to the call for actions discussed at the Third World Meeting on Terraced Landscapes, by framing the methods for mapping and inventorying terraced landscape features while supporting the understanding of their history, geography, social, and physical structures (Fig. 12.7).

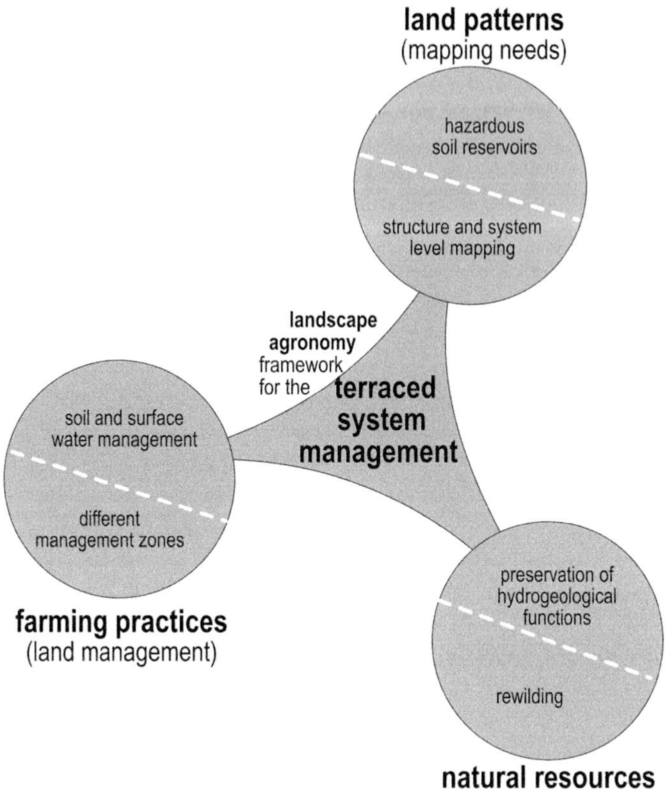

Fig. 12.7 Conceptual framework for designing management strategies of terraced systems, based on the landscape agronomy approach (Benoît et al. 2012). To ensure a suitable management, the three poles must be addressed. Each pole illustrates a couple of examples further discussed in the text

Early applications of the landscape agronomy framework concerned the characterisation of the farming system component of European agricultural landscapes (Andersen 2017; van Zanten et al. 2013) also through remote sensing techniques (Bégué et al. 2015) and the design of a simple tool for extending agronomic management on a landscape basis (Orchard and Hackney 2016). Based on the landscape agronomy framework, we stress the need to tackle the terraced system management taking into explicit account the relationships between the spatial configuration of agricultural practices and the expected and the observed impacts on soil, water, and other natural resources (Fig. 12.7). This implies to develop geographic information systems to map all the three agricultural landscape components: farming practices, land patterns, and natural resources. The underpinning rationale is to relate the single terrace field management to the landscape it belongs to. In this perspective, farming practices include both the different forms of soil and water management (at the micro- and meso-scale) and of the zoning of actual management practices (at the meso- and macro-scale). The survey of farming practices is meant to improve the evaluation of the "fragility" of land patterns (Rizzo 2009), so as to identify the most critical ones, such as hazardous soil reservoirs defined by complex management practices and located in abandoned patches. The correct mapping of land patterns and the related land management practices aims to enhance the conservation of the natural resources of terraced landscape, namely to orient the proper preservation of its hydrogeological functions (Savo et al. 2013). In summary, a landscape agronomic approach could help to better locate the terrace management actions and to relate them to the overall landscape rebuilding (Di Fazio et al. 2005; LaFevor 2014; Qiu et al. 2013). This could lead, for instance, to opt for rewilding those abandoned patches having a marginal role in the system management instead of pursuing the onerous complete restoration of a whole abandoned landscape (e.g. Rizzo et al. 2014).

12.3.3 Maintenance

The interactions between agriculture and the land have dramatically evolved during the last decades, in particular, for the steady decrease in the number of farmers and the land tenure dynamics. In plain areas, this was related to an increase of the farms size, while in slope areas, the evolution rather concerned the ageing of the local population, the arrival of newcomers, and a marked fragmentation of properties. Terraced landscapes have undergone to even more extreme changes with the loss of the system perspective due to a deepening break in the knowledge transmission of the management practices. In a landscape agronomy perspective, the proper maintenance of terrace features requires (i) to address explicitly the management practices, (ii) to zone the landscape accordingly, and (iii) to redefine the meaning of "traditional". A first limitation in a suitable terrace maintenance is the consistency of resolution between available topographic data and the terrace features, whose size is usually narrow than the most common data sources. Nevertheless, advances in remote sensing (e.g. airborne laser scanner) are enhancing the mapping capabilities, eventually allowing to overcome the difficulties to identify the characteristics of these features (e.g. Chartin et al. 2011; Rizzo et al. 2007; Sofia et al. 2014a; Sofia et al. 2016). Existing literature widely addressed the mapping of terrace features, mostly drawing the classification on the risks deriving from their abandonment (Brancucci and Paliaga 2006; Tarolli et al. 2014). Yet, a major weakness even of the most advanced studies is the poor consideration of the management practices because they are difficult to survey. The landscape agronomy framework can eventually strengthen the mapping and inventorying of terrace features by stressing in a spatially explicit way the focus on the management impacts on agri-environmental resources (Rizzo 2009). This can be useful to understand the current landscape state and to support the shared design of desired transitions (Gennai-Schott et al. 2014). Several studies contributed insofar to the description and mapping of terraced landscapes, though with increasingly isolated sectoral approaches. On the one hand, the improved computing performances of computer are boosting the application of advanced hydrological models fed on the newly available data. Cross-sectoral metrics are even facilitating the characterisation of the terrace history on very long-time spans. On the other hand, the mapping mainly resulted in the identification of patterns based on the simple geographical analysis of homogeneous spatial distribution of the surveyed attributes, though their limits depend also on underpinning processes or on emergent landscape dynamics (e.g. Farina and Belgrano 2004; Lazrak et al. 2010). In this regard, the landscape agronomy framework requires to address the spatial configuration of management practices underpinning the land patterns, so analysing the whole area as a system and explicitly addressing the relationships linking its various parts (Maigrot et al. 2004; Salem et al. 2013; Schaller et al. 2012; Sweeney et al. 2013; Thenail et al. 2000). A relevant maintenance metric could be the terrace density, following the method proposal from the ALPTER project (Varotto and Ferrarese 2008), complementary to some more adapted to Mediterranean terraced areas (Agnoletti et al. 2015; Rizzo 2009). In conclusion, the mapping and zoning of terraced landscape might lead to reconsider the definition of "traditional landscape". This is an appealing concept for studies dealing with landscape characterisation that requires, however, a complete historical approach (e.g. Antrop 1997; Leibundgut and Kohn 2014; Torquati et al. 2015; Torró 2007). Terraces are indeed long-lasting landscape features that can undergo multiple cycles of cultivation and abandonment. To inform suitable maintenance plans, it would be more relevant to frame the terrace feature mapping within a landscape level zoning of medium-/long-term land use dynamics (Fig. 12.7). This could allow to distinguish (i) *core zones* of the traditional landscape, which have been quite continuously preserved and cultivated, from (ii) *buffer zones*, periodically cultivated and abandoned following the local population dynamics, and (iii) clear *marginal areas*, marking past expansions of the terraced system afterwards definitively abandoned, for which rewilding could be the best hydrogeological and management option. Multiple data sources should be analysed to clearly distinguish buffer and marginal zone, especially in the case of field abandoned in more recent times (Bevan and Conolly 2011; Galletti et al. 2013; Rühl et al. 2005; Treacy 1987), thus benefiting of the integrative landscape agronomy framework. Altogether, informing the mapping and survey of terraced landscape on management practices could help revealing patterns to understand past land dynamics and to orient future terraced landscape policies.

12.4 Final Remarks

Since ancient times, humans have used terracing practices for agricultural activities in different environments (both hilly and mountainous areas) and regions of the world and also for mitigating soil erosion and stabilising hillslopes. The study of terraces represents a challenge for our modern society and deserves particular attention. The reasons are several: their economic, environmental, and historical–cultural implications and their hydrological functions, such as erosion control, slope stabilisation, lengthening of the rainfall concentration time, and the consequent reduction of the surface runoff. Land abandonment and the new farming and lifestyles of the young generation are seriously questioning the terraced landscapes survival. The result is a progressive increase of soil erosion and landslides that can be a serious problem if they occur in densely populated areas. Because of this, terraced landscapes need to be monitored, maintained, and preserved by taking explicitly into account the agricultural practices that shaped them. The advanced remote sensing techniques such airborne laser scanner can be useful in monitoring criticalities and provide terraces inventory map for an entire region. Also, a deeper insight into failure mechanisms can be obtained through specific geotechnical surveys that must feed agricultural management strategies at the whole terrace system level. Altogether, we suggest to adopt a landscape agronomy perspective to address consistently spatial organisation of the farming practices for the preservation of soil and other natural resources. These actions can help to overcome the critical issues related to erosion and landslides risks and, finally, to preserve the landscape identity.

References

Agnoletti M, Santoro A, Emanueli F, Maggiari G, Preti F (2012) Terracing and hydrogeological risk. A study of the environmental disaster of 25 October 2011 in Cinque Terre. In: Florens 2012 Studi e Ricerche Essays and Researches, pp 25–46

Agnoletti M, Conti L, Frezza L, Monti M, Santoro A (2015) Features analysis of dry stone walls of Tuscany (Italy). Sustainability 7:13887–13903

Andersen E (2017) The farming system component of European agricultural landscapes. Eur J Agron. Farming systems analysis and design for sustainable intensification: new methods and assessments 82, Part B, 282–291

Antrop M (1997) The concept of traditional landscapes as a base for landscape evaluation and planning. The example of Flanders Region. Landsc Urban Plan 38:105–117

Bégué A, Arvor D, Lelong C, Vintrou E, Simoes M (2015) Agricultural systems studies using remote sensing. In: Prasad ST (ed) Land resources monitoring, modeling, and mapping with remote sensing. CRC Press, pp 113–130

Benoît M, Rizzo D, Marraccini E, Moonen AC, Galli M, Lardon S, Rapey H, Thenail C, Bonari E (2012) Landscape agronomy: a new field for addressing agricultural landscape dynamics. Landsc Ecol 27:1385–1394

Bevan A, Conolly J (2011) Terraced fields and Mediterranean landscape structure: an analytical case study from Antikythera Greece. Ecol Model 222:1303–1314

Booth AM, Roering JJ, Perron JT (2009) Automated landslide mapping using spectral analysis and high-resolution topographic data: Puget sound lowlands, Washington, and Portland Hills, Oregon. Geomorphology 109:132–147

Brancucci G, Masetti M (2008) I sistemi terrazzati: un patrimonio, un rischio. Paesaggi Terrazzati Dell'arco Alpino Atlante. Marsilio, Venezia (ITA), pp 46–54

Brancucci G, Paliaga G (2006) The hazard assessment in a terraced landscape: preliminary result of the Liguria (Italy) case study in the Interreg III ALPTER Project. In: 2006 ECI Conference on Geohazards, Lillehammer, Norway

Brancucci G, Ghersi A (2018) Geodiversità dei vigneti liguri. Le relazioni tra paesaggio, suolo, vitigni e vino. EDIFIR (eds), pp 287, ISBN: 887970897X
Brancucci G, Ghersi A, Ruggiero ME (2000) Paesaggi liguri a terrazze: riflessioni per una metodologia di studio. Alinea, Firenze
Brandolini P, Cevasco A, Capolongo D, Pepe G, Lovergine F, Del Monte M (2018) Response of terraced slopes to a very intense rainfall event and relationships with land abandonment: A case study from Cinque Terre (Italy). Land Degrad Dev 29:630–642
Buisson B (2013) Vous avez dit agronomie? Agron Environ Sociétés 18:155–159
Cañas-Guerrero I, Mazarrón FR, Pou-Merina A, Calleja-Perucho C, Díaz-Rubio G (2013) Bibliometric analysis of research activity in the "agronomy" category from the Web of Science, 1997–2011. Eur J Agron 50:19–28
Cavazza L (1996) Agronomia aziendale e agronomia del territorio. Riv Agron 30:310–319
Chartin C, Bourennane H, Salvador-Blanes S, Hinschberger F, Macaire J-J (2011) Classification and mapping of anthropogenic landforms on cultivated hillslopes using DEMs and soil thickness data—example from the SW Parisian Basin, France. Geomorphology 135:8–20
COE C of E (2000) European landscape convention. In: Council of Europe treaty series
Deffontaines JP (1991) L'agronomie, science du champ. Le champ, lieu d'interdisciplinarité: de l'écophysiologie aux sciences humaines. Agronomie 11:581–591
Di Fazio S, Malaspina D, Modica G (2005) La gestione territoriale dei paesaggi agrari terrazzati tra conservazione e sviluppo. In: Atti convegno AIIA L' ingegneria agrar. Lo sviluppo sostenibile Dell'area Mediterr, Catania, pp 27–30
Duru M, Theau JP, Martin G (2015) A methodological framework to facilitate analysis of ecosystem services provided by grassland-based livestock systems. Int J Biodivers Sci Ecosyst Serv Manag 11:128–144
Farina A, Belgrano A (2004) The eco-field: a new paradigm for landscape ecology. Ecol Res 19:107–110
Galletti CS, Ridder E, Falconer SE, Fall PL (2013) Maxent modeling of ancient and modern agricultural terraces in the Troodos foothills. Cyprus Appl Geogr 39:46–56
Gennai-Schott S, Rizzo D, Sabbatini T, Marraccini E (2014) Uno studio dell'intensità dei sistemi olivicoli collinari terrazzati: il caso del Monte Pisano. In: Atti del XLIII convegno nazionale della società Italiana di Agronomia, Presented at the la sostenibilità dell'intensificazione colturale e le politiche agricole: il ruolo della ricerca agronomica, Pisa (Italy)
Inan HI, Sagris V, Devos W, Milenov P, van Oosterom P, Zevenbergen J (2010) Data model for the collaboration between land administration systems and agricultural land parcel identification systems. J Environ Manag 91:2440–2454
LaFevor MC (2014) Restoration of degraded agricultural terraces: rebuilding landscape structure and process. J Environ Manag 138:32–42
Lardon S, Moonen AC, Marraccini E, Debolini M, Galli M, Loudiyi S (2012) The territory agronomy approach in research, education and training. In: Farming systems research into the 21st century: the new dynamic. Springer, Dordrecht, pp. 257–281
Lazrak EG, Mari J-F, Benoît M (2010) Landscape regularity modelling for environmental challenges in agriculture. Landsc Ecol 25:169–183
Leibundgut C, Kohn I (2014) European traditional irrigation in transition part I: irrigation in times past—a historic land use practice across Europe. Irrig Drain 63:273–293
Levin G, Nainggolan D (2016) The significance of spatial fragmentation of land ownership for occurrence of scrubs on semi-natural grasslands. Landsc Ecol 31:2031–2044
Lin CW, Tseng CM, Tseng YH, Fei LY, Hsieh YC, Tarolli P (2013) Recognition of large scale deep-seated landslides in forest areas of Taiwan using high resolution topography. J Asian Earth Sci 62:389–400
Lo Re G, Fuller IC, Sofia G, Tarolli P (2018) High-resolution mapping of Manawatu palaeochannels. NZ Geogr 74:77–91 https://doi.org/10.1111/nzg.12186
Maigrot J, Deffontaines JP, Erard R, de Sede-Marceau MH (2004) Changes in production systems and spatial reorganisation of farming practices. Cah Agric 13:321–330

Moonen AC, Lardon S, Marraccini E, Pinto-Correia T, Rizzo D (2016) From action research to action learning—ecosystem services assessment as a learning platform for students, local land users and researchers. In: Wilcox A, Vinall S (eds) Social and technological transformation in farming systems—diverging and converging pathways. Presented at the 12th European IFSA symposium, Harper Adam University, UK, p 8

Orchard PW, Hackney B (2016) A simple tool for extending agronomic management on a landscape basis. Landsc Ecol 31:239–242

Preti F, Guastini E, Penna D, Dani A, Cassiani G, Boaga J, Deiana R, Romano N, Nasta P, Palladino M, Errico A, Giambastiani Y, Trucchi P, Tarolli P (2018) Conceptualization of water flow pathways in agricultural terraced landscapes. Land Degrad Dev 29:651–662

Qiu Z, Chen B, Takemoto K (2013) Conservation of terraced paddy fields engaged with multiple stakeholders: the case of the Noto GIAHS site in Japan. Paddy Water Environ 1–9

Rizzo D (2008) La gestione delle sistemazioni idraulico-agrarie come contributo alla tutela del paesaggio agrario terrazzato toscano. ARSIA, Pisa

Rizzo D (2009) Landscape-agronomic analysis in the agro-environmental fragility assessment of a terraced landscape. Ph.D. thesis, Scuola Superiore Sant'Anna, Pisa

Rizzo D, Galli M, Sabbatini T, Bonari E (2007) Terraced landscapes characterization. Developing a methodology to map and analyze the agricultural management impacts (Monte Pisano, Italy). Rev Int Géomat 17:431–447

Rizzo D, Marraccini E, Lardon S, Rapey H, Debolini M, Benoît M, Thenail C (2013) Farming systems designing landscapes: land management units at the interface between agronomy and geography. Geogr Tidsskr-Dan J Geogr 113:71–86

Rizzo D, Mari JF, Marraccini E, Lazrak EG (2014) Agricultural landscape segmentation: a stochastic method to map heterogeneous variables. In: Presented at the 1st IALE-Europe thematic workshop: advances in spatial typologies: how to move from concepts to practice? Lisboa, PRT

Rühl J, Pasta S, La Mantia T (2005) A method for the study of secondary succession processes in terraced old fields: the case study Pantelleria Island (Canale di Sicilia). For Riv Selvic Ed Ecol For 2:388–398

Salem AB, Majdoub R, M'Sadak Y, Khlifi S (2013) Importance of the Meskat system and its landscape insertion through the olive groves of Sousse Region (Tunisian Sahel)

Savo V, Caneva G, McClatchey W, Reedy D, Salvati L (2013) Combining environmental factors and agriculturalists' observations of environmental changes in the traditional terrace system of the Amalfi Coast (Southern Italy), Ambio, pp 1–14

Schaller N, Lazrak EG, Martin P, Mari J-F, Aubry C, Benoît M (2012) Combining farmers' decision rules and landscape stochastic regularities for landscape modelling. Landsc Ecol 27:433–446

Sofia G, Marinello F, Tarolli P (2014a) A new landscape metric for the identification of terraced sites: the slope local length of auto-correlation (SLLAC). ISPRS J Photogramm Remote Sens 96:123–133

Sofia G, Prosdocimi M, Dalla Fontana G, Tarolli P (2014b) Modification of artificial drainage networks during the past half-century: evidence and effects in a reclamation area in the Veneto floodplain (Italy). Anthropocene 6:48–62

Sofia G, Bailly J-S, Chehata N, Tarolli P, Levavasseur F (2016) Comparison of pleiades and LiDAR digital elevation models for terraces detection in farmlands. IEEE J Sel Topi Appl Earth Obs and Remote Sens 9(4):1567–1576

Sweeney S, Steigerwald DG, Davenport F, Eakin H (2013) Mexican maize production: evolving organizational and spatial structures since 1980. Appl Geogr 39:78–92

Tarboton DG (1997) A new method for the determination of flow directions and upslope areas in grid digital elevation models. Water Resour Res 33:309–319

Tarolli P (2014) High-resolution topography for understanding Earth surface processes: opportunities and challenges. Geomorphology 216:295–312

Tarolli P, Preti F, Romano N (2014) Terraced landscapes: from an old best practice to a potential hazard for soil degradation due to land abandonment. Anthropocene 6:10–25

Tarolli P, Sofia G, Calligaro S, Prosdocimi M, Preti F, Dalla Fontana G (2015) Vineyards in terraced landscapes: new opportunities from lidar data. Land Degrad Dev 26:92–102

Thenail C, Le Cœur D, Baudry J (2000) Relationships between field boundaries, farming systems and landscape: consequences on biodiversity pattern in agrarian landscapes. In: 4th European symposium on European farming and rural systems research and extension: environmental, agricultural and socio-economic issues, Volos, Association for farming system research and extension, European group, pp 63–76

Torquati B, Giacchè G, Venanzi S (2015) Economic analysis of the traditional cultural vineyard landscapes in Italy. J Rural Stud 39:122–132

Torró J (2007) Terrasses irrigades a les muntanyes valencianes, Les transformacions de la colonització cristiana. In: Estudiar i gestionar el paisatge històric medieval. Universitat de Lleida, pp 81–143

Treacy J (1987) Building and rebuilding agricultural terraces in the Colca Valley of Peru. In: Yearbook, proceedings of the conference of Latin Americanist Geographers, pp 51–57

Vallebona C, Pellegrino E, Frumento P, Bonari E (2015) Temporal trends in extreme rainfall intensity and erosivity in the Mediterranean region: a case study in southern Tuscany Italy. Clim Change 128:139–151

van Zanten BT, Verburg PH, Espinosa M, Gomez-y-Paloma S, Galimberti G, Kantelhardt, J, Kapfer, M, Lefebvre, M, Manrique R, Piorr A, Raggi M, Schaller L, Targetti S, Zasada I, Viaggi D (2013) European agricultural landscapes, common agricultural policy and ecosystem services: a review. Agron Sustain Dev 1–17

Varotto M, Ferrarese F (2008) Mapping and geographical classification of terraced landscapes: problems and proposals. In: Scaramellini G, Varotto M (eds) Terraced landscapes of the Alps. Atlas, Marsilio, pp 38–45

Chapter 13
Health, Seeds, Diversity and Terraces

Salvatore Ceccarelli

Abstract Modern plant breeding has moved towards uniformity, while the increasing demand for nutritious and safe food would require the maintenance and enhancement of biodiversity to respond to climate changes, to improve resilience at farm level and to improve health through a diversified diet. Thus, a change in the way new varieties are produced is necessary, and this is offered by participatory plant breeding, which combines modern science with farmers' knowledge and emphasizes specific adaptation. This is particularly relevant for remote, difficult to access agricultural landscapes such as terraced agriculture. Yemen, a typical country with large areas covered by terraces, offers an example that participatory plant breeding can be successfully implemented even in these challenging situations: in a three years programme, new varieties of barley and lentil, two key food crops in Yemen, were obtained. A methodology, which can be even more suitable to terraced agriculture, is evolutionary plant breeding through which farmers can manage independently a large and evolving genetic diversity. This allows them to quickly respond to climate changes and associated new pests, to be the owner of their own seed, to diversify their agricultural systems and increase their resilience and, more importantly, to improve their nutritional status with a more diversified diet without depending on external inputs.

13.1 Nutrition and Diseases

Farming and livestock production, along with the food industry that transforms and transports the products to where they are consumed, not only does not make available to the world nutritious and healthy food but at the same time has negative impacts on the environment and on the most vulnerable people (Sukhdev et al. 2016).

S. Ceccarelli (✉)
Rete Semi Rurali, Scandicci, Italy
e-mail: ceccarelli.salvatore83@gmail.com

M. Varotto et al. (eds.), *World Terraced Landscapes: History, Environment, Quality of Life*, Environmental History 9,
https://doi.org/10.1007/978-3-319-96815-5_13

The food systems are currently the cause of 60% of terrestrial biodiversity loss, 24% of greenhouse gas emissions, 33% of land degradation and 61% of the decline in commercial fish stocks (Hajer et al. 2016).

The reduced crop diversity and the consequent homogenization of food sources around the world on the one hand are reducing the capacity to cope with climate change and on the other hand are reducing our immunitary defences (von Hertzen et al. 2011) thus undermining human health around the world.

According to the report on global nutrition, malnutrition and diet are by far the major risk factors for diseases at the global level (International Food Policy Research Institute 2016).

Currently, about 800 million people worldwide suffer from hunger, two billion are malnourished, and another about two billion are overweight or obese. People suffering from diabetes are so many that if they live in the same country, that country would be the third most populated country in the world after China and India: the number of people with diabetes has quadrupled between 1980 and 2014 (Krug 2016), and in 2012, diabetes only caused a half million deaths (WHO 2015) with a global cost of 1.31 trillion dollars (Bommer et al. 2017).

13.2 Climate Change and Health

Climate change represents an intriguing research problem because, firstly, of the uncertainty of the expected changes (Nelson et al. 2009; Trenberth et al. 2015), which makes difficult predicting the increase in temperature and the decrease in rainfall anywhere on the planet with an acceptable degree of accuracy, and, secondly, because the decrease in rainfall and the increase in temperature are likely to be different according to, among others, elevation and slope. Therefore, agroecologies such as the terraces, which represent the dominant agricultural landscape in countries such as Yemen, Nepal, Bhutan and are also widespread in Ethiopia, Eritrea, Vietnam, China and in a number of countries in South America, are particularly vulnerable because of the large differences in elevation, slope and exposure. Therefore, particularly in these areas, plant breeding programmes to improve crop adaptation to climate change are likely to face a moving target and probably a different target in different areas (Ceccarelli 2014a, 2017). Thirdly, adaptation to climate change implies also adaptation (resistance or tolerance) to new insect pests and diseases, which have been shown to have altered their latitudinal ranges in response to global warming (Bebbe et al. 2013).

Climate change is also expected to have an impact on crops with a direct consequence on health: a simulation study showed that the increase in CO_2 is expected to decrease the content of iron and zinc in crops such as rice, bread wheat, maize, soybeans, field peas and sorghum (Myers et al. 2014). This is a serious problem as the deficiency of these two microelements is already causing the death of 63 million people annually. According to the World Health Organization

(WHO), there are two billion people with anaemia in the world and half of the anaemia is due to iron deficiency (WHO, UNICEF, UNU 2001).

Therefore, adapting crops to climate change represents a complex research objective.

13.3 Can We Have Both Cheap and Healthy Food?

It seems that now we must resign to a choice between these two options that reflect the dilemma between feeding and nourishing. However, there is nothing preventing us to strive for affordable healthy food.

In an attempt to get out of this dilemma, the concept of smart food is emerging, as the type of food that is good to the consumer, to the planet and to the farmer. Is good to the consumer because derived from crops rich in antioxidants, proteins, vitamins and micronutrients such as iron, calcium and zinc and are easy to digest; they do not contain gluten and prevent cancer, diabetes and cardiovascular disease; are good to the planet because are more resistant than others to high temperatures and drought and therefore are able to adapt to climate change and need less water; are good to farmers because increasing agrobiodiversity does increase the resilience of the farm, are easy to grow and can open up new markets. Examples of smart foods are those derived from legumes and from grains such as sorghum and millet, common in Asia and Africa. Sorghum has been recently defined as the new quinoa (http://www.icrisat.org/smartfood/).

Even if so far we have discussed only health and food, implicitly, we have been discussing about seed, because all our food derives from seed and our health depends largely from food; therefore, the seeds are at the root of many of the current problems.

13.4 Where the Seed Comes from?

Plant breeding is the science that produces new crop varieties, many of which give the food that ends up on our table. For millennia, it has been done by farmers and only in the last hundred plus years has been done by researchers in research centres or research stations.

During the millennia before modern plant breeding began, farmers were moving around with seeds and livestock, and because neither were uniform, they could gradually adapt to different climates, soils and uses. Whenever farmers settled, they continued to improve crops and livestock. In the case of crops, the way they did it can still be seen today in a number of countries and consists of selecting the best plants that give the seed to be used for the following season. Therefore, the selection was done in the same place where the crops were grown and the seed of the selected plants was *mixed* before planting. This process was highly location-specific in the sense that each farmer did it independently from other farmers and for his/her own conditions of soil, climate and uses. What we call ancient, old, heirloom varieties originated through this process (Ceccarelli 2017) which generated diversity both within and between farmers' fields.

With time, modern plant breeding took a different approach: being done in a research station, moved the selection away from the place where the crop was grown, thus creating a gap between the "selection environment" and the "target environment". It also moved the selection away from the people who did it for millennia, thus ignoring all the knowledge generated by that process.

As the target environment is actually represented by several locations outside the research station, it is more appropriate to talk about a "target population of environments". Hence the problem of whether the research station is representative of any of the locations it is supposed to serve. This is a particularly serious problem because, to the best of my knowledge, there are no research stations located on terraces, and one may wonder how relevant could be the research—the selection—in the specific case of plant breeding, done in a station situated in an entirely different environment, not only in a geographical sense but in a agroecological sense.

The problem of addressing a number of heterogeneous environments outside the research station was solved by smoothing the differences between the target environments with external (mostly chemical) inputs, namely fertilizers, pesticides and irrigation water. In such a way, those environments became from an agronomic point of view very similar even if geographically distant and therefore one or few varieties could be grown across all of them: these varieties were defined as "widely adapted" and the breeding philosophy that produced them "wide adaptation".

The Green Revolution (Baranski 2015) adopted the "wide adaptation" philosophy in the 1960s wheat programme in India avoiding, in the short term, the incipient danger of a famine, but causing, on the long term, penalties such as the leaching into the groundwater of fertilizers residues due to the overuse of fertilizers above the amount that plants can utilize (Good et al. 2011), the water shortage, the emergence of pesticide resistance (Gassmann et al. 2014), the increase in the population of harmful insects (Lu et al. 2013), the bypassing of farmers in marginal areas (Baranski 2015) and the loss of crop diversity by displacing or even replacing landraces (Frison et al. 2011).

As a result of this change in breeding philosophy, there has been (1) a progressive emphasis on genetic uniformity both in the self-pollinated crops (such as wheat, barley and rice) and in cross-pollinated crops (such as corn), and in the latter through the use of hybrids, and (2) the use of production per unit area as the predominant breeding objective, with the result that the quality progressively declined: in fact, globally, crops contain today less protein (−4%), less iron (−19%) and less zinc (−5%) than in the 1960s (De Fries et al. 2015). In USA, while grain yield of bread wheat has increased over time, the concentrations of copper, iron, magnesium, manganese, phosphorus, selenium and zinc have decreased (Garvin et al. 2006; Murphy et al. 2008). A similar trend has been observed in fruits and vegetables (Davis 2009).

The two developments of modern plant breeding mentioned earlier, together with (1) a growing concentration of the seed and of the pesticides markets in the hands of few large corporations, and (2) a similar concentration in few hands of the food industry, have had, have and, in a scenario of business as usual, will continue to have some negative effects on our health. The increasing uniformity of what is grown inevitably entails increasing uniformity of what we eat, and this has been put in relation with a reduction of our immunitary defence system and the consequent rise of a whole range of diseases including cancers (Khamsi 2015). Also, since modern varieties, particularly cereals, are generally less nutritious, we must eat more to meet the daily requirements, thus contributing to the increase, now endemic, of obesity.

Because food is derived from seeds, it is at the way in which the seeds are produced and made available to farmers that we have to look for the solution to environmental problems including climate change and to our and future generations' health. One solution is to change the way we select new varieties by moving back the process in farmers' fields and by making farmers equal partners in the selection process, in a model known as participatory plant breeding (Ceccarelli et al. 2009). This genetic improvement model has several advantages such as an increase in agrobiodiversity, reduction of chemical inputs because it adapts crops to the environment rather than changing the environment, a higher benefit/cost ratio (Mustafa et al. 2006) and finally the recognition that farmers can play a key role in plant breeding by combining their traditional knowledge with that of the scientists (Halewood et al. 2007; Ceccarelli et al. 2000, 2009). This model of plant breeding has been implemented in a number of countries, in different agroecologies and with various crops (Ceccarelli 2015) including the terraced agriculture in Yemen (Ceccarelli et al. 2003).

13.5 The Case of Yemen

The project, which allowed implementing participatory plant breeding (PPB) on the terraced agriculture of Yemen, was supported by the then System-Wide Program of Participatory Research and Gender Analysis (PRGA, later dismantled). The project was implemented in the Kuhlan Affar area, a steep mountain slope that descends from about 3000 m asl to about 800 m asl towards Wadi Sharis and addressed the terraced mountain slopes that range from 1700 m asl to 2800 m asl approximately, where 90% of the agriculture is located. The area is supported by traditional methods of water harvesting mainly terracing of mountain slopes. Most farming families still grow landraces and save part of their harvest as seed source for the subsequent year (Fig. 13.1).

The villages of the research area are very small in terms of population numbers. The villages of Kuhlan Affar are in Hajjah province, 123 km northwest of the capital Sana'a. The study area lies within the two districts of Sharis and Kuhlan in Hajjah province, which is located in the western escarpments of Yemen. At the time the

Fig. 13.1 A typical village in Yemen with the terraces in the background. *Photo* S. Ceccarelli

project was implemented, the total population of this province was estimated at about 1.5 million, which represented 7.8% of the total population of Yemen, and was growing at a rate of 3% annually. They produced about 5% of the total agricultural crop production of the country. Most people of Hajjah province worked in agriculture and cattle breeding. The total agricultural area in Hajjah province was estimated at about 124,600 ha, of which 36% or 46,000 ha is predominantly cultivated terraces and Wadi banks, and rangelands cover about 63% of the province or 78,000 ha. The area is famous for its coffee beans, fruit and cereals production. Tobacco and palm trees are also common in the plains. Kuhlan Affar is a remote area, on mountains, where living conditions and access to cities are difficult, and was chosen because the province represented the traditional dry lands farming systems in the country's northwestern highlands; it was a typical example of areas neglected by agricultural research, and the area was characterized by subsistence agriculture.

The size of the terraces varies, mostly in relation to the slope—the steeper the slope the smaller the terrace. Each farming family usually owns more than one terrace, with an average farm size of only about 1.4 ha; usually, only one crop is planted in each terrace, but it is not uncommon to see terraces divided between lentil and barley or between sorghum and faba bean or even between all four crops. Agriculture is mainly rainfed with an annual average rainfall of 300–500 mm, falling in two seasons: March to April and August to September. It is the principal economic activity in the area and engages 80% of the population (Aw-Hassan et al. 2000). The most important crops are sorghum, wheat, lentil, barley, dry peas, maize, millet, beans, fenugreek, coffee and qat (Ceccarelli et al. 2003). Mainly, local varieties dominate in these farming systems, and women and men farmers save part of their harvests as seed for next year planting and sometimes exchange it with neighbours under the assumption that this will improve productivity, but the seed quality is generally poor. For these reasons, we started a PPB programme in collaboration with the Agricultural Research and Extension Authority (AREA). Women started to be involved gradually into the PPB programme, especially when men farmers started to gain confidence in the project.

Three villages (Hasn Azam, Beit Al-Wali, and Al-Ashmor) were selected by the local breeders based on the importance of barley and lentil cropped in the area. The project was discussed with farmers in these villages through meetings where the objectives of collaborative research and its potential benefits for rural communities were discussed, and the responsibilities in terms of project implementation and evaluation defined.

The implementation of the project was challenging because we did not have any previous experience of working in the limited physical space offered by terraces.

The participatory barley and lentil selection in the Kuhlan Affar areas was conducted for three years with the objectives of:

(1) testing the methodology in remote locations characterized by traditional agricultural systems and difficult environments
(2) identifying improved cultivars of barley and lentil.

The initial experiments were conducted in the three villages in the Kuhlan Affar area mentioned earlier and in the research station of the Agricultural Research and Extension Authority (AREA) at Al-Erra, near Sana'a. In each of the four locations, the trial consisted of the same fifty genotypes in both barley and lentil. The 50 barleys included six landraces, collected from different areas in the Northern Highlands of Yemen and obtained from the national Gene bank of Yemen, and improved lines from the Arab Centre for Studies in Arid Dry Lands (ACSAD). The 50 lentil entries included 15 local land races, also obtained from the national Gene bank of Yemen, and 35 entries from the International Centre for Agricultural Research in the Dry Areas (ICARDA) lentil breeding programme. In both crops, one local cultivar was used as a common check in each location.

Planting of the barley and lentil trials occurred in June 1999 in Bit Al–Wali (BA) and Hasn Azam (HA), and in July in Al-Ashmor (AA) and Al-Erra (AE) research station (Fig. 13.2). Plots consisted of four rows 2.5 m long and 25 cm apart. The experimental design was the randomized complete block design with two replications hosted in two adjacent terraces. The farmers' cultural practices were followed. Both planting and harvesting were organized by the AREA researchers and done manually by the host farmer and his family. Planting was done in furrows opened by one-stilted plough pulled by a donkey in the direction of the maximum length of the terrace. This was actually suggested by the farmers and resulted in the plots being oriented as the surrounding farmer's crop. Harvesting was done by hand.

Fig. 13.2 Research station at Al-Erra and the village of Al-Ashmor at about 3000 m elevation with the participatory barley and lentil breeding experiments. *Photo* S. Ceccarelli

At the end of the first year, the farmers selected 19, 16 and 21 barley lines in Hasn Azam, Bit Al-Wali and Al-Ashmor, respectively. In lentil, the number of lines selected was 23 in both Hasn Azam and Bit Al-Wali and 21 in Al-Ashmor, respectively. During the second year, the selected lines were evaluated in the same location in which they had been selected, in two replications and in plots of 10 rows at 25 cm distance and 5 m long. The experimental design, field layout, cultural practices, planting and harvesting were as described for the first-year trials.

At the end of the second year, six barley lines were selected in each of the three locations and were tested for a third year in the three villages. The total number of different lines was 12 including 2 of the six landraces. Only one line was commonly selected in all three villages, three lines were common to two villages, and the others two were unique to a specific village. In lentil, the number of lines selected in the second year was 7 in Hasn Azam, 8 in Al-Ashmor and 11 in Bit Al-Wali. The total number of different lines was 17 out of the initial 50 with 6 lines common to two locations. The 17 lines included 8 landraces (or 53% of those present in the first year) and 9 breeding lines (or 26% of those present in the first year). All trials were planted in two replications and in plots of 10 rows at 25 cm distance and 5 m long.

The third-year trials were sufficiently small for both replications to be accommodated on the same terrace.

As this was the first time farmers (both men and women) were involved in evaluating a relatively a large number of lines, the evaluation was initially done through consensus by the group of farmers and resulted for each plot in either discarding or selecting. In the second and third years, the selection procedure was changed at the request of the farmers, since they felt more confident in their individual opinion. This eventually allowed to disaggregate the data according to men and women preferences.

The three years of participatory plant breeding in Yemen ended with the identification of two high yielding barley varieties and three high yielding lentil varieties, which were adopted and cultivated by most of the farmers in locations where in the past centralized and non-participatory breeding was not capable of introducing any new variety (Ceccarelli 2002). As a consequence, there were seed production skills emerging among the farmers which were translated into a functional and efficient seed production system.

There were much more differences between farmers' and breeder's selection in the first and second year when the diversity was higher than in the third year when the number of lines was nearly 1/10th of the initial population. This was particularly true in barley where also genotype x locations interactions played a greater role than in lentil. This suggests that if farmers do participate in the selection process during the initial phases of a breeding programme, the differences in selection by farmers and breeder may determine the final outcome of a participatory breeding programme as compared with a non-participatory programme. An additional implication is that participatory programmes based on a small number of lines, such a participatory variety selection (PVS), are neither likely to exploit the full potential of farmer participation nor can be taken as example of lack of differences in selection criteria of the various participants.

This work demonstrates that with the participation of farmers, it was possible to implement a research programme in remote and difficult to access areas where conventional research did not have any impact. This demonstration affected the policy-makers to the point that participatory research has become part of the strategy of agricultural research in Yemen.

13.6 Evolutionary–Participatory Plant Breeding (EPB)

There are several other examples of successful PPB programmes, but despite these successes, PPB has a weakness in requiring the collaboration of a research institute to provide breeding materials and technical support such as experimental designs and statistical analysis. Therefore, the sustainability of a participatory programme depends on the long-term commitment of a research institution, and this is the main weakness of the PPB because it is not possible to count on the participation of an institution on a lasting basis.

An interesting alternative is offered by evolutionary (participatory) plant breeding—participatory is in parenthesis because, though desirable, the participation of an institution is not indispensable. The idea is not new as it was proposed back in 1956 (Suneson 1956). The method consists in planting in farmers' field's mixtures of many different genotypes of the same crop, or populations built using early segregating generations, namely materials obtained from crosses. Mixtures and populations will be planted and harvested year after year, and due to the natural crossing (higher in cross-pollinated and lower in self-pollinated crops), the genetic composition of the seed that is harvested is never the same as the genetic composition of the seed that was planted. In other words, the population evolves to become progressively better adapted to the environment (soil type, soil fertility, agronomic practices including organic systems, rainfall, temperature) in which is grown. As the climatic conditions vary from one year to the next, the genetic makeup of the population will fluctuate, but if the tendency is towards hotter and drier climatic conditions as expected in view of climate changes, the genotypes better adapted to those conditions will gradually become more frequent (Ceccarelli 2014b).

An evolutionary population, which can be made by the farmers themselves by buying and mixing seed of as many different varieties (including hybrids) of a given crop, can be used by the farmers (and by the researchers if they are willing to participate) as a source of genetic diversity from which to select. When this is done, it is expected that, based on selection theory (Falconer 1981), response to selection will increase because of the large population size of an evolutionary population leading therefore to a greater selection efficiency.

This has been done in Italy (data not published) using a zucchini (summer squash) evolutionary population obtained by letting 11 commercial hybrids to freely intercross. After only two cycles of visual selection, as in the case of tomato as described in Campanelli et al. (2015), the farmer selected two varieties, differing in colour, yielding as much as the commercial hybrids. He has already started selling the two new varieties in local markets.

Evolutionary populations of different crops (Fig. 13.3) are currently grown by farmers in Jordan, Ethiopia (as part of the Bioversity International project "Strengthening cultivar diversity of barley and durum wheat to manage climate-related risks and foster food and nutritional security in marginal areas of Ethiopia" supported by GIZ), Iran, Italy, France, Portugal and India for cereal crops (maize, barley, bread and durum wheat and rice), grain legumes (common bean) and horticultural crops (tomato and summer squash). Farmers growing these populations report higher yields, lower weed infestation and disease presence and lower insect damages. The use of pesticides has consequently been reduced.

Because of their continuous evolving, evolutionary populations cannot be patented or protected by IP. According to the Commission Implementing Decision of 18 March 2014 pursuant to Council Directive 66/402/EEC, in Europe, it is currently possible to market experimentally heterogeneous materials of wheat, maize, oats and barley up to 31 December 2018 (Official Journal of the European Union 2014).

Fig. 13.3 An evolutionary population of bread wheat (left) and one of zucchini (right). *Photos* S. Ceccarelli, at the left; courtesy of Dr. Campanelli on the right

Iranian farmers growing an evolutionary population of wheat have marketed the bread obtained from the flour of the evolutionary population in local artisanal bakeries. The bread can be consumed also by customers intolerant to gluten (Rahmanian et al. 2014). Farmers growing wheat evolutionary populations in France and Italy confirmed that creating mixtures brings not only greater yield stability but also greater aroma and quality to the bread (Fig. 13.4).

Thus, evolutionary (participatory) plant breeding, being a relatively inexpensive and highly dynamic strategy to adapt crops to a number of combinations of both abiotic and biotic stresses and to organic agriculture, seems to be a suitable method to generate, directly in farmers' hands, the varieties that will feed the current and the future populations. Indeed, experimental evidence shows that with evolutionary breeding it is possible to combine high yield and stability (Raggi et al. 2017).

Combining seed saving with evolution and bringing back the control of seed production in the hands of farmers can produce better and more diversified varieties that can contribute to help millions of farmers to reduce the dependence from external inputs and the vulnerability to disease, insects and climate change and ultimately contribute to food security and food safety for all. Being simpler to implement and to manage, evolutionary plant breeding seems particularly suited to terraced agriculture.

Participatory plant breeding and evolutionary plant breeding, while benefiting from advances in molecular genetics, reconcile increased production of more readily available and accessible food, with increased agrobiodiversity while maintaining the evolutionary potential of our crops needed to cope with climate change.

Fig. 13.4 Traditional bread making in Iran with the flour of a bread wheat evolutionary population (left) and a shop selling the same bread (right). *Photos* courtesy of Ms. Maede Salimi

13.7 Conclusions

In discussing the global problems including the pandemic of obesity and diabetes, seldom it is recognized that the solution of these problems requires a change in the way seed is produced, because seed is related to all these problems. Conventional plant breeding conducted by large private seed companies needs to generate profit and is difficult to change it from the current emphasis on wide adaptation supported by a consolidation of the seed industry (Howard 2009; Fuglie et al. 2011) to an emphasis on specific adaptation. This could be conveniently done, to some an extent, by small seed companies, but mostly by public breeding such as the breeding programmes conducted by CGIAR using their large germplasm collections amounting to about 710,000 seed samples (http://www.cgiar.org/consortium-news/genebanks-investing-in-biodiversity-for-future-generations/) which include all the most important staple food crops.

However, there are three reasons to be worried about the future of seed. First is the increasing trend towards public–private collaboration, which is leading to the creation of private–public breeding activities with some parts of the public breeding programmes executed by large seed companies which derive royalties from the final products; second is the transfer of former top managers of some of the largest seed companied into top-management positions in the CGIAR and vice versa; and third is the increasing role of private foundations' support to public research (Martens and Seitz 2015). All this is made worse by the progressive consolidation of the seed market (MacDonald 2017). These three recent developments raise questions on weather in a not too distant future we may witness a, at least partial, privatization of the CGIAR gene banks. Whether this will happen or not, the evolutionary populations may play two important roles: firstly, in the hands of developing countries may represent a continuous, independent from CGIAR centres and not patentable source of better adapted genetic material for their breeding programme as an addition to or a replacement for the genetic material they usually receive from the CGIAR; secondly, in the hands of the farmers, and being non-patentable for their continuing evolving nature, they will remain as publicly available genetic resources. Once the farmers have the seed, they have the solution (Gilbert 2016).

Acknowledgements This work would not have been possible without the collaboration of men and women in the three villages in the Kuhlan Affar area. Dr. S. Grando, Dr. M. Martini and Dr. A. Aw-Hassan, then barley breeder, gender specialist and socio-economist at ICARDA, respectively, gave an important scientific support, and Mr. A. Luft, cereal breeder and head of AREA gene bank, was instrumental in making this work possible.

This work was supported by a small grant from the System-Wide Program on participatory Research and Gender Analysis (SW-PRGA).

References

Aw-Hassan A, Alsanabani M, Bamatraf A (2000) Impacts of land tenure and other socio-economic factors on mountain terrace maintenance in Yemen. In: CAPRI working paper no. 3. CGIAR systemwide program on property rights and collective action. IFPRI

Baranski MR (2015) Wide adaptation of green revolution wheat: international roots and the Indian context of a new plant breeding ideal 1960–1970. Stud Hist Philos Biol Biomed Sci 50:41–50

Bebbe DP, Ramotowski MAT, Gurr SJ (2013) Crop pests and pathogens move polewards in a warming world. Nat Clim Chang 3:985–988

Bommer C, Heesemann E, Sagalova V, Manne-Goehler J, Atun R, Bärnighausen T, Vollmer S (2017) The global economic burden of diabetes in adults aged 20–79 years: a cost-of-illness study. Lancet Diabetes Endocrinol 5(6):423–430

Campanelli G, Acciarri N, Campion B, Delvecchio S, Leteo F, Fusari F, Angelini P, Ceccarelli S (2015) Participatory tomato breeding for organic conditions in italy. Euphytica 204(1):179–197

Ceccarelli S (2002) Village based participatory breeding in the mountain slopes of yemen, Report of a PRGA small grant for April 2001–March 2002

Ceccarelli S (2014a) Drought. In: Jackson M, Ford-Lloyd BV, Parry ML (eds) Plant genetic resources and climate change. CABI International, pp 221–235

Ceccarelli S (2014b) GMO, organic agriculture and breeding for sustainability. Sustainability 6:4273–4286

Ceccarelli S (2015) Seeds of future. In: Laboratorio expo: the many faces of sustainability. Giangiacomo Feltrinelli Editore, pp 177–197

Ceccarelli S (2017) Increasing plant breeding efficiency through evolutionary-participatory programs. In: Pilu R, Gavazzi G (eds) More food: road to survival, Bentham Science Publishers, pp 22–47

Ceccarelli S, Grando S, Tutwiler R, Baha J, Martini AM, Salahieh H, Goodchild A, Michael M (2000) A methodological study on participatory barley breeding. I Selection Phase. Euphytica 111:91–104

Ceccarelli S, Grando S, Martini M, Lutf A (2003) Participatory barley and lentil breeding in Yemen. Arab World Agribus 19(5):19–20

Ceccarelli S, Guimaraes EP, Weltzien E (eds) (2009) Plant breeding and farmer participation, FAO

Davis DR (2009) Declining fruit and vegetable nutrient composition: what is the evidence? HortScience 44(1):15–19

De Fries R, Fanzo J, Remans R, Palm C, Wood S, Anderman TL (2015) Metrics for land-scarce agriculture. Science 349(6245):238–240

Falconer DS (1981) Introduction to quantitative genetics, 2nd edn. Longmann Group Ltd

Frison AE, Cherfas J, Hodgkin T (2011) Agricultural biodiversity is essential for a sustainable improvement in food and nutrition security. Sustainability 3:238–253

Fuglie KO, Heisey PW, King JL, Pray CE, Day-Rubenstein K, Schimmelpfennig D, Wang SL, Karmarkar-Deshmukh R (2011) Research investments and market structure in the food processing, agricultural input, and biofuel industries worldwide. ERR-130, U.S. Dept. of Agriculture, Econ. Res. Serv

Garvin DF, Welch RM, Finley JW (2006) Historical shifts in the seed mineral micronutrient concentration of US hard red winter wheat germplasm. J Sci Food Agric 86(13):2213–2220

Gassmann AJ, Petzold-Maxwell JL, Keweshan RS, Dunbar MW (2014) Field-evolved resistance to bt maize by western corn rootworm. PLoS ONE 6(7):e22629. https://doi.org/10.1371/journal.pone.0022629

Gilbert N (2016) Frugal farming. Nature 533:308–310

Good AG, Beatty PH (2011) Fertilizing nature: a tragedy of excess in the commons. PLoS Biol 9: e1001124. https://doi.org/10.1371/journal.pbio.1001124

Hajer M, Westhoek H, Ingram J, Van Berkum S, Özay L (2016) UNEP, Food systems and natural resources. A Report of the working group on food systems of the international resource panel

Halewood M, Deupmann P, Sthapit B, Vernooy R, Ceccarelli S (2007) Participatory plant breeding to promote farmers' rights. Bioversity international. http://www.icrisat.org/smartfood/. Accessed 7 Mar 2017
Howard PH (2009) Visualizing consolidation in the global seed industry: 1996–2008. Sustainability 4:1266–1287. https://doi.org/10.3390/su1041266
International Food Policy Research Institute (2016) Global nutrition report 2016: from promise to impact: ending malnutrition by 2030
Khamsi R (2015) A gut feeling about immunity. Nat Med 21:674–676
Krug EG (2016) Trends in diabetes: sounding the alarm. The Lancet 387(10027):1485–1486
Lu Y, Wu K, Jiang Y, Xia B, Li P, Feng H, Wyckhuys KAG, Guo Y (2013) Mirid bug outbreaks in multiple crops correlated with wide-scale adoption of Bt cotton in China. Science 328: 1151–1154
MacDonald JM (2017) Mergers and competition in seed and agricultural chemical markets. Amber waves
Martens J, Seitz K (2015) Philanthropic power and development, Who shapes the agenda? Brot für die welt/global policy forum/MISEREOR
Murphy KM, Reeves PG, Jones SS (2008) Relationship between yield and mineral nutrient concentrations in historical and modern spring wheat cultivars. Euphytica 163:381–390
Mustafa Y, Grando S, Ceccarelli S (2006) Assessing the benefits and costs of participatory and conventional barley breeding programs in Syria, ICARDA
Myers SS, Zanobetti A, Kloog I, Huybers P, Leakey ADB, Bloom AJ, Carlisle E, Dietterich LH, Fitzgerald G, Hasegawa T, Holbrook NM, Nelson RL, Ottman MJ, Raboy V, Sakai H, Sartor KA, Schwartz J, Seneweera S, Tausz M, Usui Y (2014) Increasing CO2 threatens human nutrition. Nature 510:139–142
Nelson GC, Rosegrant MW, Koo J, Robertson R, Sulser T, Zhu T, Ringler C, Msangi S, Palazzo A, Batka M, Magalhaes M, Valmonte-Santos R, Ewing M, Lee D (2009) Climate change: impact on agriculture and costs of adaptation. International Food Policy Research Institute
Official Journal of the European Union (2014) Commission implementing decision, L 147/79 of 10 March 2014. http://eur-lex.europa.eu/legal-content/EN/TXT/?uri=CELEX%3A32014D0150
Raggi L, Ciancaleoni S, Torricelli R, Terzi V, Ceccarelli S, Negri V (2017) Evolutionary breeding for sustainable agriculture: selection and multi-environment evaluation of barley populations and lines. Field Crop Res 204:76–88
Rahmanian M, Salimi M, Razavi K, Haghparast R, Ceccarelli S (2014) Living gene banks in farmers fields. Farming Matters 12–15
Sukhdev P, May P, Müller A (2016) Fix food metrics. Nature 540:33–34
Suneson CA (1956) An evolutionary plant breeding method. Agron J 48:188–191
Trenberth KE, Fasullo JT, Shepherd TG (2015) Attribution of climate extreme events. Nat Clim Chang 5:725–730. https://doi.org/10.1038/nclimate2657
von Hertzen S et al (2011) Natural immunity: biodiversity loss and inflammatory diseases are two global megatrends that might be related. EMBO Rep 12:1089–1093
WHO (2015) Fact sheet no 310: the top ten causes of death. World Health Organization
WHO, UNICEF, UNU (2001) Iron deficiency anaemia: assessment, prevention and control, a guide for programme managershttp://www.who.int/nutrition/publications/micronutrients/anaemia_iron_deficiency/WHO_NHD_01.3/en/index.html

Chapter 14
Comparative Studies on Pattern and Ecosystem Services of the Traditional Rice Agricultural Landscapes in East Asia

Yuanmei Jiao, Toshiya Okuro, Kazuhiko Takeuchi, Luohui Liang and Xuan Gao

Abstract The traditional agricultural landscape is a multifunctional geographic object, which can provide multiple ecosystem services for human beings due to complex interactions between components, patterns, processes, and dynamics. Taking the Satoyama landscape in Japan and the Hani terrace landscape in southwestern China as objects, this paper reviews and compares their patterns and multiple ecosystem services. The results indicate that both landscapes are composed of similar elements, including forests, villages, wet rice paddies or terraces, grasslands, streams, ponds, and irrigation ditches. However, they differ in distributing area, spatial pattern, ecosystem services, and socio-ecological pressures. The main elements of the traditional Satoyama landscape are secondary forests and small, gently sloping, rice paddies. Its spatial pattern is a heterogeneous mosaic of seminatural ecosystems, and the prevailing ecological process is the flow of organic fertilizer. As the landscape has been abandoned because of various social pressures (the aging farming population, urbanization, and economic globalization), its current ecological service is mainly cultural—providing Japan with a symbol of traditional rural lifestyle. The Hani terraces are still a vigorous, living landscape, which provides all kinds of ecological services for local people and tourists. This life in the landscape stems from its reciprocal effects, feedbacks among the vertically ordered components of natural and seminatural forests, villages, and huge, steeply sloping rice terraces. Because they are traditional agricultural landscapes,

Y. Jiao (✉) · X. Gao
School of Tourism and Geography Science, Yunnan Normal University,
Kunming 650500, P.R. China
e-mail: ymjiao@sina.com

T. Okuro · K. Takeuchi
Graduate School of Agricultural and Life Sciences,
The University of Tokyo, Tokyo 113-8657, Japan

L. Liang
Institute for Sustainability and Peace, The United Nations University,
Tokyo 150-8925, Japan

M. Varotto et al. (eds.), *World Terraced Landscapes: History, Environment, Quality of Life*, Environmental History 9,
https://doi.org/10.1007/978-3-319-96815-5_14

both the Hani terraces and the Satoyama are facing continuous pressures from social and economic development. Learning the efficient, adaptive management strategies from Satoyama can help Hani rice terraces meet the challenges.

14.1 Introduction

Agricultural lands comprise about 40% of global terrestrial area (FAO 2009). Although its primary services are the provision of food, fiber, and fuel, agriculture—as an ecosystem directly managed by human beings—plays a unique role in both supplying and relying on ecosystem services (Swinton et al. 2007; MA 2005; Zhang et al. 2007). The links within the agricultural landscape, namely the landscape pattern and process, impact the ecosystem services provided to society (Matson et al. 1997). Thus, it is essential to assess agriculture's ecosystem services from a landscape perspective (Tscharntke et al. 2005).

Rice is the most important irrigated crop in the world (FAO 2009). Due to the long history of rice cultivation in Asia, there are various kinds of traditional rice paddy landscapes that provide multiple ecosystem services to local people. Taking as study objects two traditional rice paddy/terrace landscapes in eastern Asia, the Satoyama in Japan and the Hani terrace landscape in southwest China, this article summarizes and analyzes the two areas' characteristics and ecosystem services. The study's aims are: (1) to compare the landscapes' pattern and ecosystem services, (2) to analyze the target and nontarget ecosystem services within and outside the landscapes, (3) to discuss the landscapes' current status and the challenges they are facing, and (4) to find lessons learned from Satoyama that may benefit the sustainability of the Hani rice terraces.

14.2 Components and Features of the Satoyama and the Hani Terrace Landscape

The traditional Satoyama landscape is a mosaic of secondary forests (also called Satoyama forests), wet rice paddies, cultivated fields, grasslands, streams, ponds, and irrigation ditches surrounding a Japanese farming village—the entire landscape necessary to supplying the needs of a community (Fukamachi et al. 2001; Kobori and Primac 2003; Takeuchi et al. 2003). Because the Satoyama landscape is the traditional rural/agriculture landscape of Japan, its features exist on a national scale (Table 14.1), and the two most famous Satoyama landscapes are the rice paddies located in Noto Peninsula and Sado Island (Fig. 14.1).

The Hani terrace landscapes are composed of forests, ponds, villages, rice terraces inundated all year, and dry fields. They are located in the Honghe Hani and Yi Autonomous Prefecture, in the southeast part of Yunnan Province, southwestern

Table 14.1 Components and features of the Satoyama and the Hani terrace landscape

Indicator	Satoyama landscape	Hani terrace landscape
Location	23°–45°N, 125°–142°E	22.5°–23.5°N, 100°–103°E
Spatial scale	National/Japan	County/Yuanyang, Lvchun, Jinping, Honghe
Area (ha)	6,000,000–9,000,000[a]	1,112,300
Terrain	From low-relief mountains to hills, lowlands, and valley bottoms	From high-relief mountain peak to deep carved river valley, distributed across the whole mountainsides
Climate	Temperate monsoon	Subtropical monsoon and vertical climate
Total annual precipitation (mm)	1000–2000	700–2400
Average temperature (°C)	10	17
Elevation range (m-a.s.l.)	175–195	105–2940
History	Yayoi era (300 B.C. ~ A.D. 300)	Tang dynasty (A.D. 618 ~ 903)
Landscape composition	Satoyama woodlands, other crop fields, grasslands, wet rice paddies, villages, streams, reservoirs, artificial ponds, irrigation ditches	Natural and secondary forests, year-round inundated rice terraces, dry fields, grasslands, villages, streams, numerous artificial small terraced ponds, and irrigation ditches
Area proportion of landscape composition	Forests: paddies: other crop fields = 1:1:1 at village scale[b]	Forests: terraces: dry lands: other land uses = 3:1:2:1 at county scale (Yuanyang)
Current status	Abandoned and underused since 1960s	Mostly are kept in original status

[a]Cited from Takeuchi et al. (2003, p. 46)
[b]Cited from Washitani (2001)

China (Fig. 14.1). People of various races, with the Hani people being the main ethnic group, have maintained this spectacular agricultural landscape for over 1300 years (Table 14.1). Compared to the Satoyama landscape, the Hani terrace landscapes are concentrated in this specific bio-cultural region. They are a unique traditional agricultural land use system in the high-relief mountainous region with a subtropical monsoon climate.

Both the Satoyama and Hani landscapes are traditional subsistence farming systems, which provide a bundle of ecosystem services (Takeuchi 2010; Jiao et al. 2012), including provisioning services (species used as food sources, timber, medicines, and other useful products); regulating services (flood control, climate stabilization); supporting services (soil formation, water purification); and cultural services (aesthetic or recreational assets, such as ecotourism attractions, providing tangible and intangible benefits) (Kremen and Ostfeld 2005). In addition,

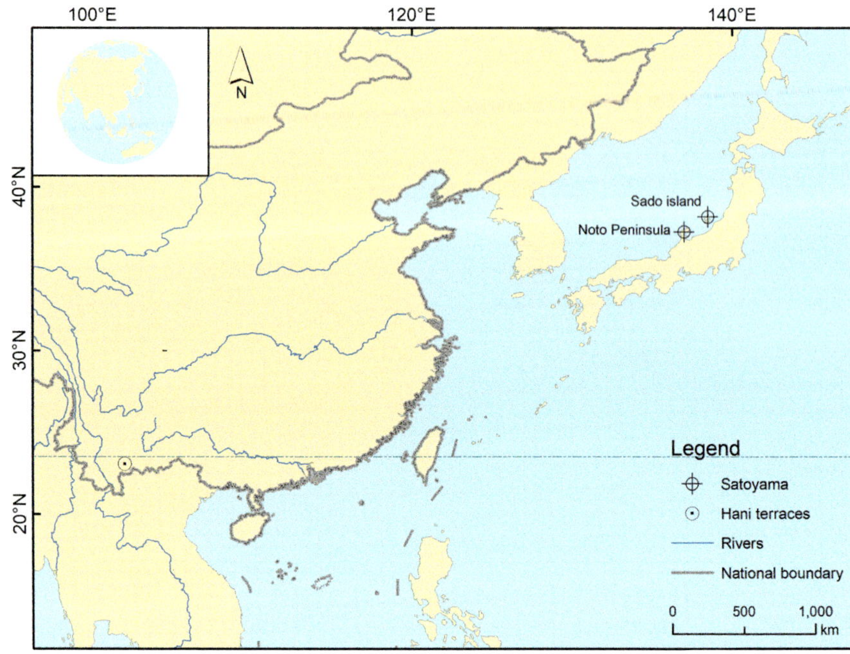

Fig. 14.1 Location of the Satoyama in Japan and the Hani terrace landscape in SW China

both landscapes are facing serious challenges on local and global scales, although these challenges differ in natural and social respects. For example, owing to economic globalization and the aging of local farmers, the Satoyama landscape was underused or abandoned, which caused a decrease in bio-cultural diversity. On the contrary, the Hani terrace landscape is under high pressure to develop economically because of the poverty of local farmers and underdeveloped socioeconomic conditions. In this circumstance, it may be argued that the Hani terrace landscape is in the early overuse stage that the Satoyama landscape faced before the 1960s.

14.3 Spatial Pattern and Ecological Process in the Satoyama and the Hani Terrace Landscape

Landscape pattern—the spatial arrangement of ecosystems—can influence the horizontal and vertical flows of materials, such as water, sediments, or nutrients (Peterjohn and Correll 1984), and other important ecological processes, such as net primary production (Turner 1989). Therefore, landscape pattern can affect the spatial distribution and delivery of ecosystem services. The center of the Satoyama landscape is the settlement surrounded by small, gently sloping rice paddies, and large areas of secondary forests (Fig. 14.2a). All the ecosystems or elements in the

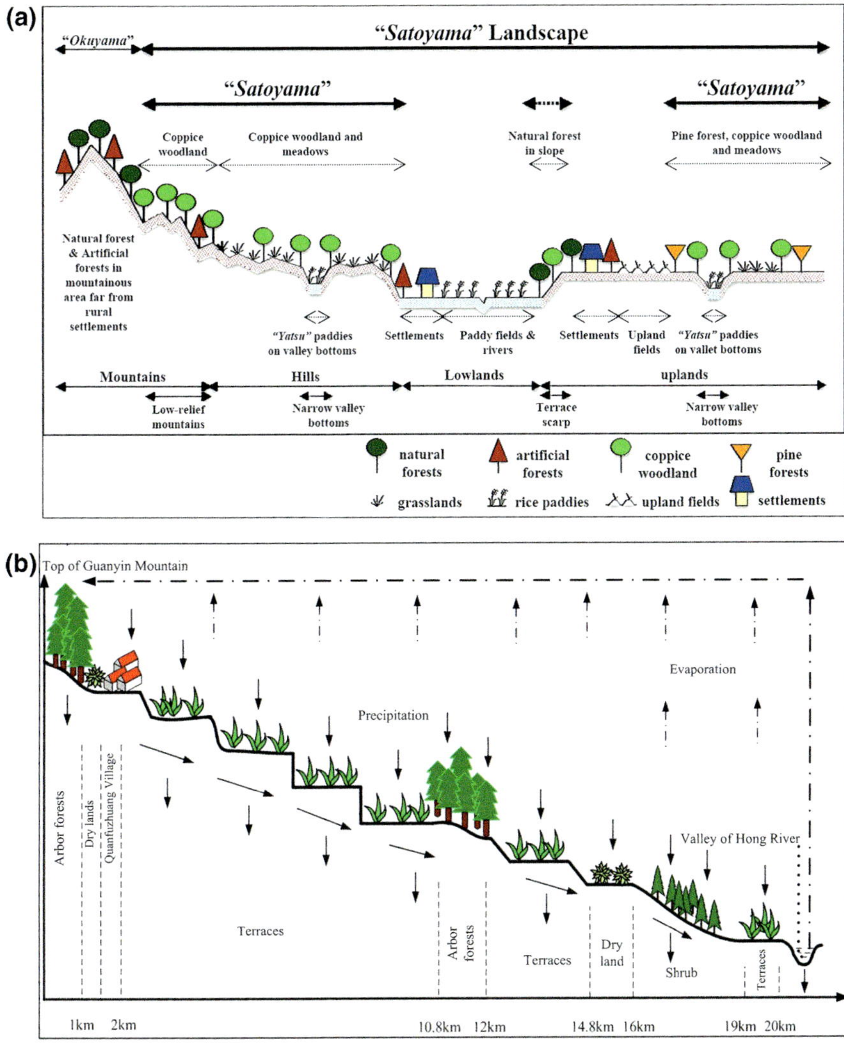

Fig. 14.2 Spatial structure from high mountains to river valley with strongly connected elements in **a** the Satoyama landscape (after Yamamoto 2001, from Takeuchi et al. 2003) and **b** the Hani terrace landscape

Satoyama landscape were once strongly connected to each other through the agricultural land use system (Takeuchi et al. 2003), mainly via organic fertilizers, such as manure, fodder, ash, and forest litter.

The Hani terrace landscape stretches across the whole mountain slope, with the natural forests on the mountaintops as one major landscape element (Fig. 14.2b). The forest-village-terraces structure along the Hani landscape's slope forms an efficient resource circulation system. The water from the forests runs through an irrigation network to reach the hamlets first, followed by the terraces. Then it continues downslope into the river valley (Jiao et al. 2012). This flow of water and nutrients in the Hani landscape is important to the provision of ecosystem services, since those services are controlled by, and normally characterized by, complex ecological processes and functions that sustain and improve human life (Daily 1997).

14.4 Ecosystem Services of the Satoyama and the Hani Terrace Landscape

Both the Satoyama and the Hani terrace landscape are mosaics of well-connected ecosystems, including forests, human settlements, rice paddies/terraces, and water areas. Each landscape element has substantial components, which have multiple ecosystem services that are directly or indirectly related to each other. These ecosystem services are divided into "within" and "outside" categories, due to the transfer of biological resources/flows, which spatially separates on-site and off-site ecosystem services (Guo et al. 2000). The ecosystem services are also classified into "target or desirable" and "nontarget" categories because, for example, provisioning services, including food, fiber, timber, and other subsistence materials, are the target ecosystem services farmers pursue, while others, especially regulating and supporting services, are the nontarget categories, which mainly benefit the public. The ecosystem services provided by the two traditional agricultural landscapes are summarized in Table 14.2.

According to Fukamachi et al. (2001), Washitani (2001), and Takeuchi et al. (2003), each element in the traditional Satoyama has its special functions or target ecosystem services related to the traditional daily life of Japanese people. For example, coppice forests were traditionally used to produce fuelwood and charcoal; to feed cattle and horses; to collect chestnuts, young shoots of ferns, herbs, and mushrooms as foods; and to gather litter in winter and green manure in summer to fertilize rice paddies or crop fields. The grasslands provided fodder for livestock, straw for homes, and organic fertilizer for rice planting. All the other landscape components provided special products or goods for local peoples, directly or indirectly (Table 14.2). Due to the continuous utilization and management of coppicing, mowing, irrigating, and harvesting over centuries, the Satoyama landscape became a heterogeneous mosaic of habitats, which consisted of a succession series of forests, grasslands, water areas, and rice paddies. In addition, collecting organic fertilizer from forests and grasslands to fertilize rice paddies consolidated the relationship between landscape elements, thus generating a conglomerate of regulating and supporting services (Table 14.2). The landscape has also generated many kinds of cultural services, such as nonmaterial spiritual and religious benefits, recreation, aesthetic values, and the cultural heritage of historic sites, including sacred woodlands, shrines, temples (Fukamachi et al. 2011).

Table 14.2 Ecosystem services of the traditional Satoyama landscape and the Hani terrace landscape

	Satoyama landscape				Hani Terrace landscape			
Landscape elements	Components	Desirable ES within landscape	Desirable ES for marketing	Nontarget ES outside landscape	Components	Desirable ES within landscape	Desirable ES for marketing	Nontarget ES without landscape
Forest	Coppice, fuelwood, secondary forest, fir forest, pine forest	P: litterfall, green leaf manure, firewood for heating and light, timbers, poles for house and fence construction, bamboo shoots, wild fruits and mushroom for food; medical plant C: recreation	P: timbers	P: seminatural habitats for species S: nutrient cycling, carbon sequestration, phyto-remediation R: climate and flood regulation C: recreation	Natural forest, natural sacred forest, natural timber-charcoal forest, seminatural Caoguo forest, cultivated tea plantation	P: freshwater for irrigation and drinking, firewood for heating and light, timbers, poles for house and fence, litterfall, bamboo shoots, wild animals, fruits and mushroom; medical plant C: culture inheritance	P: Caoguo (large cardamom) and tea	P: freshwater, seminatural habitats for species S: nutrient cycling, carbon sequestration R: climate and flood C: recreation
Cropland	(1) Paddy field (2) rain-fed (3) dry land	P: rice, straw, wheat	P: rice	P: small water area or wetland for organisms R: climate and flood, alleviation of nonpoint source pollution	Small sized rice field full of water, levee made by soil, river valley, channel system, pathway system	P: rice, straw, fishes and other aquatic animals for meat, semiaquatic and aquatic plants for food C: culture inheritance	P: rice, fishes R: nutrient retention	P: freshwater, wetland for organisms R: climate and flood C: recreation
Water area	River, reservoir	P: freshwater, fishes		P: freshwater, wetland for organisms	Numerous terraced ponds along river	P: freshwater R: flood and drought regulation		P: freshwater, numerous

(continued)

Table 14.2 (continued)

	Satoyama landscape				Hani Terrace landscape			
Landscape elements	Components	Desirable ES within landscape	Desirable ES for marketing	Nontarget ES outside landscape	Components	Desirable ES within landscape	Desirable ES for marketing	Nontarget ES without landscape
	and soil bank	R: flood and drought regulation		R: climate	valley, river, small reservoir			wetland for organisms R: climate
Village or settlement	House and livestock corral, shrine, road, vegetable land, bamboo, chestnut tree, dung yard, seedbed	P: meat, vegetable, chestnut, organic fertilizer R: manage and maintain the whole landscape S: cultured soil formation C: cultural center of residents	P: meat and vegetable	R: multiple utilities of natural resources C: recreation and tourism	House and livestock corral, amusement place, road, drinking water pool, wastewater pool, vegetable land, bamboo or fruit tree or sparse tree	P: meat, vegetables, fruits, organic fertilizer R: manage and maintain the whole landscape S: cultured soil formation C: cultural center of residents, recreation	P: meat and vegetable	R: multiple utilities of natural resources C: ethnic culture, recreation, and tourism
Grassland	Burning and mowing grassland	P: fodder for livestock, straw for house, fertilizer		P: habitats for species S: nutrient cycling, C: recreation	Natural grassland or newly planted grain to green land	P: land for grazing, straw for house, fertilizer		P: habitats for species S: nutrient cycling

P: provisioning services; R: regulating services; S: supporting services; C: cultural services (modified from Jiao et al. 2014)

Compared to the Satoyama landscape, the Hani terrace landscape is different in the following aspects.

1. It encompasses not only secondary forests, but also natural forests, tea gardens, and huge, steeply sloping rice terraces inundated all year.
2. The current Hani's daily life is mainly dependent on terrace and forest target provisioning services—such as the rice, fish, and other aquatic animals from terraces, edible hygrophytes from terrace levees and semiaquatic patches, firewood, timber, edible wild mushrooms, plants, insects, and animals from forests.
3. Due to the limited water supply for planting rice in high mountains with a subtropical monsoon climate, the Hani people have developed an excellent irrigation system and water conservation strategies to regulate water flow. During the regulating process, the Hani people have realized the target ecosystem services of water conservation and also the significant nontarget regulating and supporting services for macro-climate, hydrological cycle, soil formation, and nutrient cycling.
4. The Hani people believe many natural things, such as mountains, forests, rivers, terraces, and rice, all have a divine owner who must be respected. Therefore, the Hani terrace landscape provides many types of cultural services. Additionally, the Hani terraces have produced a highly valued creation service for both national and international societies, after the spectacular landscape was publicized by Yann Layma, a French photographer, from 1988 to 1993. His work made the Hani terrace landscape a world-famous tourist destination.

14.5 Challenges in the Satoyama and the Hani Terrace Landscape

According to the *Statistical Handbook of Japan 2011*, because of the highly developed economy and the aging of farmers, Japan's cultivated acreage shrank year after year from 6.09 million ha in 1961 to 4.59 million ha in 2010. The most common cause for the decrease was cultivation abandonment, accounting for approximately 44.0% of all cases. As one of the consequences of land abandonment, in fiscal year 2009, the self-sufficiency rate of all foods, except rice and vegetables, was highly dependent on imports from foreign countries (Statistics Bureau 2011). A similar phenomenon also happened in lumber production. Therefore, abandoning the Satoyama landscape and importing ecosystem services led to the biodiversity and ecosystem services crises in Japan, due to the loss of human-dominated seminatural habitats. In addition, high risks of climate change correlate to decreased ecosystem services in many ways (Table 14.3).

As for the Hani terrace landscape, although now it is still keeping its original status, economic and social development strategies are putting continuous pressures on its ecosystem services (Table 14.4). For example, the wonderful beauty of

Table 14.3 Biodiversity and ecosystem services crises in the Satoyama (summarized from report of *the third national biodiversity strategy of Japan*—JG 2007)

Crises	Drivers	Indirect pressure	Ecosystem services	Responses
First crisis Overuse: species and habitat degradation	Excessive human activities: industrialized monoculture, coniferous plantations, urbanization	Social demands due to rapid economic growth	People want to get efficient production of large amount of ecosystem services	Habitat restoration strategic environmental assessment
Second crisis Underuse (Satoyama issue: degradation of Satochi-Satoyama)	Insufficient level of management: abandonment of secondary forests, low-profitable rice terraces and other croplands	Social aging problem and shortage of rural labor after rapid economic growth period	Dependence upon imported foreign ecosystem services including foods, timber, and energy	Sustainable use of local resources; management of abandoned plantations; large-scale wildlife management
Third crisis Ecosystem disturbances	Artificially introduced alien species, chemical contaminations	Homogeneity of vast, intensively managed areas Decreasing of natural enemies for alien species	Degradation of regulating and supporting services	Monitoring and management of invasive species population; concentrated management scheme for prioritized targets
Climate change crisis Huge potential for species extinction and ecosystem collapse	Loss of estuary and seashore habitats, high risks of extreme climate	Population explosion, global scale degradation of natural vegetation	Degradation of regulating and supporting services	Enhancement of monitoring; developing adaptation methods

terraces attracts numerous national and international tourists. Thus, tourism has become a pillar industry in Yuanyang. Nevertheless, the conflicts are very serious between farmers and tourism companies because the beautiful scenery is created and maintained by farmers, but the marketed recreation value only benefits the company and local government. Additionally, large numbers of tourists consume lots of wild, edible plants, as well as upstream water, thus exhausting some wild species and causing water shortages for irrigation in downstream areas. Therefore, the trade-offs among different ecosystem services and human well-being have become difficult risks.

At present, the Hani terrace landscape is facing many challenges, such as severe droughts caused by global climate change, landslides and terrace collapse caused by

Table 14.4 Challenges faced by the Hani terrace landscape

Items	Challenges
Loss of rice landraces	In 1980, there were 195 local and 47 wild rice landraces, respectively; in 2008, only 48 local rice landraces were recorded. Tropical and subtropical areas were planted with hybrid rice varieties (Jiao et al. 2011)
Drought	March to May 2005, a "37-year return drought" occurred, and about 1200 ha of terraces suffered from the drought (Zuo 2005). April 2010, the Hani terrace landscape suffered another 100-year drought, and about 3300 ha of terraces were damaged (Ren 2010)
Changes of forests	Due to the implication of policies and economic development strategies, such as grain to green and the poverty elimination project, many croplands, and tea gardens are converted to plant *Alnus nepalensis*. *Amomum tsao-ko Crevost et Lemaire* and *Radix Isatidis* are planted under the forest canopy. Rubber trees are planted in the southern part. *Eupatorium adenophorum Spreng* has invaded in bare land, grassland, and the forest fringe area
Change of rice terraces	The rice terraces located in the dry-hot valley have been changed into tropic croplands for bananas, etc. Almost all the main ditches are consolidated in Quanfuzhuang River basin (about 10% of the total). Only the three paddy levees owned by one family were consolidated in 2005
Landslides	All were relatively small landslides, averaging 30 m in width, 20 m in length, and 3–5 m in depth
Social pressures	Extreme poverty of local farmers, migrating labor of younger generations, and conflicts among stakeholders including farmers, tourism companies, and local governments

a vulnerable environment of steep slopes and friable metamorphic rocks, loss of rice landraces caused by the expansion of hybrid rice and application of modern agricultural technology, poverty of local peoples, loss of traditional ecological knowledge, off-farm labor migration caused by social and economic factors, and conflicts among stakeholders in tourism development.

Although originally both the Satoyama and the Hani terrace landscape had many target and nontarget ecosystem services, these have changed, and will continue to change, with the pressures from both the inside and the outside worlds. Adaptation of sustainable management is essential to conserving the landscapes and keeping their multiple functions and multi-services (Takeuchi et al. 2003).

14.6 Lessons from the Satoyama Crises to Benefit the Development of the Hani Terrace Landscape

Compared to the three crises of the Satoyama landscape, the Hani terrace landscape is still managed in a traditional, rice-farming way. It provides multiple ecosystem services because most Hani people, including younger generations, still depend on the terrace landscape to survive, and the holders of traditional knowledge are passing their oral legacy to new generations, although few young residents know all of it.

Currently, many strategies have been proposed and implemented to revitalize the Satoyama landscape, such as the "Satoyama Initiative," the "Sub-global assessment of Satoyama and Satoumi in Japan," "The 10th Conference of the Parties (COP 10) to the Convention on Biological Diversity (CBD)," and related strategies. It should be noted that these strategies' most essential idea is to find dynamic, balanced interactions between human and nature in the long term, and to rebuild a sustainable, low-carbon, resource-circulating, and nature-harmonious society with an eye toward resilience enhancement (Takeuchi 2010). In 2011, two pilot sites of the Satoyama landscape were designated as Globally Important Agricultural Heritage Systems (GIAHS). One of them is "Sado's Satoyama in harmony with Japanese crested ibis." At present, valuable actions of environment-friendly farming are being carried out on Sado Island in the Niigata Prefecture. Through winter flooding of rice paddies, together with organic farming and fish ladders, the Satoyama has been restored to provide a suitable habitat for many aquatic species, such as crested ibises, loaches, and other fish. The rice produced there is certified as "Creating villages coexisting with crested ibises," and it is then traded at higher prices than ordinary rice. In this system, ecosystem services and human well-being are integrated into a multifunctional landscape. The Sado success can be used as a revitalizing model for rural development in Japan.

Some lessons can be learned from the Satoyama evolution to help manage the Hani terrace landscape.

1. The Satoyama was abandoned because of aging farmers as a result of decreasing birth rates and younger generations migrating to cities seeking higher income. Now, the Hani terrace landscape is facing the same problems of losing the younger generation. In the near future, the Hani landscape will be facing terrace abandonment. Therefore, determining how to increase the farmers' income and human well-being is vital to maintaining the whole landscape. The brand-certified rice produced by this area should be traded at higher prices via specific policies. Efficient management of local resources is the most important strategy to conserve biodiversity as well as ecological services.
2. The consolidation of irrigation and drainage systems, paddy levees, and the banks of both river and pond has diminished habitat quality and caused fragmentation by cutting off the flow of water and nutrients in the Satoyama landscape. Thus, the application of agricultural technologies in the Hani terrace landscape should consider a new kind of building material, rather than concrete, which will preserve flow connectivity and the seminatural habitats for the conservation of biodiversity and ecological services.
3. To combat global climate change, the Hani rice terraces should also enhance monitoring and develop corresponding adaptation methods from both scientific and indigenous ways. New strategies and policies should also be put into practice to protect the rich rice landraces and endangered species.

In 2010, a report on the background of the "Satoyama Initiative" compared cases from many countries, including Kenya, Tanzania, Argentina, Mexico, Sri Lanka, Cambodia, Bangladesh, India, Spain, Australia. In this report, the Satoyama landscape was defined as a "socio-ecological production landscape," a name which highlights the various physical structures, management techniques, and governance systems that characterize these landscapes, the benefits they provide, or the threats they currently face. In this way, the "Satoyama Initiative" can reorganize the importance of these landscapes for ecosystem services and human well-being, as well as promote reconstructing mechanisms for managing such landscapes internationally (Takeuchi 2010). The Hani terrace landscape should learn from these international cases, such as the Satoyama crises, to avoid the negative effects of modernization and globalization and instead cultivate a sustainable path to development.

Acknowledgements This study was funded by the Environment Research and Technology Development Fund (E-0902) of Japan's Ministry of the Environment, and by the National Natural Science Foundation of China (grant numbers 40401022, 41271203, 41761115). The authors also thank for Anton Pijl and the other anonymous reviewer for their constructive comments, and the editors for their editing of the manuscript.

References

Daily G (ed) (1997) Nature's services: societal dependence on natural ecosystems. Island Press, Washington, DC

FAO (Food and Agriculture Organization) (2009) http://www.fao.org/statistics/en. Accessed 15 Apr 2018

Fukamachi K, Oku H, Nakashizuka T (2001) The change of a Satoyama landscape and its causality in Kamiseya, Kyoto Prefecture, Japan between 1970 and 1995. Landsc Ecol 16:703–717

Fukamachi K, Miki Y, Oku H, Miyoshi I (2011) The biocultural link: isolated trees and hedges in Satoyama landscapes indicate a strong connection between biodiversity and local cultural features. Landsc Ecol Eng 7(2):195–206

Guo ZW, Xiao XM, Li DM (2000) An assessment of ecosystem services: water flow regulation and hydroelectric power production. Ecol Appl 10:925–936

JG (Japanese Government) (2007) The third national biodiversity strategy of Japan. http://www.env.go.jp/en/focus/attach/071210-e.pdf. Accessed 20 Oct 2017

Jiao YM, Li X, Liang LH, Takeuchi K, Okuro T, Zhang D, Sun L (2012) Indigenous ecological knowledge and natural resource management in the cultural landscape of China's Hani terraces. Ecol Res 27(2):247–263

Jiao YM, Liang LH, Okuro T, Takeuchi K (2014) Ecosystem service and biodiversity of traditional agricultural landscapes: a case study of Hani terraces in SW China. In: Hong S, Bogaert J, Min Q (eds) Biocultural landscape: changing knowledge and value. Springer, pp 81–88

Kobori H, Primac R (2003) Participatory conservation approaches for Satoyama, the traditional forest and agricultural landscape of Japan. Ambio 32(4):307–311

Kremen C, Ostfeld RS (2005) A call to ecologists: measuring, analyzing, and managing ecosystem services. Front Ecol Environ 3:540–548

MA (Millennium Ecosystem Assessment) (2005) Ecosystem and human well-being: synthesis. Island Press, Washington DC

Matson PA, Parton WZ, Power AG, Swift MJ (1997) Agricultural intensification and ecosystem properties. Science 277:504–509

Peterjohn WT, Correll DL (1984) Nutrient dynamics in an agricultural watershed: observations on the role of a riparian forest. Ecology 65(5):1466–1475

Ren RG (2010) Fifty thousands mu Hani terraced fields were dried by high temperature and no rain. Chuncheng Evening News, April 17. (In Chinese)

Statistics Bureau (2011) Statistical handbook of Japan. http://www.stat.go.jp/english/data/handbook/pdf/2011all.pdf. Accessed 23 Apr 2018

Swinton SM, Lupi F, Roberson GP, Hamilton SK (2007) Ecosystem services and agriculture: cultivating agricultural ecosystems for diverse benefits. Ecol Econ 64:245–252

Takeuchi K (2010) Rebuilding the relationship between people and nature: the Satoyama initiative. Ecol Res 25(5):891–897

Takeuchi K, Brown RD, Washitani I, Tsunekawa A, Yokohari M (eds) (2003) Satoyama: the traditional rural landscape of Japan. Springer, Tokyo

Tscharntke T, Klein AM, Kruess A, Steffan-Dewenter I, Thies C (2005) Landscape perspectives on agricultural intensification and biodiversity-ecosystem service management. Ecol Lett 8:857–874

Turner MG (1989) Landscape ecology: the effect of pattern on process. Annu Rev Ecol Syst 20:171–197

Washitani I (2001) Traditional sustainable ecosystem “SATOYAMA” and biodiversity crisis in Japan conservation ecological perspective. Glob Environ Res 5(2):119–133

Yamamoto S (2001) Studies on the effect of changes in rural landscape structure on secondary forest plants in Japanese rural areas. Bull Natl Inst Agroenviron Sci 20:1–105

Zhang W, Ricketts TH, Kremen C, Carney K, Swinton SM (2007) Ecosystem services and dis-services to agriculture. Ecol Econ 64:253–260

Zuo XJ (2005) The first drying crack of Hani terraced fields within hundred years. Beijing Youth Daily, June 1. (In Chinese)

Chapter 15
Terraced Lands: From Put in Place to Put in Memory

Ada Acovitsióti-Hameau

Abstract The terracing of sloping, rocky, and hilly land is man's answer to the physical environment's potentialities and constraints. It sets up singular territories where nature and culture are tightly linked. The places of action and life thus created report a high level of symbiosis between inhabitants/users and used time/ spaces. This symbiosis defines the quality of the areas where humans live, the *écoumène* as explained by Augustin Berque (Berque in Écoumène, Introduction à l'étude des milieux humains. Belin, Paris, 2000). Acquainted to a put in writing of the relief, these arrangements tangibly mark the way users—permanent, seasonal, or passers-by—perceive the territory. Expressed by tracing and by artifacts widely executed in dry stone, this "writing on the soil" conforms and orders space and establishes a network of relationships concerning family; neighborhood; production; cooperation; and structuring societies, land property, and technical systems. Functional efficiency and aesthetics create another register of shared perceptions, which gives meaning to forms of the landscape, to constructed works, and to modes of building, conferring an identity recorded by the collective memoirs. These perpetual comings and goings established between the territory's reality and its representation, this fertile *trajectivity*—to paraphrase Gérard Chouquer—between the material and the mental, justify and legitimize attention to and care for terraced lands because, beyond the universe of forms and production of goods, these sets help us appreciate lifestyles and think about man's way of being in the world.

15.1 "Marks" on the Lands

The terracing of sloping, rocky, and hilly land is man's answer to the natural environment's potentialities and constraints for earning a profitable living. These arrangements matter among the objective evidences of the adaptation reactions

A. Acovitsióti-Hameau (✉)
Société Scientifique Internationale Pour L'étude Pluridisciplinaire de La Pierre Sèche, Brignoles, France
e-mail: aser2@wanadoo.fr; contact@pierreseche-international.org

M. Varotto et al. (eds.), *World Terraced Landscapes: History, Environment, Quality of Life*, Environmental History 9,
https://doi.org/10.1007/978-3-319-96815-5_15

between a given territory and its permanent, seasonal, or passing-by users. If terracing hillsides is a model of the relief, this action also mirrors writing. As with all graphic acts, terracing is seen by results that "leave traces" on or "mark out" the ground that insert forms and affix signs in the physical environment. The latter, in its turn, is not simply a medium for expressing oneself. The existing space is an active, dynamic entity, which influences the choice of disposals and brings about the intervention of the scripters. So, combined tightly with the ground, land terracing participates in the constitution and maintenance of viable, fertile, and sustainable spaces, while simultaneously clearing up and spreading information on the statuses and functioning of lands. In the long term, land terracing becomes familiar, is a representative element of lifestyles, and reveals group and individual identities. For these reasons, the value of terraces are founded on explicitly perceptible motivations, but also on implicitly felt ones. It is through this double aspect, material and symbolic, that we propose to analyze the strong, current tendency in favor of the preservation and development of this territorial equipment.

15.2 Terraces, Landscapes, Societies

Several mechanisms are at work during territorial arrangement and, consequently, during the implementation of terraces. They correspond to a "taking of possession of the territory" (Ambrosi 1990), this "conquest" being made gradually and respecting the place's physical, social, and economic particularities. This process requires consciousness of the place's filiation: good knowledge of the succession of aspects and statuses, which link together to include the past in the realizations of the present and the perspectives for the future. The clarity and acceptance of this filiation protect the quality (for local products or for links among groups and among individuals) of new disposals brought by economic and social changes (Acovitsióti-Hameau 2002). The case of Vilafranca (València province, Spain) is a good example (Fig. 15.1).

The town is as much industrious (spinning and cloth factories) as tourist (mountain holidays), and since about 2005, it has increasingly promoted and cared for its territory's peasant features. The urban core rises in the middle, framed by cultivated lowland and grazed highland, both of which are compartmented and terraced. This landscape is looked after by the will to maintain a modest agro-pastoral economy that stimulates hiking and agritourism while taking advantage of the market thus created. This development plan allows partial preservation of vernacular territorial arrangements. The project is forward-looking, and its realization is still in progress despite disagreements on its contents. At Vilafranca (and in all other similar cases), natural land and built works compose *in fine* a "whole," a single inseparable set. This vision gives sense to the concerned

Fig. 15.1 Town of Vilafranca (València Region, Spain) between its mountain and its countryside (*Photo* ASER Association)

space and makes it a self-identity landscape for people who occupy and use it and a cultural landscape (asking for attention and care) for all outer observers.

Passing from the concept of space to the concept of landscape is an evolution dependent on human intervention. This intervention requires in-depth knowledge of the territory via diligent attendance and in the long term, followed by a detailed transmission of collected information and proven sensations. In this approach, the individual and collective feeling remains decisive. Landscape is based on perceptions. In other words: "the transformation of the physical environment by man is the expression of his cultural representation of this environment" (Blanchemanche 1990). To the geographical, primary notion, are added factors issuing from history, sociology/ethnology, art, and aesthetics. To physical, temporal, and social realities are added their representations: images and concepts based on these realities. The geographer Gerard Chouquer (Chouquer 2001) believes the realities and their representations to be equal, each of them depending on the other to be conceived, achieved, and explained. This equality shows itself especially in terms of dynamics. There is a continual interpenetration between these groups of powers, giving landscapes a "fuzzy and evolutionary" character.

Perpetually transformed, landscapes thus appear always new in themselves but also already known because of the multiple inheritances that they incorporate and that common users accept. The obliteration of these inheritances would empty the landscapes of their anthropological substance, of their quality of "place," of space where we can decipher "the registrations of the social link" (Augé 2010). For instance, in the mid-season mountain pastures of the Montegrosso village (Liguria, Italy), the gradual conversion of the former farms and sheepfolds to residences of a holiday resort did not erase the memory or the festivities bound to pastoralism. This change also did not limit the handling of lands and buildings by indigenous families, despite the reduction of common land and the disappearance of local herds (Acovitsióti-Hameau 2010). Since the 2000s, the landscape of small farmhouses, tidied up in a three district, terraced environment supported by an herb forehead (Fig. 15.2), is partially maintained by strong personal and collective memories and identical considerations attached to the place.

The notion of "place" is important for understanding the effects of the mechanisms transforming spaces. We can even say that these transformations aim exclusively at producing places: spaces qualified by their specific properties and by their material and immaterial functions. Soil composition, land configuration, the extent of the sight, the sound atmosphere, the penetration or lack of winds, the light, the exhalations, the cold, or the heat make their character. The issued products, the displayed activities, the attached rites, and the referring imagination make personalized entities—"markers" for country planning, for everyday travels, for relationships between individuals and among neighboring groups. As with the territory, the place is not simply the space (Brochot and

Fig. 15.2 Mid-seasons' hamlet of Case Fascei (Montegrosso-Pian-Latte, Liguria, Italy) among its pasture terraces (*Photo* ASER Association)

de la Soudière 2010). Precise in its location, defined in its temporality (the "place" can also be the convenient moment), and—for these reasons—having a name, the place is often personified as concealing a "spirit" or a "genius" (*genius loci*); the spirit of a "place" is not attached to a "space," which has flexible outlines and a moldable morphology. Places (or *topoi*) are thus assimilated to beings and are our "social body": an extension of ourselves in the environment (Leroi-Gourhan 1943). This beautiful metaphor reminds us of the ceaseless comings and goings between reality and perception—the perpetual interaction between the physical territory and its felt, historic, social, and patrimonial contexts. This complex process ends by developing local cultures, which are made and unmade in groups and periods. Each of these stages or concrete moments corresponds to a time/space, combining the objective space with the material, ideal, and sensory elements of each instant into a *concretude* (Chouquer 2001). All these instants are different and succeed each other with the complicity of the land users.

15.3 Country Planning: Lifestyles

Implementing terraced lands mobilizes the concepts and mechanisms we have described above. Terracing participates in global country planning, prioritizing rural lands and, most of all, those having an agro-sylvo-pastoral production. Their spread can be considerable, and though other operations are necessary for their creation, the most visible technique used is dry-stone walling. This technique absorbs a large part of the materials excavated from freshly turned soils and supports the drainage and aeration of terraced plots. It maintains the place's unity and coherence because it is largely manual and craft (few operators, no special tools, empirical apprenticeship despite theoretical books that have circulated since the nineteenth century and are multiplied today). The material (the stone) and the technique (building with no mortar) call up to performance, and this factor encourages the use of terraces and dry-stone construction. Additionally, dry-stone walling, just as land terracing, shows—beyond knowledge, skill, and ingenuity—the sensibility of the body and of the spirit via adaptation to contexts and milieus.

Finally, expressed by drawings and artifacts largely executed in dry stone, the "writing" of terraces arranges and orders space by showing usage rules (cultures, irrigation, storages, passages, etc.) and by inciting relevant behaviors (cyclic presence/absence of users, joint arrangements and help for productions, development of leisure activities and user-friendliness, etc.). These functions strengthen every link: inter-spatial (matching land location to the plans of traffic, etc.); inter-seasonal (matching cultures, wood cuts, animals' passes, hunting calendars, etc.); intergenerational (transmission of land ownership, uses, features of the soil, water circulation, climate, etc.); and inter-community (drainages, road networks, trade, fairs and other festivities, etc.). There is a whole set of relationships concerning family, neighborhood, production, and cooperation that structures societies, property, and technical systems.

Similarly, if this "putting in writing" brings order, it also draws attention to activities and behaviors and thereby perpetuates them. Signaling arrangements (wall crowning, cairns, marks), hunting devices (seats, recesses, low walls), and pastoral markers (signals forbidding or allowing entry in pasture land, signals for huts and water supplies) bear indications for possible actions, misadvised or authorized, and instructions of good behavior for habitual or occasional users and for passers-by. All of them should recognize these signals or—at least—note them and leave them in place.

Efficiency and aesthetics create another register of shared perceptions, giving immaterial qualities to landscape patterns, constructed works, and modes of building—conferring an identity recorded by collective memoirs. For instance, people of Banyuls keep cultivating their vineyard in the "old way" (Fig. 15.3) and promote their products by exalting the configuration of the relief carrying the plants: steep the slope, strong the wine!

Feelings of neatness and strength from terraced spaces are emphasized as the view embraces big areas without caring about details. These panoramic views increase the "misleading clarity that confer the height and a faraway vision" (Urbain 2010). Examined near the ground or on a small scale, the same arrangements give a different image. It is because the vertiginous masks the profusion of the structures on the ground. These structures often have a modest appearance and visibility, but their existence is essential for the arrangement, the stability, and the functioning of the terraced lands' implementation. Most of the devices for road networks, water

Fig. 15.3 Steep terraced slopes of the Banyuls—Collioures vineyard, Roussillon, France (*Photo* ASER Association)

Fig. 15.4 Terraced high mountain around Arés, Maestrat, Spain (*Photo* ASER Association)

streaming, storage, and rest are in this class. Even important constructions, such as huts for laborers, are not always recognizable in the overviews, which display "staged" spaces, the shown images being conditioned to impress (Fig. 15.4). Walking through these spaces allows people to discover their "secrets." These are often revealed to be treasures of ingenuity, demonstrating that places, functions, and building techniques fuse to one another and are attached to social points and imaginary items to create personalized spatial unities.

15.4 What Is Important? The System

Carefully examining terraced lands and dry-stone settings shows their complexity and diversity. The smallest details reveal lifestyles and bear witness to the builders' intelligence. Global country planning presupposes many operations and realizations, all essential and complementary: the breaking-up and drainage of lands; the shaping of plots; water courses and distribution; the partition and demarcation of cultivated, free, and wooded land; the implementation of roads and paths; the construction of utilitarian premises for peasants and foresters; and the establishment of pastoral parks and buildings. Access ramps and flying staircases through terraces, specific places for fruit trees such as olive trees (Fig. 15.5), a sun-bathed wall in front of which the dessert grapes mature, a "niche" to keep provisions cool or a hollow in the wall to tidy up tools, ropes, and flocks' bells are all elements indispensable for the system.

Fig. 15.5 Olive trees' protections in Lesvos, Greece (*Photo* ASER Association)

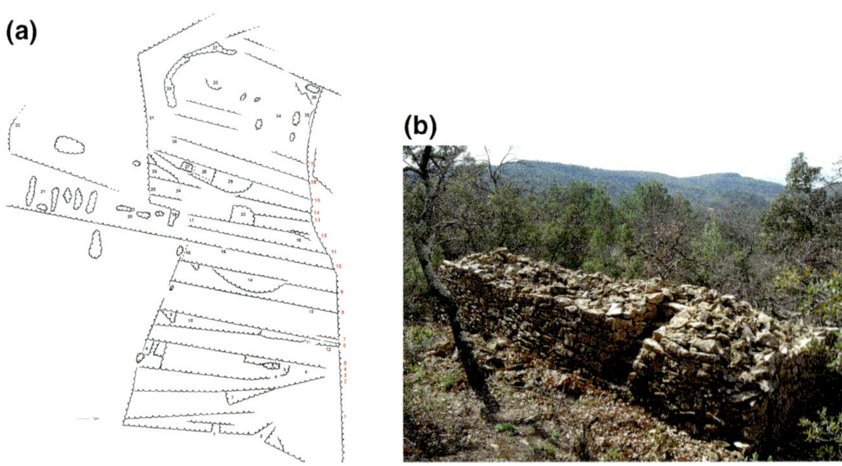

Fig. 15.6 Detailed country planning of a modest slope, Var, Provence, France (*Photo* ASER Association)

The planning affecting a hillside in the Var department's hinterland (Provence, France) shows the complicity and complementarity between the voluminous and the restricted, the mass and the detail. The set consists of terraces, stone heaps, and enclosure walls, and it covers the whole slope and the little plateau forming its summit (Fig. 15.6).

It is bare, thankless land, where stone appears everywhere and where one wonders if culture, even minor, is possible. However, the little, local farming community made this area fruitful, cultivating it annually and driving flocks to pasture, without neglecting the leisurely but also food-supplying activities of hunting and gathering. Functional since at least the middle of the nineteenth century, the area was gradually abandoned after World War II. Modest hunting and the grazing of small family herds subsisted until the early 2000s. Some years ago, the area became public, and the municipality intends to preserve all land accommodations, even if there is no precise plan of action yet.

Preliminary studies (topographies, ethnographical surveys, and archival research) are now finished, and the district periodically accepts guided visits, which discuss dry-stone walling and the lives of former rural people, whose hard labor is illustrated by built devices. Hunters and gatherers continue crossing the hillside. Despite the banality of the constructions (or, perhaps, because of it), the feeling of owning an exceptional patrimonial place makes its way among the state's and inhabitants' representatives. The concept of "lands of stone" and of "food giving stones" (Godefroid 2014) insinuates itself in the community's agricultural and forest heritage. If the researchers and the authorities have a real responsibility for this evolution, this insinuation would not be possible without the sense (the spirit maybe?) of the place itself, which, apparently, awakens recollections of lifestyles practiced by well-known and intimate ancestors. Because memory can take hold of this sense, the place becomes part of the cultural heritage in a credible and practicable way.

The case presented above is generalizable for most of the terraced territories that are being recovered. For instance, recovering cultivation in several terraced areas of Majorca (Baleares, Spain) is intended to support alternative tourism touching the hinterland and the autumn/winter season and absorbing traditional agro-pastoral products, which are transformed and packaged in an innovative way (proposed forms and quantities, packaging types, various lots for tasting, etc.). This initiative also maintains farmers on these lands who, without such measures, would have abandoned them. So, numerous vernacular arrangements on the hillsides are rebuilt and reused, either because they welcome and supply circuits of agritourism or because the desire to develop internal wealth allows (and values) maintaining or returning to rural life. In Italy, "recoveries" or "stubborn forms of return" to rural life on terraced lands seem to have success, not due to the "fétichisation" of chosen places, but from their rehabilitating manners of living and reports to the environment that the society understands and adopts (Varotto 2008). In all contexts where innovations remain attentive to their heritages and are moderately achieved, social (new populations) and economic (new productions) transformations that accompany them do not upset the traditional vocations of places (dominant activities suitable to each land unit) or all people/territories (networks of traffic, mutual aid, and exchanges). These conditions are indispensable for the survival of terraced territories; the high level of symbiosis between inhabitants/users and time/spaces defines, not only the quality, but also the proper essence of their being. For this reason, "putting in place" and "putting in memory" terraced lands are processes which cannot be dissociated from the cause-and-effect of natural, economic, human, and identity progressions.

The tendency to put forward the indivisibility of the place, the products, the human groups, the social relationships, and the perceptions of all kinds is visible in the inscription of the dry-stone fashioned landscapes on the UNESCO world heritage lists. Among a dozen so distinguished sites between 1996 and 2014, two-thirds include, or consist completely, of agricultural terraces, and half of those show vine growing as the key activity, followed by olive growing and rice growing. All are qualified as "cultural landscapes." The holistic approach is clear in the arguments' discourse, where social links, practices, and customs take as much place as the purely material data. Arguments in favor of a set of terraced rice fields on Bali in 2012 even give primacy to a local philosophical system (the Subak), which simultaneously manages land holding and use, means of production, social implications of organizations, and distribution of goods. In the arguments for the partial inscription of a vineyard in Burgundy, France, (2014), popular know-how about geomorphology and microclimates plays a role as important as academic data. The focus of examining and judging dry-stone landscapes seems to be the idea that proven environmental, social, and cultural practices work in favor of the parallel self-fulfillment of people and lands. They also help to discover or assert mentalities, ways of action, and identical or similar interests, which strengthen understanding of the other and promote collaborations. Terraced territories are—apparently—a good laboratory to test these benefits.

References

Acovitsióti-Hameau A (2002) Constructions en pierre sèche en Méditerranée: pour une reconversion en souplesse. In: Bulletin de l'Association des géographes Français (BAGF-Géographies): pp 332–343

Acovitsióti-Hameau A (2010) Une montagne à vivre et à partager: le territoire de Montegrosso-Pian-Latte en Ligurie. In: Tzortzis S, Delestre X (eds) Archéologie de la montagne européenne, Actes du colloque de gap (septembre 2008). Errance et Centre Camille Jullian, Paris et Aix-en-Provence: pp 97–107

Ambrosi A (1990) L'archittetura in pietra a secco: costruzione, progetto, tipologie. In: Ambrosi A, Degano C, Zaccaria DA (eds) Architettura in pietra a secco, Actes du 1er séminaire international sur la pierre sèche—Noci-Alberobello, septembre 1987, Schena, Brindisi, pp 17–84

Augé M (2010) Retour sur les 'non-lieux'. Communications 87:171–177

Berque A (2000) Écoumène, Introduction à l'étude des milieux humains. Belin, Paris

Blanchemanche Ph (1990) Bâtisseurs de paysages: Terrassement, épierrement et petite hydraulique agricole en Europe (XVIIe-XIXe siècle). M.S.H, Paris

Brochot A, de la Soudière M (eds) (2010) Autour du lieu. In: Communications 87

Chouquer G (2001) Nature, environnement et paysage au carrefour des théories. Études rurales 157–158:235–252

Godefroid G (2014) Les pierres qui nourrissent. Chasse, élevage et agriculture 'de la pierre' à Perremenguier Sainte-Anastasie-sur-Issole. Cahier de l'ASER 18:37–47

Leroi-Gourhan A (1943) L'homme et la matière. Albin Michel, Paris
Urbain J-D (2010) Lieux, liens, légendes. In: Brochot A, de la Soudière M (eds) Autour du lieu. Communications 87: 99–107
Varotto M (2008) Conclusions. Vers une redécouverte des "paysages intermédiaires". In: Scaramellini G, Varotto M (eds) Paysages en terrasses des Alpes Paysages en terrasses des Alpes, Projet ALPTER. Marsilio, Venise, pp 112–117

Chapter 16
Economic Analysis of the Traditional Cultural Terraced Olive-Growing Landscape and Participatory Planning Process

Biancamaria Torquati, Lucio Cecchini, Sonia Venanzi and Giulia Giacchè

Abstract Terraced landscapes are increasingly considered as valuable cultural, social, and environmental systems. However, we could attribute the loss of most of these landscapes and their tangible and intangible heritages to the abandonment or degradation of these areas that could be valorised and protected. The main goals of this study are providing a methodology and tools to analyse traditional terraced landscapes integrating spatial analysis with socio-economic analysis and suggesting operational or political proposals to reduce the abandonment of agricultural terraces. The selected study area is the olive-growing area in the municipality of Trevi (Umbria Region, Italy). This study's innovative contribution is its methodology composed of two main steps: construction of an integrated and open database followed by launch of a participatory planning process. The results show that the main weakness of the conservation of the olive-growing landscape is the profitability level of olive oil production and sale. Some measures need to be implemented to address these issues, and a joint public and private effort is required. On the one hand, public institutions should provide direct funding and incentives, and on the other, consumers should be willing to pay more for extra-virgin olive oil with landscaped value. Moreover, nomination for inscription into UNESCO's World

B. Torquati (✉) · L. Cecchini · S. Venanzi
Department of Agricultural, Food and Environmental Sciences, University of Perugia, Perugia, Italy
e-mail: bianca.torquati@unipg.it

L. Cecchini
e-mail: luciocecchini89@gmail.com

S. Venanzi
e-mail: sonia.venanzi@gmail.com

G. Giacchè
Mixed Research Unit Spaces and Societies (ESO), University of Rennes, Rennes, France
e-mail: giacche.giulia13@gmail.com

M. Varotto et al. (eds.), *World Terraced Landscapes: History, Environment, Quality of Life*, Environmental History 9,
https://doi.org/10.1007/978-3-319-96815-5_16

Heritage List as a cultural landscape, advocated by local institutions, could generate useful synergies to implement efficient collective marketing policies, which are particularly demanded by olive growers.

16.1 Introduction

Terraced landscapes exist all over the world. They have been studied widely both due to their considerable landscape value, the delicate environmental balances they help safeguard, and the increasingly neglected state they have fallen into (de Graaff et al. 2008; Duarte et al. 2008; Barbera et al. 2010; Kieninger et al. 2013; LaFevor 2014; Tarolli et al. 2014; Agnoletti et al. 2015; Arnáez et al. 2015; Torquati et al. 2015; Bonardi and Varotto 2016; Fukamachi 2016; Ridder et al. 2016; Ferro-Vázquez et al. 2017).

Terraced landscapes are increasingly considered as valuable cultural, social, and environmental systems to be protected and enhanced. In cultural terms, their value lies in the related traditional farming practices and the presence of historical hydraulic-agricultural works, while in social terms, they represent a source of livelihood. Environmentally, terraced landscapes mean a low degree of agricultural mechanization and an extremely limited option for external inputs. However, policymakers are having trouble identifying clear and, above all, coordinated strategies for their conservation.

In Italy, several initiatives foreground the protection of terraced landscapes. They range from international cooperation (e.g. International Terraced Landscapes Alliance) to landscape catalogues (e.g. Italian Historical Rural Landscapes: Agnoletti 2012), from the establishment of local landscape observatories (e.g. Canale del Brenta) to local spatial planning (e.g. *Regolamento urbanistico di Radda in Chianti* [Urban planning by the municipality of Radda in Chianti]), and from measures towards rural development programmes (e.g. support for non-productive investments) to listing of the location as a UNESCO World Heritage Site (e.g. Cinque Terre and Costiera Amalfitana).

In recent years, new approaches to the study of traditional landscapes have been proposed. These approaches are mainly based on integrated and interdisciplinary methodologies (Barbera and Biasi 2011), and they often rely on the application of the Geographical Information System (GIS) (Cullotta and Barbera 2011) . GIS technology has allowed scholars to investigate the historical evolution of traditional landscapes by comparing multi-temporal maps; produce a series of ecological indices underpinning landscape metrics and quantitatively measuring their structure and function (Bender et al. 2005; Vizzari 2011; Modica et al. 2012; Sklenicka et al. 2017); integrate local knowledge with knowledge of experts in conservation-oriented landscape planning (Torquati et al. 2011); and characterize the spatial structure of farms to provide a more accurate picture of the real structure of the agriculture sector.

Moreover, it offers useful information for designing policies to improve competitiveness in remote rural areas (Colombo and Perujo-Villanueva 2017).

To our knowledge, no study has combined spatial information on traditional terraced landscapes with socio-economic information on farms on which they are located.

In this context, this study aims to provide a methodology to analyse traditional terraced landscapes that can integrate spatial analysis with socio-economic analysis, by focusing on the interrelationship between farms, terraced landscapes, and economically and socially sustainable intervention strategies. Additionally, this study tries to identify public and private interventions that can help prevent the abandonment of marginal terraced olive growing, since, despite a favourable economic context, it still is vulnerable.

The study area is represented by the olive-growing area in the municipality of Trevi (Umbria region, Italy) (Fig. 16.1a), which has been chosen for two reasons. First, modern olive groves coexist in this area with traditional terraced olive groves, which is typical of contemporary arboriculture in foothills and mountainous areas (Barbera and Biasi 2011) (Fig. 16.1b). Second, the local extremely dynamic olive economy is supported by active institutional initiatives. These initiatives include, among others, cultural events organized by the association *Città dell'Olio* (Oil Towns); protected designation of origin (PDO) certification; historical routes linked to Saint Francis; registration in the National Registry of Rural Landscapes of Historical Interest, of farming practices and of traditional knowledge; presence of several hectares of organically grown olive groves; and nomination for inscription into UNESCO's World Heritage List.

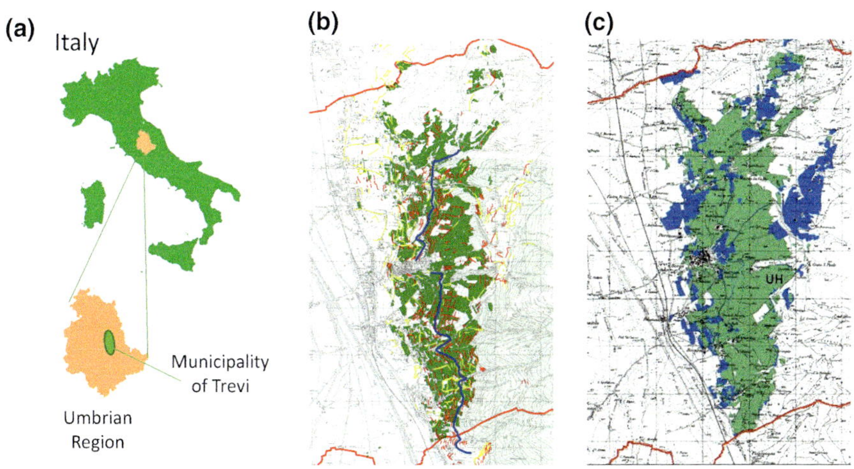

Fig. 16.1 **a** Geography location of municipality of Trevi (Italy). **b** Olive-growing area in 2012 with the position of dry-stone walls (coloured in red), slopes (coloured in yellow), and level line of 400 m above sea level (coloured in blue). **c** Olive-growing areas that disappeared between 1955 and 2012 (coloured in blue). *Source* **a** our elaboration; **b** elaboration of Leonardo Laureti

The remainder of the paper is structured as follows. We begin by describing the general aspects of the study area, focusing on the importance of the traditional cultural olive-growing landscape and its current situation. In the next section, we outline the cartographic sources and methodology used for data analysis. We then present the results and discuss implications for decision drawing our conclusions in the last part of the paper.

16.2 Case Study: The Traditional Cultural Landscape of Olive Groves in Trevi (Italy)

16.2.1 Historical Overview of the Study Area

The olive-growing area in the municipality of Trevi is located between 250–300 and 550–600 m above the sea level. In this area, olive trees grow near the lower thermal limit. Olive groves that cover the foothills are not suitable for intercropping due to their planting density, and as a result, a unique and homogenous landscape can be seen all along the Umbrian valley (Bevilacqua 2012).

In this area, as well as in the entire Umbria Region, appearance of the olive tree was not spontaneous, and its introduction by inhabitants goes back to the Etruscan era. Historical evidence of olive tree cultivation in the Trevi area is offered by the discovery of a Roman sandstone mill. In the Middle Ages, olive tree cultivation in the area suffered a setback, and olive trees were grown within the city walls—or close to them—in small plots protected by stone walls, the so-called *Chiuse* (Bevilacqua 2012). Simultaneously, the number of pastures increased, and animal fats replaced olive oil.

Around the fifteenth century, planting of olive trees was resumed by local people to the detriment of the woods on mountain slopes. The olive-growing area significantly increased between the late eighteenth century and the nineteenth century, following the laws issued by the Papal State, which had jurisdiction over these territories. New plantings were subsidized, and local people were even allowed to uproot the wooded area on the condition that land with high slopes would be properly terraced. This engineering work, resulting from secular efforts, still characterizes the whole area. When the slope is moderate, plots are flanked by grassy embankments (*ciglioni*). At higher altitudes and steep slopes, semi-circular walls (the so-called *lunette*) and terraces protected by dry-stone walls have frequently been built (Bevilacqua 2012). These architectural landscape elements are sometimes mixed with rocky outcrops that have often been incorporated into the structure of the dry-stone walls.

This area has clear landscape value: olive trees dwell here as if they were on the top of a tower, and from a distance, they seem to climb into the woods, creating a pleasant colour contrast. Against the backdrop of the centenary olive trees, some watchtowers and many farmhouses with dovecotes still stand out, and beautiful little rural Romanesque churches and old painted and frescoed wayside shrines (typical of local tradition) can be found in the area.

In the eighteenth and nineteenth centuries, olive trees were multiplied by excising ovules from the mother plant stump. The scarcity of stumps prevented local people from making a qualitative selection based on olive production and on adaptability to the specific climate and soil of the area. Instead, such a selection occurred over the years thanks to the combined action of adverse climatic events and farming, resulting in the current high level of biodiversity of these olive groves. In the municipality of Trevi, modern olive growing based on regular planting distances is a recent phenomenon (olive trees are only 40–50 years old), and it has marginally affected some plots at lower altitudes. A diachronic analysis of land use from 1955 to 2012 shows that the olive-growing area decreased from 1266 ha in 1955 to 961 ha in 2012, meaning an overall decrease of 24%.

The reduction of the olive-growing area affected the entire territory devoted to olive cultivation; the phenomenon was particularly serious in the valley floor because of urbanization and in inaccessible mountain areas that were abandoned and colonized by forest cenoses (Fig. 16.1c). It is important to note that in some areas of the municipality of Trevi, especially in the Northeast, new olive groves have been planted.

Evaluation of the landscape's integrity level[1] confirms the reduction of the olive-growing area in the western part of the plain and the mountain margins to the East. It also clearly highlights that the olive grove belt of Trevi, which distinguishes the landscape of the eastern slope of the Umbrian Valley, shows high invariability in agricultural uses. This traditional landscape integrity is an asset to the area, and its conservation plays a crucial role not only from an agricultural, ecological, and cultural perspective but also in terms of promoting the local and regional value of olive oil production.

16.2.2 Critical Issues in the Traditional Cultural Olive-Growing Landscape

There is reduction of the olive-growing areas, especially in marginal olive groves that are increasingly at risk of abandonment. In our study area, an olive grove can be classified as marginal based on different parameters: plant density (presence of less than 204 plants in the olive grove is considered marginal); land development (presence of terraces and *lunette*); slope (when the slope is higher than 25%, the

[1]Landscape integrity level was assessed using a particular methodology (in Italian Valutazione Storico-Ambientale, VASA), that is, through evaluation of the dynamics and integrity of the traditional agricultural landscape. VASA evaluations are required by Italian law to register the area in the National Registry of Rural Landscapes of Historical Interest, of farming practices, and of traditional knowledge (Art. 4 and Annexes 2 and 3 of Ministerial Decree no. 17070 of 19/11/2012).

area can be scarcely accessed by mechanical vehicles); and altitude (above 500 m, the soil and climatic conditions are considered unfavourable). In the municipality of Trevi, the dividing line between marginal and non-marginal olive groves can be set at 400 m above the sea level. This limit is roughly located on the old road of the so-called *Strada delle Selvette* to the North and the so-called *Strada dei Condotti* (or *dell'Acquedotto medievale*) to the South. This imaginary line also allows one to identify and divide older and newer olive groves according to their cultivation methods.

Older olive groves partly date back to Roman times and were later expanded by the Benedictines; their irregular planting distance follows the mountainous and upland terrain. On the valley-facing side, plots are delimited mainly by grassy embankments. Newer olive groves were planted instead, thanks to the benefits offered under the Papal State, in mostly purposely deforested areas with quite steep slopes and stony soil. The planting distance is quite regular, and stepped terraces vary according to the steepness of the slope. Towards the valley, plots are often protected by grassy embankments or dry-stone walls that sometimes turn into semi-circular walls (*lunette*). To date, their preservation status varies.

Moreover, some old peri-urban olive groves are located in the surroundings of Trevi, sometimes in areas with steep slopes (e.g. *Costa di Santa Caterina*). The planting distance follows the ground patterns, while plots are highly terraced, and steps are often very narrow. As seen above, towards the valley, olive groves are protected by variously preserved dry-stone walls and grassy embankments.

This particular landscape of terraced olive groves expresses, more than anything else, the diversity of traditional Umbrian landscapes shaped by the history of local communities.

For the municipality of Trevi, it represents a fundamental source of income from olive oil production, synergy with the historical and cultural heritage, local food and wine traditions, slow tourism, and so on. At the same time, the area shows several problems related to the management of terraced olive groves and the fragility of this form of agriculture. In the light of that concrete proposals and actions for the recovery and conservation of terraced olive-growing landscapes are needed. Such proposals also require identifying and prioritizing intervention areas, as well as establishing a consortium of stakeholders able to make the appropriate interventions.

16.2.3 Structural and Technical Data of Olive Groves in the Municipality of Trevi

Out of the 961 ha of olive groves mapped in 2012, only 683 were surveyed during the last Agricultural Census in 2010 (Table 16.1), which means a further loss of olive-growing hectares. Moreover, according to data from the Agricultural Census, in 2010, 412 out of the total 529 agricultural enterprises based in the municipality

Table 16.1 Structural data of agriculture in the municipality of Trevi

	1982	1990	2000	2010
Farms (number)	707	691	700	529
Farms (ha)	5628	5262	5219	5170
Farms with olive groves (number)	365	380	445	412
Farms with olive groves (hectares reserved for olive groves)	605	601	615	683
Average farm size (ha)	7.96	7.62	7.46	9.77
Average olive grove size (ha)	1.66	1.58	1.38	1.66

Source ISTAT, Agricultural Census 2010

Table 16.2 Agricultural land use: Olive groves in Trevi (whole municipality)

Farm activities	2013 (ha)
Olive grove	675.81
Wood	209.02
Pasture	69.20
Cereal	42.79
Tare and uncultivated area	25.22
Alfalfa	21.32
Vineyard	11.23
Legumes	7.30
Truffle	6.34
Oilseeds	1.88
Farmhouse and annex	1.21
Annual forage	1.18
Orchard	1.15
Horticultural crop	1.00
Vegetable garden	0.09
Short rotation forestry	0.07
Total	1074.81

Source Our elaboration on SIAN database, 2013

(78%) were olive producers and olive-growing cooperatives. Moreover, 360 (87%) out of the total 412 olive-growing holdings surveyed in 2010 Agricultural Census are listed in the National Agricultural Information System database in 2013 (in Italian, Sistema Informativo Agricolo Nazionale, SIAN). This database includes all the farms that have applied for payments under the Common Agricultural Policy (CAP).

However, the olive-growing area in 2013 (675 ha, SIAN database) corresponds to 99% of the 2010 Agricultural Census area (683 ha). The 675 ha of olive groves are part of a wider agricultural production context and account for 63% of the total

Table 16.3 Olive groves in the municipality of Trevi (technical data)

Information	2013
Number of farms	387
Utilised agricultural area (ha)	1074
Total utilised agricultural area (ha)	1968
Cadastral area (ha)	2014
Number of farms without olive groves	27
Number of farms with olive groves	360
Olive cultivation area (ha)	675
Total olive cultivation area (ha)	806
Cadastral olive cultivation area (ha)	833
Introduction and/or support of Integrated Production Method (Measure 212_A) (ha)	223
Introduction and/or support of Organic Production Method (Measure 212_B) (ha)	168
Ground cover to prevent carry-over of pollutants from soil to water (Measure 212_H) (ha)	4
Total (Agri-Environmental Measure 212) (ha)	395
Natural handicap payments to farmers in mountainous and other areas (Measure 211) (ha)	228

Source SIAN database 2013

utilised area (Table 16.2). The remaining area is occupied mainly by woods (209 ha), pasture and fodder crops (90 ha), cereals (43 ha), and vineyards (11 ha).

Therefore, the socio-economic characterization of the study area is strongly influenced by the presence of olive-growing holdings: 360 out of the 387 agricultural enterprises located in the municipality grow olive groves. As regards production techniques, integrated production (223 ha) and organic production (168 ha) methods are applied the most. Furthermore, olive-producing holdings benefit from payments to farmers in mountainous areas or other areas with handicaps (a total of 228 ha; Table 16.3).

16.3 Materials and Methods

The methodology adopted was structured into two steps: (i) the conception and construction of a database suitable for planning and managing the Traditional Agricultural Landscape (TAL) and (ii) launch of the participatory process.

16.3.1 Devising an Operative Database for Planning and Management of Olive Groves in the Municipality of Trevi

A database was developed for the planning and management of olive groves in the municipality of Trevi at the 'reference parcels' level, in which the minimum unit of cultivation is defined as a continuous land surface geographically bounded within a plot with a single land use.[2]

The database was built by aggregating SIAN data in a series of ad hoc processed map themes. The following information has been attributed to each land parcel:

1. Number of olive trees calculated through a GIS zonal analysis;
2. Average elevation of each olive grove polygon calculated through GIS zonal statistics from Digital Elevation Model (DEM);
3. Average slope of each olive grove polygon calculated through GIS zonal statistics from DEM;
4. Average inter-visibility of each olive grove (land parcel) from lookouts at the historical centre of Trevi or from other significant viewpoints along the main trekking route in the area, the 'Sentiero degli Ulivi' (Path of Olive trees);
5. Classification of the area (mountainous or disadvantaged) through GIS processing of data from the database of crop polygons and regulation maps;
6. Presence of slope embankments without stone walls as identified by the Numerical regional Technical Cartography (in Italian Carta Tecnica Regionale Numerica, CTRN) scale, 1:5000;
7. Presence of dry-stone walls through GIS zonal analysis from the regional CTRN scale, 1:5000, and from the database of particle polygons;
8. Estimate of the volume of dry-stone walls per polygon, value estimated at 161 cubic metres per hectare;
9. Size of polygons falling within the area of the Olive Grove Park (*Parco degli oliveti*) as indicated in the Urban Master Plan (in Italian Piano Regolatore Generale, PRG) of the municipality of Trevi;
10. Classification of olive groves according to the number of trees per surface unit as marginal (trees/hectare < 204), traditional (trees/hectare between 204 and 400), or intensive (trees/hectare > 400);
11. Labour use (hours per tree) for the three types of olive groves (marginal olive grove = 1.36; traditional olive grove = 1.11; intensive olive grove = 0.80); data obtained from direct interviews;
12. Olive oil production (kilograms per tree) for the three types of olive groves (marginal olive grove = 1.40; traditional olive grove = 1.40; intensive olive grove = 1.52); data obtained from direct interviews;

[2]This definition was enforced by Council Regulation (EC) No. 1593/2000 for the identification of agricultural parcels when carrying out administrative checks on the areas declared by farmers.

13. Cost of olive oil production in olive groves without dry-stone walls (euros per tree) for the three types of olive groves (marginal olive grove = 24; traditional olive grove = 20; intensive olive grove = 15); data obtained from direct interviews;
14. Cost of olive oil production in olive groves with dry-stone walls (euros per tree) for the two types of olive groves with dry-stone walls (marginal olive grove = 28.30; traditional olive grove = 22); data obtained from direct interviews;
15. Cost of the restoration of dry-stone walls, estimated at €400/cubic metres;
16. Carbon dioxide equivalent (CO_2eq) emissions based on the number of trees per hectare; data estimated assuming CO_2eq emissions in its tenth year, *leccino* olive *cultivar*, planting distance of 5.5 × 5.5, and in dry conditions: tCO_2eq/hectare = 1.419, where tCO_2eq is total CO_2eq;
17. CO_2eq sequestration based on the number of trees per hectare; data estimated assuming CO_2eq sequestration in the tenth year, *leccino* olive *cultivar*, planting distance of 5.5 × 5.5, and in dry conditions: tCO_2eq/hectare = 2.03; and
18. Revenue from hypothetical carbon credits for long-lasting wood products: €60/tCO_2 (Source: project LIFE07 ENV/IT/000388).

In particular, the following base maps were used: (1) geographic database of 2006 cadastral polygons, transformed into Gauss-Boaga coordinate system (East fuse, Datum: Roma 1940), for an extension comprising all the cadastral sheets that intersect the limits of the study area; (2) alphanumeric database of 2013 crops recorded in SIAN for all the cadastral sheets that intersect the limits of the study area; and (3) alphanumeric database of the 2011, 2012, and 2013 actions and measures under Rural Development Plan (RDP) recorded in SIAN for all the cadastral sheets that intersect the limits of the study area.

Given the unavailability of an updated SIAN cartography and time misalignment between the SIAN database (2013) and cadastral maps (2006), resulting in a mismatch between SIAN and RDP data and cadastral polygons (no geometries of cartographic limits for each sub-polygon crop were available), further action was required. In order to generate a dataset of SIAN and RDP information, the 2013 SIAN data records have been associated with cadastral polygons in MS Access with a relationship of one → many, which associated each sub-particle row of the SIAN data with a 2006 parcel polygon, using a unique join key.

The association of one → many has been carried out exclusively on alphanumeric records, irrespective of their geometry, in order to link each parcel record to one or more crop records corresponding to the same cadastral polygon. Results are encouraging, since the match equals to 92%.

The economic analysis of production costs and revenue was carried out through nine direct interviews (three for each type of olive grove), which allowed us to assess for each type of olive growing: labour use (hours per tree), olive oil production (kilograms per tree), olive oil production costs in olives groves without dry-stone walls (euros per tree) and with dry-stone walls (euros per tree), and cost of restoration of dry-stone walls (euros/mc). In particular, dry-stone wall restoration costs were estimated using information provided by two local agronomists, while the CO_2eq balance (tCO_2eq/hectare = 1419) was derived from existing literature (Proietti et al. 2014). Finally, the classification of olive groves based on 'market propensity' was carried out, considering the ownership structure and annual consumption units satisfied by the production, where one consumption unit has been estimated as 12 kg of extra-virgin olive oil per year.

16.3.2 Participatory Planning Process

A participatory planning process was launched in 2015, and five meetings were organized involving 64 participants (farmers, technicians, and public institutions). The first step of the participatory process was the creation of an advanced Web-based content management system. Map data and spatial information have been published and shared on the website in order to make them available in the decision-making processes. Information and data were shared first with the technicians of the municipality of Trevi and with then with staff of the local Mountain Community, who have taken an active part in the research unit. At a later stage, stakeholders were involved, and they participated in the three meetings held in the town hall of Trevi.

A computer-based system was developed to provide all the parties involved in participatory planning a virtual working table, thereby making the participatory governance hypotheses accessible at all levels of stakeholders, from the operational level of farmers to the decision-making level of local public administrators. The working method is based on the construction of thematic maps that can be consulted directly on the website, which provides the opportunity to build queries on proposed topics, make reports, upload photographs, and add personal reflections and comments on project-action proposals.

The main objectives were to (1) search for and build a form of territorial cooperation aimed at integrating environmental and landscape improvement actions in agricultural practices; (2) search for and build new forms of rural governance as a vehicle for enhancing farmers' environmental commitments, and as a way of developing responsibility in local communities for the implementation of landscape policies; (3) disseminate the importance of protecting the historical olive-growing landscape and continuing the cultivation of olives against abandonment, also the light of the new Directives adopted in the context of the CAP 2014–2020 and aimed at protecting the environment by introducing specific targeted measures to promote linkages between tourism and agriculture; and (4) highlight potential benefits to be gained by the application of olive groves in the municipality of Trevi or part thereof for UNESCO World Heritage Site status.

Actions to be taken were discussed with the stakeholders (technicians of the municipality of Trevi and the Mountain Community, farmers, representatives of trade unions, and workers of social cooperatives) and then submitted in the form of a questionnaire during the last two meetings with only farmers. Actions were divided into two groups. The first group of actions analyses how a farmer may contribute to the conservation of the olive-growing landscape in the municipality of Trevi by considering the technical and economic characteristics of his/her olive-growing farm. The second group investigates the extent to which a farmer considers it useful to become involved in collective initiatives aimed at the conservation of the olive-growing landscape and rural development of the area. For each action, respondents specified their level of agreement or disagreement on a five-point Likert scale: strongly disagree (0), disagree (1), neutral (2), agree (3), and strongly agree (4).

In more detail, the first group of actions to be taken at farm level has been organized as follows: (1a) reducing production costs; (1b) increasing the selling price of olive oil; (1c) increasing oil yield; (1d) expanding integrated farming production method; (1e) expanding organic olive-growing production method; (1f) establishing a contribution for traditional/marginal olive growing; (1g) introducing premiums for carbon credits; (1h) introducing subsidies for the maintenance of dry-stone walls; (1i) activating an RDP measure to restore dry-stone walls with 100% of funding; and (1j) activating an RDP measure for the recovery of abandoned olive groves with up to 75% of funding.

As for the second group, the following actions were considered: (2a) creating a collective project for the restoration of dry-stone walls; (2b) establishing a cooperative for olive harvesting; (2c) establishing a dry-stone walling school; (2d) encouraging different forms of renting between smaller and leading olive-growing farms; (2e) encouraging active involvement of social cooperatives in the management of olive groves; and (2f) implementing efficient collective marketing policies.

16.4 Results and Discussion

The association of each olive-growing parcel with its owner, as well as the combination of spatial and economic information, allowed a detailed reconstruction of the different types of olive-growing farms within the study area.

Marginal olives groves, where the risk of abandonment is greater, cover a total area of 61 ha. Traditional olive growing which ensures lower production costs than marginal olive growing takes up the largest area, corresponding to 520 ha (77% of the total). Intensive olive groves extend over 95 ha (14% of the total). Table 16.4 illustrates the different distributions of the three types of olive growing across farms. Olive producers and olive-growing cooperatives with a single type of olive grove are less numerous: 20 farms contain only marginal olive groves, while 47 only intensive olive groves.

This means that apart from the 47 intensive olive-growing farms, the remaining 88% farms deal with daily issues related to traditional and/or marginal agriculture.

The annual average socio-economic and environmental results are presented in Table 16.5. As for labour demand, the 230,526 olive trees located in the study area require more than 245,000 h of work, while the production amounts to 328,000 kg

Table 16.4 Different types of olive growing across farms

	Farms		Olive groves	
	N.	%	N.	%
Marginal olive groves (olive trees/hectare < 204)	20	5	12	2
Traditional olive groves (olive trees/hectare between 204 and 400)	103	27	93	14
Intensive olive groves (olive trees/hectare > 400)	47	12	15	2
Marginal and traditional olive groves	41	11	96	14
Marginal and intensive olive groves	5	1	3	0
Traditional and intensive olive groves	93	24	152	23
Marginal, traditional, and intensive olive groves	51	13	305	45
Other areas	27	7	–	0
Total	387	100	676	100

Source Our elaboration

of extra-virgin olive oil for a total cost of €4.4 million (note that these are production costs, including the costs of transforming olives into extra-virgin olive oil, but also the net of dry-stone wall maintenance costs and oil packaging costs). The net sequestration of CO_2 equals to 427 tCO_2eq.

Economic sustainability, that is, net of CAP premiums, is reached when the revenue is at least equal to estimated production costs. Figures differ according to the type of olive growing. For marginal olive groves, the revenue amounts to approximately €312,000 with a production of more than 18,000 kg of olive oil (estimated selling price of €17.20/kg). For traditional olive groves, it reaches approximately €3.5 million with a production of more than 241,000 kg of olive oil (estimated selling price of €14.30/kg). Intensive olive groves produce more than 68,000 kg of olive oil, generating a revenue of approximately €675,000 (estimated selling price of €9.86/kg).

When the annual average dry-stone wall maintenance costs, estimated at €3.7/mc, are added, average extra-virgin olive oil selling prices that ensure economic sustainability of the olive groves increase. They are estimated at €20.21/kg for olive oil produced from marginal olive groves and at €15.71/kg for olive oil produced from traditional olive groves.

The current average selling price for 1 kg of unpackaged extra-virgin olive oil produced in the municipality of Trevi is €15. This price can be considered very high if compared with the price of extra-virgin olive oils marketed in Italy. This is mainly the result of PDO certification, which has played a crucial role in promoting the olive oil characteristics linked to its geographical origin, as well as its organoleptic qualities, thus making it recognizable by consumers.

Landscape protection through dry-stone wall restoration and subsequent maintenance is not a common practice among local olive growers, since they already have trouble remunerating factors of production, especially family labour force, at opportunity cost. Not even the 2007–2013 RDP measure for non-productive

Table 16.5 Socio-economic characterization of the three different types of olive growing

Technical/economic data	Utilised agricultural area (ha)	Number of olive trees	Labour force in olive growing (h)	Olive oil production (kg)	Net carbon sequestration (tCO_2eq)
Marginal olive groves (olive trees/hectare < 204)	61	13,007	17,690	18,210	24
Traditional olive groves (olive trees/hectare between 204 and 400)	520	172,425	191,391	241,395	319
Intensive olive groves (olive trees/hectare > 400)	95	45,094	36,075	68,543	83
Other areas	399	–	–	–	–
Total	1075	230,526	245,156	328,147	427

Technical/economic data	Dry-stone wall restoration costs (€)	Annual cost for olive oil production (olive groves without dry-stone walls; €)	Annual cost for dry-stone wall maintenance (€)	Annual cost for olive oil production (olive groves with dry-stone walls; €)
Marginal olive groves (olive trees/hectare < 204)	1,761,804	312,168	55,930	368,098
Traditional olive groves (olive trees/hectare between 204 and 400)	17,965,005	3,448,495	344,849	3,793,344
Intensive olive groves (olive trees/hectare > 400)	2,256,512	676,410	–	676,410
Other areas	–	–	–	–
Total	21,983,320	4,437,073	204,436	4,641,508

Source Our elaboration

investments, which provided non-repayable grants of 80% of the cost incurred, succeeded in stimulating the restoration of dry-stone walls. Yet, they can be found in around 340 ha of the olive-growing study area. More specifically, they are located in 45% of the area covered by marginal olive groves and in 50% of the area covered by traditional olive groves. With an average presence of 161 mc of dry-stone walls per hectare of olive grove, we estimated a total of 55,000 mc of dry-stone walls. With a restoration percentage of 100% and an estimated cost of €400/mc, this would mean an overall investment of €22 million.

The main weakness in the conservation of the olive-growing landscape clearly lies in the profitability level of olive oil production and sale. This adds up to farms being slowly, yet inevitably, transferred to the next generation. Despite being aware of the value and opportunities offered by the conservation of this 'cultural landscape', new generations do not share the same knowledge, skills, and commitment of their previous generation. Therefore, contractors increasingly provide farming and landscape maintenance interventions. This entails a series of issues, among which are difficulty finding appropriately trained workers and an inevitable rise in costs. Costs are therefore less and less offset by revenues, even though they are substantial, since olive oil is sold at a higher price and marketed as a high-quality and unique product.

The ownership structure is instead one of the most valuable in the conservation of the olive-growing landscape. On the one hand, the highly fragmented ownership of agricultural land, which is divided into small plots, has allowed local people to grow olive trees as a side occupation, driven by strong attachment to this species. On the other hand, the small extension has ensured rational use of resources and sufficient penetration of some rich niche markets. Furthermore, as already underlined, PDO certification has played a crucial role in making olive oil recognizable by consumers.

However, nowadays, different types of PDO olive oils are sold at reasonable prices under commercial labels in supermarkets, thus making it necessary to move beyond PDO certification and develop new ways to strengthen the relationship between products, production areas, and local people.

Possible actions and proposals formulated and discussed during meetings with stakeholders have been variously assessed by olive growers who were given the survey questionnaire. The direct survey involved 10 olive growers who together cultivate 35% of the olive-growing study area. The results, presented in Table 16.6, show that, according to respondents, conservation of the olive-growing landscape depends on the introduction of premiums for dry-stone wall maintenance and on the activation of an RDP measure for the restoration of dry-stone walls with 100% funding, both at the farm level and, on implementation of efficient collective marketing policies, at the territorial level.

At the farm level, great importance is given to the activation of an RDP measure for the recovery of abandoned olive groves with up to 75% of funding, as well as for the increase in the selling price of extra-virgin olive oil. Moreover, further efforts should be made to reduce production costs, as it would be desirable to establish a contribution for marginal and traditional olive growing and to introduce premiums for carbon credits. At the territorial level, collective projects for the restoration of dry-stone walls with the establishment of a dry-stone walling school are not considered particularly relevant.

At the farm level, less importance is given to actions related to farming and production methods and to the increase of oil yield. At the collective level, there is certain opposition to the establishment of a cooperative for olive harvesting and to the involvement of social cooperatives.

Table 16.6 Results from the participatory planning process

First question: With reference to your farm, do you think these actions are effective to ensure the olive groves landscape in the municipality of Trevi?	
Answers to the first question	Average value
Introduce bonuses for dry-stone wall maintenance	3.4
Activate an RDP measure to restore dry-stone walls with 100% public cost contribution	3.2
Activate measures improving the restoration to avoid olive grove degradation when abandoned, supporting up to 75% of the measure	2.8
Increase olive oil price	2.6
Reduce production costs	2.4
Provide a contribution for traditional/marginal olive growing	2.4
Introduce premiums for carbon credits	2.4
Improve organic production methods	1.8
Increase the olive oil yield	1.6
Improve the integrated farming production method	1.2
Second question: Regarding the olive groves area in the municipality of Trevi, do you think these actions could be useful to ensure olive-growing conservation and rural area development?	
Answers to the second question	Average value
Implement efficient collective marketing policies	3.2
Develop a collective dry-stone wall restoration project	2.4
Establish a school where it is possible to learn how to build and restore dry walls	2.4
Encourage different forms of renting between smaller olive farms and leading olive farms	1.8
Establish a cooperative for olive harvesting	1.6
Encourage social cooperatives to have an active role in olive grove management	1.4

Source Our elaboration
Note The interviewee can answer: not at all (0), little (1), rather (2), very (3), or extremely (4)

Building upon the results of the direct survey, we focused on the olive-growing area, where the risk of abandonment is greater, and we tried to identify and assess the most appropriate actions and measures to be taken. This area was extracted from the database by selecting olive-growing parcels with a density of less than 400 trees per hectare, located above 400 m above the sea level, with a slope higher than 20%, and with dry-stone walls.

Using these parameters, 42 parcels belonging to 28 farms in a total area of 31 ha were extracted. Out of the total 31 ha, 3 are occupied by marginal olive groves, while traditional olive-growing covers 28 ha (the spatial distribution of olive groves can be seen in Fig. 16.2).

It should be noted that the selected parcels belong to farms in which the extension of the olive-growing area differs greatly (from 69 to 0.37 ha). Similarly,

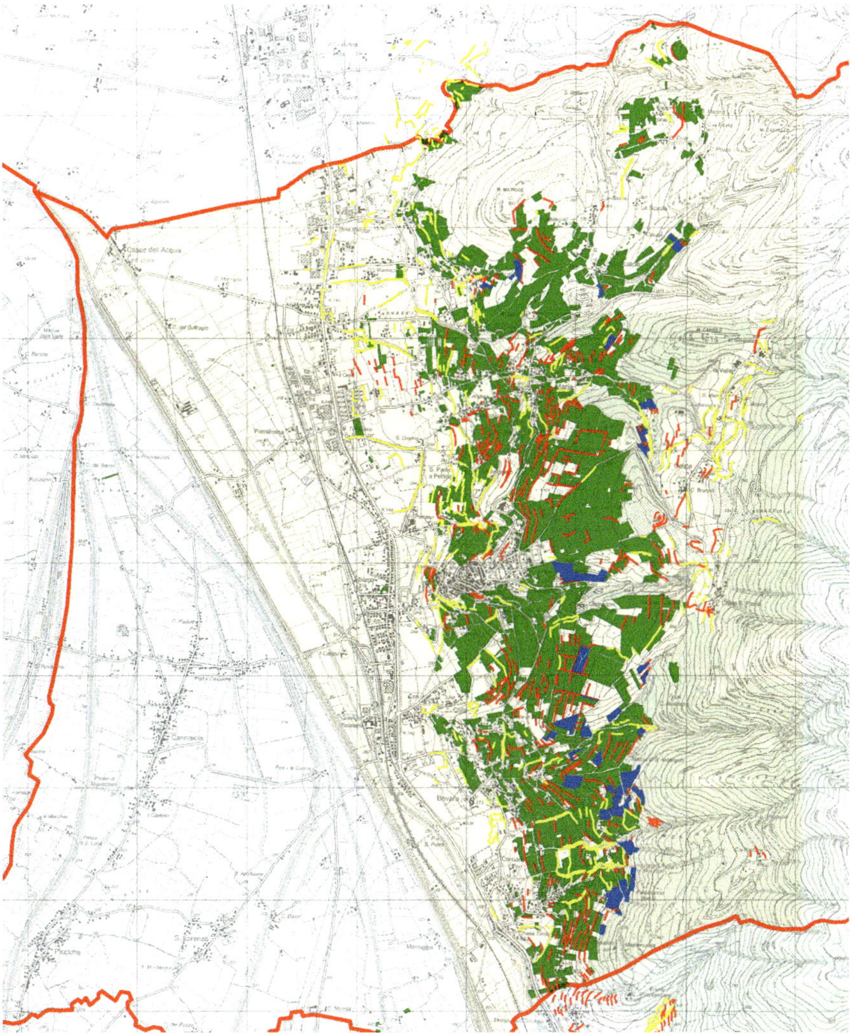

Fig. 16.2 Olive-growing area in 2012 with the position of dry-stone walls (coloured in red), slopes (coloured in yellow), and the olive-growing areas that are potentially more vulnerable to abandonment (areas in blue). *Source* Our elaboration

the weight of olive groves that are potentially more vulnerable to abandonment varies (from 1 to 100%). This confirms that olive-growing abandonment is a phenomenon that is spreading across several farms, in which it carries a different weight.

Public intervention in the area would entail an investment of approximately €1 million for the restoration of dry-stone walls, assuming a restoration percentage of 50%, plus €22,000 for their maintenance, corresponding to €1.5 for each kg of olive oil produced. Profitability would be ensured by the selling price of about €16/kg, which would mean an increase in the current selling price by 7%. Such an increase could be easily achieved through a collective marketing policy.

16.5 Conclusions

Combining spatial information on traditional terraced landscapes with socio-economic information on farms in which they are located has turned out to be an extremely useful analysis methodology to put the conservation of terraced landscapes into its economic context and assess the actions needed to protect them.

Similarly, the participatory planning process launched with the research has proven useful to identify which actions and measures would have been favourably accepted by olive growers. It also gave rise to team spirit among local stakeholders in dealing with a complex issue such as the conservation and promotion of traditional terraced landscapes.

The research shows that the vulnerability of terraced olive growing in the study area is mainly due to its lack of profitability. As mentioned by Palazzo and Aristone (2017), unavoidable trade-offs among identity and profitability ought to be implemented. Overcoming this major limitation requires a joint public and private effort. On the one hand, public institutions should provide direct funding and incentives. On the other hand, consumers should be more willing to pay a premium price for extra-virgin olive oil with landscaped value.

Numerous studies have been carried out on the subject. These studies have found a willingness to pay for attributes tied to different dimensions of agro-food production sustainability (Scarpa and Del Giudice 2004; Krystallis and Ness 2005; Tempesta et al. 2010; Moser and Raffaelli 2012; Vecchio and Annunziata 2015; Tait et al. 2016; Zhou et al. 2016).

To this end, further research should be conducted that analyses consumer preferences for extra-virgin olive oil with landscape-label-friendly certification. The estimation of the willingness to pay a premium price and identification of the central drivers of extra-virgin oil consumption with such a credence attribute could contribute to both shedding light on this under-investigated topic and supporting the farms' marketing strategies.

The collective project for the restoration of dry-stone walls could be a sound basis for setting up a pathway to cooperation in olive cultivation, which in turn would entail a reduction of production costs. Another path to promote cooperation is represented by European projects and initiatives to revitalise rural areas, such as the LEADER Programme or the Operational Groups of the European Innovation Partnerships.

The favourable economic context of our study area has highlighted that issues related to traditional terraced landscapes also affect areas where several steps have been taken at the institutional level to promote agricultural products, starting from the PDO label. The nomination for inscription into UNESCO's World Heritage List as a cultural landscape, advocated by local institutions, could generate useful synergies to implement efficient collective marketing policies, which are particularly demanded by olive growers.

The future possibility of 'selling' not only excellent olive oil, but also a whole *terroir* with its history, culture, traditions, and quality of life might persuade younger generations to remain in the area. It should be remembered that olive groves are considered a 'social crop', as olive growing is amongst the agricultural activities that create the most jobs per hectare (Colombo and Perujo-Villanueva 2017), and olive groves are recognized as a Mediterranean landscape identity (Loumou and Giourga 2003). To be included in the World Heritage List, a site should stand out for its exceptional value, be of Outstanding Universal Value, and meet some criteria, including integrity and/or authenticity of the properties, adequate protection, and management.

However, in agreement with Sklenicka et al. (2017), we would like to underline that it is essential to select protection and management methods that best correspond to both the natural conditions of the site and economic interests of farmers.

Acknowledgements The research on which this paper is based was undertaken as part of a research project titled 'Traditional agricultural landscapes in Italy: multi-disciplinary and multi-scale assessment for the development of an integrated model for landscape planning and management', co-financed by Programs of Relevant National Interest (PRIN) Grant number 2010LE4NBM_013, and as a part of a research project titled 'The quality of the landscape for socio-economic development: the case study of Trevi's historical olive groves', co-financed by Basic Research of University of Perugia. We are particularly grateful to Carlo Sportolaro for his cooperation in construction of the GIS database and for realisation of maps, and to Alvaro Paggi and Tiziana Ravagli for their cooperation in conducting the study. We would like to thank Valeria Illuminati for editing the English language in the manuscript. The authors thank the anonymous reviewers for their valuable comments on an earlier version of this article.

References

Agnoletti M (2012) The Italian historical rural landscape. Cultural values for the environment and rural development. Springer, Dordrecht

Agnoletti M, Conti L, Frezza L, Monti M, Santoro A (2015) Features analysis of dry stone walls of Tuscany (Italy). Sustainability 7(10):13887–13903. https://doi.org/10.3390/su71013887

Arnáez J, Lana-Renault N, Lasanta T, Ruiz-Flaño P, Castroviejo J (2015) Effects of farming terraces on hydrological and geomorphological processes. A review. Catena 128:122–134. https://doi.org/10.1016/j.catena.2015.01.021

Barbera G, Biasi R (2011) I paesaggi agrari tradizionali dell'albero: il significato moderno di forme d'uso del suolo del passato. Italus Hortus 18(1):23–40

Barbera G, Cullotta S, Rossi-Doria I, Rühl J, Rossi-Doria B (2010) I paesaggi a terrazze in Sicilia: metodologie per l'analisi, la tutela e la valorizzazione. ARPA Sicilia, Collana Studi e Ricerche, 7

Bender O, Boehmer HJ, Jens D, Schumacher KP (2005) Using GIS to analyse long-term cultural landscape change in Southern Germany. Landsc Urban Plan 70(1):111–125. https://doi.org/10.1016/j.landurbplan.2003.10.008

Bevilacqua P (2012) Umbria. In: Agnoletti M (ed) Paesaggi Rurali Storici. Per un catalogo nazionale, Laterza

Bonardi L, Varotto M (2016) Paesaggi terrazzati d'Italia: eredità storiche e nuove prospettive. FrancoAngeli

Colombo S, Perujo-Villanueva M (2017) Analysis of the spatial relationship between small olive farms to increase their competitiveness through cooperation. Land Use Policy 63:226–235. https://doi.org/10.1016/j.landusepol.2017.01.032

Cullotta S, Barbera G (2011) Mapping traditional cultural landscapes in the Mediterranean area using a combined multidisciplinary approach: Method and application to Mount Etna (Sicily, Italy). Landsc Urban Plan 100(1):98–108. https://doi.org/10.1016/j.landurbplan.2010.11.012

de Graaff J, Zuazo VHD, Jones N, Fleskens L (2008) Olive production systems on sloping land: prospects and scenarios. J Environ Manage 89(2):129–139

Duarte F, Jones N, Fleskens L (2008) Traditional olive orchards on sloping land: Sustainability or abandonment? J Environ Manage 89(2):86–98

Ferro-Vázquez C, Lang C, Kaal J, Stump D (2017) When is a terrace not a terrace? The importance of understanding landscape evolution in studies of terraced agriculture. J Environ Manage. https://doi.org/10.1016/j.jenvman.2017.01.036

Fukamachi K (2016) Sustainability of terraced paddy fields in traditional satoyama landscapes of Japan. J Environ Manage xxx:1–7. https://doi.org/10.1016/j.jenvman.2016.11.061

Kieninger PR, Penker M, Yamaji E (2013) Esthetic and spiritual values motivating collective action for the conservation of cultural landscape—a case study of rice terraces in Japan. Renew Agric Food Syst 28(4):364–379. https://doi.org/10.1017/S1742170512000269

Krystallis A, Ness M (2005) Consumer preferences for quality foods from a South European perspective: a conjoint analysis implementation on Greek olive oil. Int Food Agribus Manag Rev 8(2):62–91

LaFevor MC (2014) Restoration of degraded agricultural terraces: rebuilding landscape structure and process. J Environ Manage 138:32–42. https://doi.org/10.1016/j.jenvman.2013.11.019

Loumou A, Giourga C (2003) Olive groves: "The life and identity of the Mediterranean". Agric Hum Values 20:87–95. https://doi.org/10.1023/A:1022444005336

Modica G, Vizzari M, Pollino M, Fichera CR, Zoccali P, Di Fazio S (2012) Spatial-temporal analysis of the urban–rural gradient structure: an application in a Mediterranean mountainous landscape (Serra San Bruno, Italy). Earth Syst Dyn 3(2):263–279. https://doi.org/10.5194/esd-3-263-2012

Moser R, Raffaelli R (2012) Consumer preferences for sustainable production methods in apple purchasing behaviour: a non-hypothetical choice experiment. Int J Consum Stud 36(2):141–148. https://doi.org/10.1111/j.1470-6431.2011.01083.x

Palazzo AN, Aristone O (2017) Peri-urban matters. Changing Olive growing patterns in Central Italy. Sustainability 9. https://doi.org/10.3390/su9040638

Proietti S, Sdringola P, Desideri U, Zepparelli F, Brunori A, Ilarioni L, Nasini L, Regni l, Proietti P (2014) Carbon footprint of an olive tree grove. Appl Energy. https://doi.org/10.1016/j.apenergy.2014.04.019

Ridder E, Galletti CS, Fall PL, Falconer SE (2016) Economic and social activities on ancient Cypriot terraced landscapes. J Environ Manage. https://doi.org/10.1016/j.jenvman.2016.12.037

Scarpa R, Del Giudice T (2004) Market segmentation via mixed logit. Extravirgin olive oil in urban Italy. J Agric Food Ind Organ 2:1–7

Sklenicka P, Kottová B, Šálek M (2017) Success in preserving historic rural landscapes under various policy measures: Incentives, restrictions or planning? Environ Sci Policy 75:1–9. https://doi.org/10.1016/j.envsci.2017.05.010

Tait P, Saunders C, Guenther M, Rutherford P, Miller S (2016) Exploring the impacts of food label format on consumer willingness to pay for environmental sustainability: a choice experiment approach in the United Kingdom and Japan. Int Food Res J 23(4):1787–1796

Tarolli P, Preti F, Romano N (2014) Terraced landscapes: from an old best practice to a potential hazard for soil degradation due to land abandonment. Anthropocene 6:10–25. https://doi.org/10.1016/j.ancene.2014.03.002

Tempesta T, Giancristofaro RA, Corain L, Salmaso L, Tomasi D, Boatto V (2010) The importance of landscape in wine quality perception: an integrated approach using choice-based conjoint analysis and combination-based permutation tests. Food Qual Prefer 21(7):827–836

Torquati B, Vizzari M, Sportolaro C (2011) Partecipatory GIS for integrating local and expert knowledge in landscape planning. In Andreopoulos Z, Manos, Polman P, Viaggi D (eds) Agricultural and environmental informatics, governance, and management: emerging research applications, pp 378–396. Information Science Reference, Hershey, PA, USA. ISBN 978-1-60960-621-3 (hbk). ISBN 978-1-60960-622-0 (ebook)

Torquati B, Giacchè G, Venanzi S (2015) Economic analysis of the traditional cultural vineyard landscapes in Italy. J Rural Stud 39:122–132. https://doi.org/10.1016/j.jrurstud.2015.03.013

Vecchio R, Annunziata A (2015) Willingness-to-pay for sustainability-labelled chocolate: an experimental auction approach. J Clean Prod 86:335–342

Vizzari M (2011) Peri-Urban transformations in agricultural landscapes of Perugia, Italy. J Geogr Inf Syst 3:145–152. https://doi.org/10.4236/jgis.2011.32011

Zhou G, Hu W, Huang W (2016) Are consumers willing to pay more for sustainable products? A study of eco-labeled tuna steak. Sustainability 8(5):494. https://doi.org/10.3390/su8050494

Chapter 17
The Multidimensional Benefits of Terraced Landscape Regeneration: An Economic Perspective and Beyond

Luigi Fusco Girard, Antonia Gravagnuolo and Fortuna De Rosa

Abstract Terraced landscapes represent a particular type of multifunctional, historic–cultural agrarian landscapes, today at risk of abandonment due to socio-economic changes. These landscapes are an ancient example of a "circular" model in using resources, able to be productive in multiple dimensions, providing many ecosystem services to local communities and thus contributing to human well-being. Terraced landscapes have a complex value for society: sociocultural, environmental and economic values, which can become a driver of territorial regeneration, if a systemic economic–territorial perspective is adopted. This paper proposes the circular economy as a viable model of sustainable territorial development that can support terraced landscapes regeneration, exploiting their structural multifunctionality and thus enhancing the multidimensional territorial productivity. The ecosystem services' assessment framework, which includes economic, spatial, quantitative and qualitative evaluation tools, can be integrated in agri-environmental policies to make operational the "circular" paradigm of regeneration. A selection of economic tools and case studies is presented to show how circular processes can be activated in terraced landscapes, reducing costs and waste of resources, increasing multidimensional productivity, and finally attracting more investments towards a new systemic urban-rural "circular" development model.

L. Fusco Girard · F. De Rosa
Interdepartmental Research Centre in Urban Planning "A. Calza Bini", University of Naples Federico II, Naples, Italy
e-mail: girard@unina.it

F. De Rosa
e-mail: fortuna.derosa@unina.it

A. Gravagnuolo (✉)
Institute for Research on Innovation and Services for Development (IRISS), National Research Council, Naples, Italy
e-mail: a.gravagnuolo@iriss.cnr.it

M. Varotto et al. (eds.), *World Terraced Landscapes: History, Environment, Quality of Life*, Environmental History 9,
https://doi.org/10.1007/978-3-319-96815-5_17

17.1 Towards a Circular Paradigm for the Regeneration of Terraced Landscapes

Terraced landscapes represent all over the world a particular type of multifunctional, historic–cultural agrarian landscapes, today at risk of abandonment (Varotto 2009).

Terraced landscapes are an ancient example of a "circular" model in using resources, able to be productive in multiple dimensions (economic, environmental, social, aesthetic–cultural…): rainwater is channelled and conserved into cisterns, stones are reused in building dry-stone walls, and energy comes from local sources. The work of man and nature created hybrid dramatic landscapes that are a source of identity for local communities.

However, globalization, agricultural intensification and socio-economic changes turned agricultural terraces from a productive resource into an unsustainable heritage conservation cost. The beauty of terraces attracts tourists, but the high costs of maintenance and labour effort reduce their attractiveness for productive uses, generating progressive abandonment processes, multidimensional poverty, increased environmental risks and the loss of cultural resources, thus reducing community well-being.

Today terraced landscapes are geographic–economic marginal areas. Youths tend to abandon these landscapes to seek for qualified jobs in urban centres, contributing to growing urbanization costs and further impoverishment of lands in a vicious circle. Since the problem is systemic, it requires a systemic approach to transform a vicious into a circular process: which plural interconnected models are able to break this vicious circle and to enhance the resilience in these landscapes? How to integrate economic growth (e.g. growth of enterprises) with social and environmental sustainable development?

New economic models are needed to recover the multifunctionality of terraced landscapes, balancing investment cost and the costs of maintenance and management with the returns from agricultural/touristic/multifunctional productivity.

The thesis here is that the reuse of terraced landscapes can be operationally implemented in the perspective of the circular economy: of the circular city/territory model, of the circular agriculture, of circular tourism, of circular building economy.

The "circular model" is based on reduced ecological resources consumption, reuse of all available resources, no waste. It increases productivity in multiple dimensions, creating new jobs and enhancing environmental health and well-being of people.

In this paper, we explore how the circular model can be operationalized in terraced landscapes, which notion of multidimensional value can support its implementation, rebalancing costs, and which economic instruments can be adopted to turn the actual costs into economically, socially and environmentally productive investments.

17.1.1 Circular Economy and Circularization of Processes

The non-used cultural terraced landscape is a "cost" under many points of view. Its *creative and systemic functional reuse* can reduce this "cost", turning it into an investment. It becomes effective if it is incorporated in a circular economy perspective/strategy: circular city economy, circular tourism, circular urban/territorial agriculture, circularization of building sector. Here, the proposal is to adopt circular processes (applied in particular to specific "commons") through creative functional reuse that can generate new values, jobs, positive ecological impacts, using *new evaluation tools* in choices.

The circular processes are those that mimic the organization of natural systems, which are able to self-reproduce themselves, and 'support' other systems at the same time.

Circular economy is gaining increasing attention as a potential way for our society to increase prosperity, while reducing dependence on primary materials and energy (Le Moigne 2014; Ellen MacArthur Foundation 2015). At the same time, a lively debate is going on about the attractiveness of a circular economy model of production/consumption for different stakeholders and its implications for employment, growth and the environment (Rizos et al. 2015).

The linear economic development model, based on infinite resources consumption, is no longer sustainable (Costanza et al. 2014). Circular economy enriches the mainstream economics (with its traditional linear model), introducing it into a multidimensional space.

The circular economy model reduces/avoids social and environmental negative impacts, promoting economically, socially and environmentally convenient activities based on closed short loops at local level through the production and enhancement of synergies/relationships (between the single human being and the community, human beings and the environment, human beings and the landscape). This model replaces current linear models, reducing raw materials consumption, energy consumption and waste production by reusing existing assets/structures. It includes the reuse of the embedded energy and the reuse of knowledge itself.

Thus, in the circular model we can recognize:

- the reduction of materials resources—reducing the need of new land, water, energy, and thus reducing costs;
- reuse (and shared use) of existing resources with new functions;
- maintenance of existing resources (buildings, stones, walls, etc.) ensuring longer life;
- energy recovery—valorizing the embodied energy and using renewable energy sources;
- recreation of value through the use of parts of existing (ancient, historical) buildings (refurbishing/remanufacturing);
- reuse of the specific local knowledge for valorizing the identity of territories in the regeneration of the city/territory system.

Considering costs and benefits, as well as impacts and externalities, in a multidimensional and multi-actor perspective (not only economic–financial flows), circular economic models increase the overall productivity of territorial systems (Fusco Girard 2016).

The circular economy has some main characteristics:

1. It is oriented to enlarge the lifetime of resources, assigning them new functions (in a long-time perspective);
2. It is based on synergies/symbioses between actors in fostering closed loops of value creation: economic costs are reduced, and new wealth is created through the multiplication of relationships;
3. It enhances the productivity, decoupling wealth production from negative environmental impacts.
4. It takes into account/incorporates the external effects on the natural and social environment in generating economic wealth;
5. It expresses a form of co-evolutionary capitalism that makes integration of environmental, social, development goals (Porter and Kramer 2011);
6. It projects the conventional economy in a multidimensional space in which, therefore, economic, ecological and social values coexist;
7. It modifies and enriches the very notion of value towards a complex economic, ecologic and social value (*Complex Value*);
8. It modifies the choices of investment/design/planning that necessarily become systemic;
9. Technological innovation fosters innovation reducing costs/enhancing performances.

The *closed loop* is the key principle of this circular model that can be applied not only to industrial processes, but also to financing, business and management models, creating synergies between multiple actors.

The empirical evidence shows that circular economic processes are linked to the reduction of costs (investment costs, management and operating costs, environmental and sociocultural costs) first of all because of the creation of productive synergies/symbiosis (Fujita et al. 2013) between actors (professionals, entrepreneurs, policy-makers, investment funds, civil society).

The circular paradigm is assumed here not only for the economic co-evolutive growth with natural ecosystems, but also for promoting the *human* development paradigm: no "waste of people" (and not only of resources).

Circular economy is thus able not only to reduce different forms of poverty, but more in general to integrate concretely beauty, economy and fairness because it conserves the quality of the natural/cultural environment, producing new jobs and economic wealth.

Many cities and regions are shifting to a circular economy model (Amsterdam, Paris, Vienna, Rotterdam, Helsinki, Barcelona, Brussels, Peterborough…). Recent research and policy practice show how circular territorial models can be applied in the city-region (European Commission 2015; ESPON et al. 2016; Ellen MacArthur

Foundation 2017; Partnership Circular Economy 2018). A circular model can be adopted in terraced landscapes for regenerating their multifunctionality and multidimensional productivity, starting from the recognition of their complex value through a systemic landscape approach.

17.1.2 The Landscape Systemic Approach

"The 'complex systemic (urban/territory) landscape' consists of combinations of, and interaction amongst, six landscapes: natural, man-made, man-made/cultural, financial, social, and human landscape. The specific character of a city/territory, its particular identity (its attractiveness) derives from the particular intensity and reciprocal combinations of these landscapes" (Fusco Girard 2016). The landscape approach offers a unifying, holistic and relational notion/concept/idea for facing all the goals of local development.

The landscape can be considered the synthetic/holistic indicator of the sustainable, inclusive, safe, resilient city/territory. The more important challenges of our time (e.g. health, safety, climate change, migration flows, urbanization, energy, pollution, social disparities, poverty) are embedded in the landscape. The "health" of a city/territory and the human well-being can be read in the landscape. All the values/goals/interests of a society are reflected in the landscape: here, we can recognize the culture itself of a society. As the result of mutual combination of different forms of capital, for example of the expertise and skills of a territory, it shapes the comparative advantage of an area compared with others.

The landscape has thus a particular development potential. It can become key for launching a smart sustainable development model, starting from local resources to activate creative processes of circular economy through a synergistic approach, combining the touristic, economic, local productions activities with cultural heritage regeneration, with the creativity of its inhabitants.

In the landscape, perspective is intrinsically embodied also an aesthetic dimension: the landscape itself evokes this specific aesthetic perspective. The particular beauty of the landscape (connected to its shapes, colours, microclimate, light values, local materials, and life that build its "image", its soul) expresses the combination of human and natural creativity and characterizes the true identity of a territory.

The terraced landscapes are good examples of beauty all over the world: in China, Morocco, Yemen, Japan, Peru, Canary Islands, etc. In Italy, Cinque Terre, Amalfi coast, Gargano, Euganei Hills, Valtellina are well known (Scaramellini and Varotto 2008; Bonardi and Varotto 2016).

The circular processes contribute to the beauty of a system because they allow that each component is linked to the other ones in a comprehensive order and in a dynamic harmony: the visual impacts of this systemic harmony are perceived as their *beauty*.

The quality itself of the landscape depends on the density of circular processes, of symbioses, of synergies, that multiply the flow of benefits.

17.1.3 The Complex Social Value of Cultural Terraced Landscape as Common Good

Cultural terraced landscapes can be considered "common goods". Their value is not linked to a market value, but to quality of life, and thus to the satisfaction of human needs. However, terraced landscapes have also an "intrinsic value" that is linked to their capacity of supporting ecosystems, and thus life. This value is independent from human use. The value of terraced landscapes should be understood within a complex notion of value of natural and cultural resources. They contribute to human well-being and thus have a *complex social value* (Fusco Girard 1987; Fusco Girard and Nijkamp 1997; Forte and Fusco Girard 2009), which depends on their value for all stakeholders, including future generations. The complex value includes instrumental values and independent-of-use values. Cultural terraced landscapes-represents a symbol of common and shared characteristics, rooted in the history of a community, which will be demanded by future generations. The complex social value can be understood as the comprehensive flow of net multidimensional benefits coming from the existence of a particular resource. It includes:

- a use value, which depends on its localization (e.g. real estate values), state of conservation (related to costs), re-functioning possibilities (economically productive/non-productive functions), branding (attractiveness for tourism/local use);
- an independent-of-use value, which is linked to its historic–cultural significance, symbolic value for the community, local identity that it expresses/conveys, and its value for future generation.

The comprehensive flow of net benefits, coming from agricultural, touristic, ecosystems productivity, from the well-being perception of people, is the "Complex Social Value" of the landscape.

The proposal here is oriented to interpret terraced landscapes as comprehensive/complex landscapes, as a key local resource for implementing circular economic processes preserving and enhancing their "intrinsic value", turning cultural heritage/landscape into an economically attractive resource, also able to strengthen social awareness and cohesion. In this perspective, terraced landscape enhances the territory multidimensional productivity.

The following section explores which multidimensional evaluation methods and tools can be used to assess the complex value of terraced landscapes, supporting the transition towards a circular economy by producing evidence of the costs and benefits of landscape abandonment versus regeneration options.

17.2 Evaluation Tools

17.2.1 Multidimensional Impacts of Land Abandonment in Terraced Landscapes

The abandonment of cultural agrarian landscapes generates negative impacts (costs) on multiple dimensions: sociocultural, environmental and economic, undermining their Complex Social Value. Table 17.1 shows these impacts as they have been reported in the literature, which can be applied to terraced landscapes.

These multidimensional costs are avoided if we recover the circular model. They can become the benefits of reusing terraced landscapes, contributing to create new markets and jobs with high social and environmental impacts.

Table 17.1 Multidimensional impacts (costs) of land abandonment in terraced landscapes

	Negative impacts (costs) of land abandonment
Sociocultural	Loss of cultural landscape (decrease in well-being)
	Loss of cultural identity and cultural diversity
	Loss of traditional knowledge
	Increased poverty and marginality of rural areas (loss of jobs in agriculture and agri-food sector)
	Ageing of rural/mountain population
	Increased social vulnerability of local populations
	In some extreme cases, loss of human lives due to landslide or flooding events
	Decrease of food sovereignty and food security in rural and related urban areas
	Negative impacts on human health/well-being due to loss of complex environmental factors and to loss of landscape beauty
	Impoverishment of agrarian biodiversity and irreversible loss of autochthonous seeds/agrarian qualities
Environmental	Increase of environmental risk (landslide, flooding, fire)
	Increased soil erosion
	Decrease of organic soil fertility
	Reduction of fresh water in aquifers related to less retention through canalizations, storage systems and water retention from terraces
	Increased climate-changing gas emissions related to unsustainable farming
	Increased climate-changing gas emissions related to food importation
	Loss of biological diversity related to higher environmental fragmentation
	Loss of habitat for autochthonous species
	Higher vulnerability to invasive alien species
	Increased pollution, waste and energy demand in urban areas due to increased migration flows

(continued)

Table 17.1 (continued)

	Negative impacts (costs) of land abandonment
Economic	Loss of attractiveness for tourism and recreation (long term)
	Higher costs of recovery of productive lands (also for future generations) in medium–long term
	Costs of recovery from environmental extreme events (short–medium term)
	Decreased profit from food production (short term)
	Negative spillover effects of increased poverty on local economy (short–medium term)
	Loss of agricultural revenues

Here, we will focus our attention in particular on evaluation approaches and tools that can specify these multidimensional costs. Then, we analyse some economic instruments that can be used to implement a circular regeneration model in terraced landscapes, based on successful models implemented in cultural agrarian landscapes.

17.2.2 The Ecosystem Services Assessment for Operationalizing the Social Complex Value

Methods and tools applied in ecological economics (Common and Stagl 2005; Costanza et al. 2014) can be employed to assess the societal cost of land abandonment and conversion (Gaitán-Cremaschi et al. 2017).

The ecosystem services' theory has provided meaningful definitions of ecosystem (and landscape) functions and services, and evaluation tools for their assessment, which are employed to understand the complex systems of relationships between land uses and their biophysical structures, functions and processes, and how they can be related to changes in human welfare (de Groot et al. 2010; Haines-Young and Potschin 2010). The concept of ecosystem functions and services has been introduced by Costanza et al. (1997) and further explored by the 'Millennium Ecosystem Assessment' (MEA 2003; MA 2005). Ecosystem Services (ES) have been defined by MEA as "the benefits people obtain from ecosystems" (MEA 2003). They are generated by ecosystem functions, which are in turn underpinned by biophysical "supporting" structures and processes (de Groot et al. 2010). The benefits include material products of land such as food and freshwater, the regulation of natural hazards, maintenance of natural resources such as pure air and biodiversity, and cultural benefits linked to the meanings and values that people recognize to the 'landscape'.

"The Economics of Ecosystems and Biodiversity" initiative developed a global study on the economic value of ecosystem services (TEEB 2010), with a focused analysis on ecosystem services in agricultural systems (TEEB 2015).

Classifications of ecosystem services include:

- *Provisioning services* defined as the material or energy outputs from ecosystems, which include food, water and other material resources;
- *Regulation and maintenance services*, which include the category of "supporting" or habitat services (such as habitats provided to species and the maintenance of genetic diversity), defined as the services that ecosystems provide by acting as regulators e.g. regulating the quality of air and soil or by providing flood and disease control;
- *Cultural services*, which have a clear link with the landscape scale and human well-being (Chan et al. 2012; Milcu et al. 2013), defined as the immaterial benefits such as recreation and mental/physical health, well-being, aesthetic appreciation, spiritual experience and sense of place.

The link between ecosystem services, landscape and human well-being has been widely recognized in the scientific literature (Haines-Young and Potschin 2010; Iverson et al. 2014; Plieninger et al. 2014; Hicks et al. 2015). Many authors have stressed the need of integrate ES in landscape and territorial planning (de Groot et al. 2010; Hartel et al. 2014).

Key ecosystem services in terraced landscapes are identified in Table 17.2. The selection represents the ES that are more likely to be reduced as a consequence of land use change and land abandonment in terraced landscapes (Gravagnuolo 2014, 2015).

These ES can be evaluated in economic terms to assess the societal costs of their reduction or loss. The economic valuation of ES is necessary to support informed choices in the implementation of economic instruments, balancing conflicting interests between farmers and other providers of ES and the wide range of actual and potential beneficiaries.

Table 17.2 Key ecosystem multidimensional services in terraced landscapes

Categories (from MEA, TEEB)	Key ecosystem services in terraced landscapes (adapted from Gravagnuolo 2014, 2015)
Provisioning	Food
	Freshwater
Regulating and maintenance	Moderation of extreme events (landslide, flooding, fire)
	Erosion prevention and maintenance of organic topsoil
	Maintenance of agro-biodiversity
	Local climate and air quality
	Control and mitigation of Invasive Alien Species
	Habitat for species
Cultural	Conservation of local knowledge, traditional farming and building techniques
	Cultural identity, spiritual experience and sense of place
	Tourism
	Recreation and mental and physical health
	Aesthetic appreciation

17.2.3 Economic Valuation Methods of ES

The economic valuation of ecosystem services has been experimented in many cultural agrarian landscapes (Nahuelhual et al. 2014; Gravagnuolo 2015; TEEB 2015); however, it "may not always be possible or meaningful, as they often render themselves better to qualification than to quantification, such as in the case of cultural services" (Gaitán-Cremaschi et al. 2017).

The theory of Total Economic Value defines the value of a cultural and/or ecological resource as the sum of its direct and indirect use values and its independent-of-use (non-use) values (Randall 1987). Economic valuation techniques differ for the estimation of the components of value (Defra 2007). Table 17.3 shows the main evaluation methods that can be applied for ES economic valuation.

The estimation of monetary value of cultural services should be carefully approached (Chan et al. 2012; Daniel et al. 2012; Winthrop 2014; Fish et al. 2016). If we adopt the notion of Complex Social Value of terraced landscapes, the Total Economic Value is enriched including qualitative and quantitative "measures" of value in multiple dimensions.

The notion of the Complex Social Value projects the traditional economic evaluations into a multidimensional space, where the consequences of economic choices are assessed in relation also to their environmental and sociocultural

Table 17.3 Components of Total Economic Value and evaluation methods in relation with ES categories (adapted from DEFRA 2007)

Category of ES	Total Economic Value components			
	Direct use value	Indirect use value	Option value	Non-use values
Provisioning	Market-based techniques	–	Stated preference methods (contingent valuation, choice modelling), benefit transfer	–
Regulation and maintenance	–	Avoided cost method	Stated preference methods (contingent valuation, choice modelling), benefit transfer	Stated preference methods (contingent valuation, choice modelling)
Cultural	Incomes from tourism and recreation activities, revealed preference methods (travel cost, hedonic price)	–	Stated preference methods (contingent valuation, choice modelling), benefit transfer	Stated preference methods (contingent valuation, choice modelling)

impacts. The economic value of ecosystem services in terraced landscapes represents thus an acceptable (under) estimation of their Complex Social Value for society, since it can include, even if partially, future generations. It should be integrated with non-economic indicators on well-being, health, etc. as far as possible.

The economic value of ES, and particularly cultural ES, should not be taken always as a "precise" estimate, but it can be used to discuss the applicability of economic instruments for land use management to preserve the multifunctionality of cultural agrarian landscapes, and to justify public interventions (e.g. incentives, grants).

17.2.4 Evaluation of Costs and Expected Flows of Benefits

The evaluation of costs and expected flows of benefits is a fundamental evaluation step to support informed policy choices and also private investments (FAO & Global Mechanism of the UNCCD 2015).

A careful stakeholder assessment must be performed, considering who pays for landscape recovery and maintenance (typically farmers) and who benefits—and how much—from the regeneration (e.g. tourism activities, tourists, service providers, residents, community at large). Ecosystem services should be assessed at their provision location, and in the territory in which beneficiaries are located.

Costs for terraced landscape regeneration can be classified into three main categories:

- Investment costs (recovery of abandoned land)
- Landscape maintenance costs (dry-stone wall maintenance, water channels and cisterns maintenance, other necessary periodic works)
- Operating costs (costs of conducting farms such as energy, materials, administrative services, etc.)

The traditional cost–benefit analysis (CBA) should be enriched with the ES assessment and stakeholder assessment (Hein et al. 2006; Darvill and Lindo 2016) in the perspective of a richer multidimensional Community Impact Evaluation (Lichfield 1996).

Terraced landscape regeneration represents a cost at the micro-level (farm business) and a benefit at the macro-level (societal benefits from ES). Possible economic instruments should be designed on careful evaluation of ES provision costs and benefits at the farm scale (business analysis) and at the territorial scale.

In the following section, we discuss the applicability of some economic instruments in terraced landscapes, with a view to successful cases of application in different cultural agrarian landscapes.

17.3 Some Economic Instruments

17.3.1 Applicability and Impact of Subsidy Policies in Terraced Landscapes

The value of conservation of traditional agriculture can be described as the "contribution that would be socially fair to provide to farmers who preserve traditional landscapes" (Tempesta 2013). Subsides can be thus conceived as "mechanisms of societal recognitions to farmers for their historical stewardship of ecosystem services" (Gaitán-Cremaschi et al. 2017).

Agro-environmental measures (such as the EU CAP instrument) typically envisage subsidies for farmers, but the amount provided has usually weak correlation with ES. Land abandonment and change towards more intensive agricultural uses are still the largest causes of loss of traditional agrarian landscapes and related ES (Bignal and McCracken 2000; MacDonald et al. 2000; Thiene et al. 2006).

Taking into account ecosystem services in land use planning and policies "requires participatory approaches that combine multiple values and languages of valuation, as recognition of the many incommensurable values" provided by traditional farming practices (Gaitán-Cremaschi et al. 2017). The question is how governments can link the economic valuation of ES to agro-environmental policy choices.

A possibility could be to compare the average economic return of farms in terraced landscape to the actual average returns of intensive farms in close locations and use the opportunity cost as a measure of compensation to more costly agricultural practices, which in turn provide ES. This method could be socially acceptable and provide the necessary income to farmers for halting and reverse land abandonment.

However, in terraced landscapes the majority of land managers are often unprofessional farmers, who cultivate for self-consumption, and non-resident owners, who do not entrust terraces management to professional farmers. Moreover, the average properties often do not exceed the single hectare; thus, the amount of subsidy provided to the few professional farmers can be of low significance.

These factors substantially reduce the impact of agro-environmental subsidy policies in terraced landscapes, increasing the need of more innovative financing and management mechanisms that empower unprofessional farmers and foster social and technological innovation.

17.3.2 Sharing Economy Innovation in Eco-Labelling Mechanisms

Voluntary eco-labelling represents a viable economic instrument to raise awareness between farmers, residents, visitors and consumers, of the multiple benefits of preserving cultural landscapes. In terraced landscapes, eco-labelling can be challenging due to the large presence of unprofessional farmers, who lack the technological and economic means necessary to follow the standards of eco-labels.

Social innovation in eco-labelling mechanisms has been promoted by civic associations in the UNESCO World Heritage Site of the Amalfi Coast (Italy) for the recovery and regeneration of the terraced landscape.

A bottom-up project (*Effetto Costiera*) has connected small local farmers with ethical consumers groups (Ethical Purchasing Groups) in the territory surrounding the UNESCO site. Consumers' groups can buy local products from terraces directly from local farmers, creating a market for the residual products from self-consumption.

Local associations are in charge of 'labelling' the products and collect them from local farmers. All actors (farmers, intermediary associations and consumers) operate in a "network of trust" that enables sharing economy practices. The activity of the associations ensures additional income to farmers and healthy Km0 products to consumers, creating circular beneficial loops at local level.

The opportunities of the sharing economy have been exploited in the Amalfi Coast for preserving local agro-biodiversity. A crowdfunding campaign to recover the autochthonous seeds of a special variety of tomato (*Fiascone*) has been promoted in 2016, resulting in a first experimental production and the launch of an agro-business start-up for its commercialization.

Impact assessment and evaluation tools can support these initiatives by providing the necessary evidence base for upscale and replication of experimental tools, and for informing policy choices towards the operationalization of the circular economy in terraced landscapes.

17.3.3 Payments for Ecosystem Services Promoting Sustainable Farming Practices

The term payments for ecosystem services (PES) is used to describe economic schemes in which the beneficiaries, or users, of ecosystem services provide direct payment to the stewards, or providers (FAO 2011; Smith et al. 2013). PES often involve a series of payments to land managers in return for a guaranteed flow of ecosystem services (or, more specifically, for management actions likely to enhance their provision).

This mechanism is a type of subsidy that aims to protect ecosystem services by providing an economic incentive by the beneficiaries—directly and voluntarily—to land managers to adopt land use or management practices favourable to the protection of ecosystem services. According to the OECD (2010), there were already more than 300 PES or PES-like programmes in place by 2010 at national, regional and local levels. PES schemes are considered a viable model for supporting rural livelihoods integrating agri-environmental subsidies (Wynne-Jones 2013; Ingram et al. 2014).

In north-eastern France, the Vittel company (Nestlé Waters), world leader in mineral water bottling, developed a payment scheme to compensate local farmers who adopted sustainable farming practices, which reduced nitrate contamination in the aquifer, thus reducing the cost to the company for bottling water in that area.

Incentives, in the forms of contracts between the company and farmers, were structured according to the cost structure and location of the individual farms, and the link between ecosystem service (water filtration and maintenance of adequate levels of nitrate in the plant sub-root system) and management practices had been established scientifically at the sub-basin and plot level.

Despite the difficulties of balancing conflicting interests (Déprés et al. 2008), in ten years all 26 farms in the area had adopted the new farming system; 1700 ha of intensive crop cultivation were converted to more sustainable farming, and 92% of the sub-basin enhanced its capacity of provision of the ecosystem service related to freshwater. A clear indicator of success has been the request from young farmers who have taken over the family farm to enter into 30-year contracts (Perrot-Maître 2006).

The experience of Vittel shows that PES is a complex tool that requires the consideration of scientific but also social, economic, political, institutional relationships and conflict-solving ability (Perrot-Maître 2006). The ability to support farmers' income and finance the technological changes needed to shift to sustainable farming was an important element of success, but the key factor of success was not financial. Trust-building, through the creation of an intermediary institution (locally based and led by a "champion" sympathetic to the farmers' cause); the development of a long-term participatory process to identify alternative practices and a mutually acceptable set of incentives; the ability to link incentives to land tenure and debt cycle issues and to substitute the traditional technical and social support networks with new ones, was all fundamental conditions of success (Perrot-Maître 2006; Smith et al. 2013).

17.3.4 Pay for Action and Pay for Result Mechanisms

"Pay for action" and "Pay for result" mechanisms have been experimented in the "Burren Programme" on Ireland's western Atlantic coastline, to recover an area internationally recognised for its cultural and natural heritage, which includes dry-stone walls, traditional farming and pastoralism, and priority habitats listed in the Habitats Directive.

Recent years have seen the withdrawal, restructuring and reduction of farming activity, which has led to the slow degradation of priority habitats due to under-grazing, abandonment and loss of land management traditions.

The Burren LIFE project (2005–2010) started the regeneration process. Demonstrative actions were implemented involving 20 pilot farms, with the aim of recover priority habitats and the cultural built heritage and increase farmers' income (BLP 2010). Burren LIFE promoted an extensive survey to assess the willingness to pay (WTP) for maintaining limestone pavements and orchid-rich grasslands in the Burren (Burren Life Group 2010). The results were used to support the structuring of a payment scheme for Burren farmers using funds from agro-environmental policies. The "Burren Farming for Conservation Programme" (BFCP) (2010–2015) involved 160 farmers on 15,000 ha to support and incentivize farmers in maintaining and enhancing the habitats of the Burren (DAFF 2010).

The Burren Programme (DAFM 2016), started in 2016 with 200 farmers, will continue to implement solutions to help manage and protect the Burren cultural landscape. For its funding, the Department of Agriculture, Food and the Marine (DAFM) provides 1 M€ annually to farmers and the National Parks and Wildlife Service (NPWS) provides additional funding for administration. Further funding is provided by the Irish Farmers Association (Burren IFA) and Teagasc (Agriculture and Food Development Authority), both of whom were centrally involved in the original BurrenLIFE project, and from a range of other stakeholders including The Heritage Council.

The Burren experience has pioneered a novel "hybrid" approach to farming and conservation which sees farmers paid for both work undertaken (Payment for Actions) and for the delivery of agreed environmental objectives (Payment for Results).

The *Payment for Actions* is related to the co-design of an annual farm plan, which contains a list of actions with the aim of improving the site's management and conservation condition. Each work is budgeted, co-founded and carried out within the year by the farmer, with a final payment for complete and satisfactory standard of work.

The *Payment for Results* system is based on the yearly evaluation of the actions carried out by farmers. The result-based payment system allows farmers greater freedom to decide how to manage their land. The results on ecosystems conservation are evaluated annually with a 'habitat health' checklist, scored between 1 and 10 for each field. A score greater than 3 receives payment, but higher scores receive higher payments. This gives farmers the incentive to manage their fields in ways that will improve their scores and their payment as well and, at the same time, guarantees the taxpayer better value for money.

17.4 Conclusions and Discussion

Complex economic and societal factors are causing the abandonment of terraced landscapes, generating further environmental damages and a dramatic decrease of well-being in mountain marginal areas.

Abandoned terraced landscapes represent an unacceptable waste of territorial, environmental, social, economic resources, which generates negative impacts on the decay of historic towns and villages, producing a vicious circle over time. Systemic solutions should be explored, able to generate jobs and income starting from "land" as vital/living resource, which provides many ecosystem services.

Mountain, youths and development are interdependent: an inclusive development model is needed, able to take care of mountain and hill areas. The circular economy has been here proposed as a model that tends to avoid all wastes of resources, able to trigger processes of territorial regeneration. A new entrepreneurship, inclusive and responsible, able to go beyond the maximization of the economic income taking into account the production of multiple values, is certainly necessary.

In other words, the fundamental thesis of this paper is that the negative impacts of land abandonment in terraced landscapes can be avoided through the implementation of *circular processes*, thus integrating beauty, economy and fairness through a systemic landscape approach.

To implement the circular economy model in terraced landscapes, social, technological, financial and economic innovation is needed. Evaluation tools must be employed to produce evidence of the costs and flows of benefits of landscape regeneration for multiple stakeholders, also in the surrounding urban areas.

The ecosystem services assessment can be applied to evaluate costs and benefits of maintaining traditional agricultural systems (TEEB 2015). Mapping and assessment of ES in terraced landscapes provide evidence of the complex spatial, physical and intangible interrelations between economic, sociocultural and environmental landscape functions, linking them to human well-being. Policy choices should be informed by the knowledge of these relationships, to compensate local farmers for the multidimensional societal benefits they provide. The recognition of the role of farmers for the conservation of cultural agrarian landscapes (because they provide ecosystem services fundamental for human well-being) is the starting point to structure economic instruments that secure an income and a new social role to rural households, thus contributing to halt migration flows from rural to urban areas.

The circular economy model can be implemented in terraced landscapes reducing the overall costs of maintenance and enhancing the social, environmental, cultural and economic productivity. Spatial proximity becomes relational proximity, creating synergies between many actors, that reduces costs.

Spatial relationships can be extended to the urban-rural ensemble, considering ES at their provision location and in the larger urban-rural areas of benefit (Plieninger et al. 2013; Nahuelhual et al. 2014). Urban and rural areas are mutually interdependent: rural villages and cities/metropolitan areas are two ends of a human settlements continuum (United Nations 2015). Ecosystem functions in the rural landscape support human well-being in urban areas by providing services such as local food, freshwater, air purification, cultural identity and other supporting services needed for human life (TEEB 2015). Thus, economic and financing mechanisms can be structured at the city-region level, implementing healthy "City Region Food Systems" as proposed by FAO (2015).

Large investments are required to recover disused landscapes, with lower and longer-term economic returns compared to areas of intensive agriculture. This reduces the attractiveness of the investment in the traditional economy. The circular economy, adopting the approach of ecological economics, changes the economic perspective taking into account the multidimensional benefits of landscape reuse as productive factors comparable to the economic return on investment, creating new market demands (willingness to pay) and thus new jobs for agro-environmental businesses.

Cooperative entrepreneurship in abandoned terraced landscapes, also in a systemic perspective with more productive plane agricultural lands, is an example of social responsibility, because it generated jobs, income and environmental preservation. Certainly, this is not the only solution to regenerate terraced landscapes. However, it should not be overlooked, considering that the civil/solidarity economy is structurally a form of circular economy, where close loops of giving-receiving-giving back are enhanced, exploiting spatial and relational proximity. Civil, solidarity, sharing economy models are all forms of circular economy that consider intrinsic values (and not only instrumental ones), the long term (future generations) and not only the short term of return on investment, intangible values and not only the economic-quantitative ones.

Case studies show that it is possible to foster social, business, financing and institutional innovation through public-private-social cooperation. In some cases, the role of third sector as intermediary has enhanced sharing/solidarity local economies introducing short chains and loops between production and consumption in internal areas and in urban-rural areas, able to valorize small-scale activities, family/community led. The "well-living" (Acosta 2010; Becchetti et al. 2017) can enhance the attractiveness of terraced landscapes, where beauty, health and proximity (spatial and relational) are key factors of quality of life.

This "soft" perspective can produce "hard" consequences on the economic dimension, influencing consumers' choices, policy choices and localization choices of businesses and residents. The economic instruments analysed can be implemented to foster innovation towards the transition to a circular model of development in terraced landscapes and attract new investments.

Financing cultural agrarian landscapes can be particularly challenging. According to FAO (2015), "Considering the variety of direct and indirect costs and benefits provided by restoration of mosaic landscapes, a mix of investors and financing instruments will be required for effective financing landscape restoration". Social impact investment funds and community-led investment models such as crowdfunding could be promoted for financing terraced landscapes regeneration, linking the social and environmental impacts with economic performances. Dewees et al. (2011) identify four main categories of investors depending on their goals:

(1) traditional investors that seek a financial return;
(2) social investors have other goals besides earning a return on their investment and are willing to accept higher risks;
(3) conservation investors use their capital to protect or restore a specific landscape, habitat or species;
(4) impact investors, who mix the approaches of the previous categories.

The attractiveness of investments is here linked to the possibility of reducing costs and enhances the multifunctionality and productivity of terraces. The evaluation of the complex social value of terraced landscape, using the methodologies of ES assessment, is necessary to provide evidence to many actors of the expected mutual flows of benefits of investments in regeneration.

Innovative financing mechanisms have been widely experimented in urban regeneration practice (Dalberg Global Development Advisors 2014), and thus, they could be transferred and adapted to terraced landscape regeneration.

As shown in the case studies, an important prerequisite for the success of economic instruments is the participation of the public actor as promoter, validator and/or co-financing subject, that ensures feasibility and a strong process based on transparency, trust, sharing and collaboration.

To improve the potential of funding for cultural resources, communication and *awareness raising* processes are needed, because from this depends the availability of financing. A second action is linked to the *accessibility of terraces*, which can encourage or discourage the availability of financing, to variable degrees when it comes to private, public or civic/social bodies. The third perspective refers to the *reduction of functional re-adaptation costs through technology and/or public subsidy*.

These three different actions have different significance when it comes to private financing (e.g. Trust, Foundations, Patrons), public bodies (national, regional, local: incentives, grants, tax reduction or tax exemption) or social impact funds/third sector bodies (e.g. non-profit organizations, businesses cooperatives, hybrid businesses and impact finance funds).

The above case studies demonstrate that a "circular paradigm", based on circular financing, business, management and institutional models, is feasible. It can enable regeneration creating new values, new attractiveness, new markets and jobs from cultural and ecological resources in terraced landscapes.

Thus, adopting the circular paradigm, terraced landscapes can become models of landscape-led regeneration for rural areas, contributing to build inclusive, safe, resilient and sustainable human settlements and to the overall *human* development.

References

Acosta A (2010) El Buen Vivir en el camino del post-desarrollo. Una lectura desde la Constitución de Montecristi. *Policy Paper* 9, vol 9

Becchetti L, Bruni L, Zamagni S (eds) (2017) Le città del ben-vivere. Il manifesto programmatico dell'economia civile per le amministrazioni locali (forthcoming)

Bignal EM, McCracken DI (2000) The nature conservation value of European traditional farming systems. Environ Rev 8:149–171. https://doi.org/10.1139/er-8-3-149

BLP (2010) The Burren LIFE project. Farming for conservation in the Burren. Final Report. Carron, Ireland

Bonardi L, Varotto M (2016) Paesaggi terrazzati d'Italia. Eredità storiche e nuove prospettive. Franco Angeli

Chan KMA, Satterfield T, Goldstein J (2012) Rethinking ecosystem services to better address and navigate cultural values. Ecol Econ. https://doi.org/10.1016/j.ecolecon.2011.11.011

Common M, Stagl S (2005) Ecological economics. An introduction. Cambridge University Press, Cambridge

Costanza R, Cumberland JH, Daly H, Goodland R, Norgaard RB, Kubiszewski I, Franco C (2014) An introduction to ecological economics, 2nd edn. CRC Press

Costanza R, D'Arge R, De Groot R, Farber S, Grasso M, Hannon B, Limburg K et al (1997) The value of the world's ecosystem services and natural capital. Nature 387:253–260. https://doi.org/10.1038/387253a0

DAFF (2010) Terms and conditions of the Burren farming for conservation programme implemented by the minister for agriculture, fisheries and food under the provisions of article 68.1(a)(i) of Council Regulating (EC) 73/2009 April 2010

DAFM (2016) The Burren Programme—Terms and conditions. Tranche 2

Dalberg Global Development Advisors (2014) Innovative financing for development: scalable business models that produce economic, social, and environmental outcomes

Daniel TC, Muhar A, Arnberger A, Aznar O, Boyd JW, Chan KMA, Costanza R et al (2012) Contributions of cultural services to the ecosystem services agenda. Proc Natl Acad Sci 109:8812–8819. https://doi.org/10.1073/pnas.1114773109

Darvill R, Lindo Z (2016) The inclusion of stakeholders and cultural ecosystem services in land management trade-off decisions using an ecosystem services approach. Landsc Ecol 31:533–545. https://doi.org/10.1007/s10980-015-0260-y

Defra (2007) An introductory guide to valuing ecosystem services. Forestry 68

Déprés C, Grolleau G, Mzoughi N (2008) Contracting for environmental property rights: the case of Vittel. Economica 75:412–434. https://doi.org/10.1111/j.1468-0335.2007.00620.x

Dewees P, Place F, Scherr SJ, Buss C, Ajayi OC, Buck LE, Elson D et al (2011) Investing in trees and landscape restoration in Africa. What, where, and how. Washington, DC

Ellen MacArthur Foundation (2015) Towards a circular economy: business rationale for an accelerated transition

Ellen MacArthur Foundation (2017) Cities in the circular economy: an initial exploration

ESPON, Interact, Interreg Europe, and URBACT (2016) Pathways to a circular economy in cities and regions. Lille, France

European Commission (2015) Closing the loop – an EU action plan for the circular economy, vol 2. Brussels, 2.12.2015 COM(2015) 614 final

FAO (2011) Payments for ecosystem services and food security

FAO (2015) Food for the Cities Programme—Building sustainable and resilient city region food systems

FAO & Global Mechanism of the UNCCD (2015) Sustainable financing for forest and landscape restoration: Opportunities, challenges and the way forward. Italy, Rome

Fish R, Church A, Winter M (2016) Conceptualising cultural ecosystem services: a novel framework for research and critical engagement. Ecosyst Serv, 1–10. https://doi.org/10.1016/j.ecoser.2016.09.002

Forte F, Fusco Girard L (2009) Creativity and new architectural assets: the complex value of beauty. Int J Sustain Dev 12:160–191

Fujita T, Satoshi O, Dong L, Minoru F (2013) Eco-industrial development as a circularization policy framework toward sustainable industrial cities. Lesson and suggestions from the eco town program in Japan. BDC. Boll Del Centr Calza Bini 13:35–52. https://doi.org/10.6092/2284-4732/2449

Fusco Girard L (1987) Risorse architettoniche e culturali: valutazioni e strategie di conservazione. Franco Angeli

Fusco Girard L (2016) The city and the territory system: towards the "new humanism" paradigm. Agric Agric Sci Procedia 8:542–551. https://doi.org/10.1016/j.aaspro.2016.02.070

Fusco Girard L, Nijkamp P (1997). Le valutazioni per lo sviluppo sostenibile della città e del territorio. Franco Angeli, Milano, Italy

Gaitán-Cremaschi D, Palomo I, Baraibar Molina S, De Groot R, Gómez-Baggethun E (2017) Applicability of economic instruments for protecting ecosystem services from cultural agrarian landscapes in Doñana, SW Spain. Land Use Policy 61:185–195. https://doi.org/10.1016/j.landusepol.2016.11.011

Gravagnuolo A (2014) Una proposta metodologica per la valutazione dei landscape services nel paesaggio culturale terrazzato. BDC. Boll Del Cent Calza Bini 14(II):367–385. https://doi.org/10.6092/2284-4732/2932

Gravagnuolo A (2015) La valutazione dei paesaggi culturali. Approcci e strumenti per la tutela e valorizzazione dei sistemi terrazzati. University of Naples

De Groot RS, Alkemade R, Braat L, Hein L, Willemen L (2010) Challenges in integrating the concept of ecosystem services and values in landscape planning, management and decision making. Ecol Complex 7:260–272. https://doi.org/10.1016/j.ecocom.2009.10.006

Haines-Young R, Potschin M (2010) The links between biodiversity, ecosystem services and human well-being. In: Raffaelli DG, Frid CLJ (eds) Ecosystem ecology: a new synthesis, pp 110–139. Cambridge University Press, Cambridge

Hartel T, Fischer J, Câmpeanu C, Milcu AI, Hanspach J, Fazey I (2014) The importance of ecosystem services for rural inhabitants in a changing cultural landscape in Romania. Ecol Soc, 19. https://doi.org/10.5751/es-06333-190242

Hein L, Van Koppen K, De Groot RS, Van Ierland EC (2006) Spatial scales, stakeholders and the valuation of ecosystem services. Ecol Econ 57:209–228. https://doi.org/10.1016/j.ecolecon.2005.04.005

Hicks C, Cinner JE, Stoeckl N, Mcclanahan TR (2015) Linking ecosystem services and human-values theory. Conserv Biol 29:1471–1480. https://doi.org/10.1111/cobi.12550

Ingram JC, Wilkie D, Clements T, Balas Mcnab R, Nelson F, Hogan Baur E, Sachedina HT, Dean Peterson D, Andrew C, Foley H (2014) Evidence of payments for ecosystem services as a mechanism for supporting biodiversity conservation and rural livelihoods. Ecosyst Serv 7:10–21. https://doi.org/10.1016/j.ecoser.2013.12.003

Iverson L, Echeverria C, Nahuelhual L, Luque S (2014) Ecosystem services in changing landscapes: an introduction. Landsc Ecol 29:181–186. https://doi.org/10.1007/s10980-014-9993-2

Lichfield N (1996) Community impact evaluation: principles and practice. Routledge

MA (2005) Ecosystems and human-well-being: wetlands and waters synthesis. World Resources Institute, Washington, DC

MacDonald D, Crabtree JR, Wiesinger G, Dax T, Stamou N, Fleury P, Gutierrez Lazpita J, Gibon A (2000) Agricultural abandonment in mountain areas of Europe: Environmental consequences and policy response. J Environ Manage 59:47–69. https://doi.org/10.1006/jema.1999.0335

MEA (2003) Ecosystems and human well-being: a framework for assessment

Milcu AI, Hanspach J, Abson D, Fischer J (2013) Cultural ecosystem services: a literature review and prospects for future research. Ecol Soc 18:art.44. https://doi.org/10.5751/es-05790-180344

Le Moigne R (2014) L'économie circulaire. Comment la mettre en œuvre dans l'entreprise grâce à la reverse supply chain? Dunod

Nahuelhual L, Carmona A, Laterra P, Barrena J, Aguayo M (2014) A mapping approach to assess intangible cultural ecosystem services: The case of agriculture heritage in Southern Chile. Ecol Ind 40:90–101. https://doi.org/10.1016/j.ecolind.2014.01.005

OECD (2010) Paying for biodiversity: enhancing the cost-effectiveness of payments for ecosystem services. Paris

Partnership Circular Economy (2018) Urban Agenda for the EU. Draft action plan

Perrot-Maître D (2006) The Vittel payments for ecosystem services: a "perfect" PES case? Environment. London, UK

Plieninger T, Dijks S, Oteros-Rozas E, Bieling C (2013) Assessing, mapping, and quantifying cultural ecosystem services at community level. Land Use Policy 33:118–129. https://doi.org/10.1016/j.landusepol.2012.12.013

Plieninger T, Van der Horst D, Schleyer C, Bieling C (2014) Sustaining ecosystem services in cultural landscapes. Ecol Soc. https://doi.org/10.5751/ES-06159-190259
Porter ME, Kramer MR (2011) Creating shared value. Harvard Bus Rev 89:62–77
Randall A (1987) Total Economic Value as a basis for policy. Trans Am Fish Soc 116:325–335. https://doi.org/10.1577/1548-8659(1987)116%3c325:TEVAAB%3e2.0.CO;2
Rizos V, Behrens A, Kafyeke T, Hirschnitz-Garbers M, Ioannou A (2015) The circular economy: barriers and opportunities for SMEs. 412. CEPS Working Document
Scaramellini G, Varotto M (eds) (2008) Paesaggi terrazzati dell'arco alpino. Atlante, Marsilio
Smith S, Rowcroft P, Everard M, Couldrick L, Reed M, Rogers H, Quick T, Eves C, White C (2013) Payments for ecosystem services: a best practice guide. London
TEEB (2010) In Pushpam Kumar (ed) The economics of ecosystems and biodiversity ecological and economic foundations. Earthscan, London and Washington
TEEB (2015) TEEB for Agriculture & Food: an interim report. Geneva, Switzerland
Tempesta T (2013) Il valore economico dei paesaggi tradizionali. Lavorare il paesaggio. Terrazzamenti e muri a secco tra identità e valori. Albiano (TN)
Thiene M, Bazzani GM, Tempesta T (2006) Le conseguenze della riforma di Politica Agricola Comunitaria sul paesaggio rurale. Econ Dirit Agroaliment 11:79–95
United Nations (2015) Habitat III issue papers 10—Urban-rural linkages. New York
Varotto M (ed) (2009) Nuove Terre Alte. Dalla "montagna che scompare" alla "montagna che vive". Riv del Club Alpino Italiano, 70–73
Winthrop RH (2014) The strange case of cultural services: limits of the ecosystem services paradigm. Ecol Econ 108:208–214. https://doi.org/10.1016/j.ecolecon.2014.10.005
Wynne-Jones S (2013) Connecting payments for ecosystem services and agri-environment regulation: An analysis of the Welsh Glastir Scheme. J Rural Stud 31:77–86. https://doi.org/10.1016/j.jrurstud.2013.01.004

Chapter 18
The Challenge of Tourism in Terraced Landscapes

Theano S. Terkenli, Benedetta Castiglioni and Margherita Cisani

Abstract Terraced landscapes contain and produce natural and cultural values which tend to be highly relevant to tourism. The variable interactions developing between the tourism industry and local agricultural systems sustaining terraced cultures may have different consequences on both local development and tourism trends. After a brief theoretical introduction into landscape–tourism interrelationships, this chapter addresses the long array of circumstances and consequences of tourism development in terraced landscapes. With the aid of a series of diagrams which serve as its analytical framework, the chapter lays out and discusses the empirical circumstances and types of challenges stemming from different types of tourism in such landscapes, in terms of polarities: (a) types of landscape uses from the demand side of tourism, (b) impacts of tourism on the agricultural system of these landscapes and (c) impacts of tourism on the socio-economic system of terraced landscapes. Both risks and opportunities incurred by tourism impacts on visited landscapes are especially pressing in the case of terraced landscapes, running the full range from most negative (i.e. destruction) to most positive (i.e. rejuvenation) possible consequences. Even though the diagrams used in this chapter present mass and mild tourism as polarities, they serve as a basis for elaboration on the attractions and expectations of tourists and local communities, in such cases, and on landscape-related tourism consequences on local agricultural and broader local/regional socio-economic systems. They also allow for some conclusive remarks on the environmental, economic and social/cultural sustainability of terraced landscape tourism, in the context of broader local/regional development, while laying the ground for further analysis.

T. S. Terkenli (✉)
University of the Aegean, Mytilene, Greece
e-mail: t.terkenli@aegean.gr

B. Castiglioni · M. Cisani
Department of Historical and Geographic Sciences and the Ancient World, University of Padova, Padua, Italy
e-mail: etta.castiglioni@unipd.it

M. Cisani
e-mail: margherita.cisani@unipd.it

M. Varotto et al. (eds.), *World Terraced Landscapes: History, Environment, Quality of Life*, Environmental History 9,
https://doi.org/10.1007/978-3-319-96815-5_18

18.1 Introduction

The significance of landscape to the variety of tourist experiences at a visited destination is well-established and considered paramount. Statistical information indicates an overwhelming visitor preference, in the post-war era, primarily for coastal, sea–sand–sun (3S) destinations, during the most accommodating time of year (i.e. summer), and secondarily (excluding various forms of urban tourism) for other types of rural and natural environments, such as mountains, forests, lakes or terraced landscapes (UNWTO 2011). All types of landscapes and places may potentially hold interest for some type of visitor, for purposes of consumption of goods, services, activities, experiences, etc. Contextualized, as well as overarching, leisure and tourism experiences increasingly inform and substantiate new types of landscapes of tourism and leisure. Thus, globalizing rates and patterns of mobility and consumption, generally speaking, necessitate renewed and more in-depth investigation as to the sites and attractions sought by visitors and to the role of landscape in visitor experiences (Aitchison et al. 2000; Terkenli 2014). Such issues are further complicated by the enormous current proliferation of a broad range of alternative and special-interest forms of tourism or leisure, variably (and often, intricately) connected to the visited landscape—a field that begs for further and more concerted inquiry (Hall et al. 2014; Hall and Page 2014).

In this study, we focus on such variability and contingency in tourist/visitor experience, in different types of terraced landscape destinations. The objective of our study is to explore, analyse and discuss the variable ways in which terraced landscapes cater to and are impacted by various types of tourism/recreational/leisure activities. Towards this goal, we propose analytical diagrams of the interrelationships that develop through tourism uses of terraced landscapes, in terms of: (a) attractions and expectations of tourists and local communities, (b) tourism impacts on the agricultural system which creates the terraces and (c) landscape-related tourism consequences on the broader local/regional socio-economic system.

We begin with a brief discussion of the interrelationships that develop between tourism/leisure activities and visited landscapes, in general; then, we proceed to explore this set of interrelationships in the case of terraced landscapes, focusing on when and how a terraced landscape becomes a tourist attraction, and, finally, we turn to a discussion on the direct and indirect impacts of the latter activities on the terraced landscapes, in the context of broader local/regional development.

18.2 Tourism and Landscape: A Brief Theoretical Staging

The centrality of the sightseeing experience to tourism (Urry 1990, 1995), coupled with the definition of the landscape itself (Council of Europe 2000), attests to the fact that there may not be tourism without landscape, and no landscape is such without its viewer/observer (in a broader sense). We ought to clarify, at the outset, our use of the terms tourist versus tourism, in conjunction with the concepts of landscape, activity, development, etc.—an area of much confusion in tourism literature. The noun tourism is intended to show the processes through which a landscape, activity, development, etc., are shaped in order to serve purposes of tourism (also, i.e. landscapes of tourism), whereas the noun/adjective tourist is used to indicate the ways/reasons/processes, in which such activities, landscapes, etc., are substantiated or used via the phenomenon of tourism. Simply put, tourism landscapes imply the ways in which these landscapes are produced, whereas tourist landscapes imply the ways in which these landscapes are consumed. We also employ the term visitor, in order to encompass all types of such intended expropriation of points, areas or sites of appeal, inciting such interest, from a broad range of parties: tourists, excursionists, day-trippers, explorers, recreationists, etc.

As a geographical medium conceived and appropriated through the senses and the power of cognition and symbolism, landscape represents the first and most enduring medium of contact between tourist and prospective or consumed place of travel; through acquired photographs, it becomes a traveller's lasting memoir (Terkenli 2014). The tourism industry markets images and discourses about landscapes, through representations of cultural signs, on the basis of which the tourist, through processes of experiential reinterpretation of the sign, may assess the sight and validate the meanings of the visited landscape, within the predominant discourse (MacCannell 1992; Terkenli 2014). On the basis of the tourist's involvement in local life, tourist activity and experience in the landscape may range from pure leisure activity (with no involvement in local life), to change from everyday habitual activity (with minimal involvement), to activity bearing new experiences, to experimental activity (with substantial involvement) and finally to experiential activity (with significant involvement in local life) (Cohen 1979; Urry 1990). Thus, the connection between landscape and tourism is not restricted either to the representational/performative or to the essentially geographical/physical nature of the travel experience. It extends to the pleasure sought in the experience, a component of tourism that has become much more central and predominant in the historical evolution of twentieth-century forms of tourism (Löfgren 1999; Rose 1996) and acknowledged through theories of emotion and affect, as well as more-than-representational geographies of human–landscape interaction (Terkenli 2014; Lorimer 2005; Crouch 1999).

On the one hand, then, the impact of the contribution of the landscape to the tourist experience and tourism business is inadvertently positive, albeit variably multifold (Terkenli 2014; Carmichael 1998). On the other hand, tourism impacts on the landscape are also highly variegated and multifold, but run the whole range from most positive to most negative. The tourism scientific literature abounds in examples of deterioration of tourism destinations through certain types of tourism development, both in economic terms (economic monocultures, increasing tourism dependency and local underdevelopment, etc.) and in all sorts of non-economic terms (visual clutter, cultural impoverishment, social degradation, environmental deterioration, etc.) (Pearce 1995; Tsartas 1996). Where tourism development, growth and expansion are unchecked, the landscape often loses its previous character and becomes infested by an inundation of incongruous and out-of-plan construction, urban, peri-urban or suburban sprawl and all types of visual pollution. In some cases, tourism development in rural communities imparts gentrification processes associated with property market impacts and the risk of exclusion of locals (housing, farming activities, everyday leisure places, etc.), especially farmers and low-income people. In other cases, rural landscape management and overall stewardship are abandoned, in favour of either more amenable and profitable economic activities (i.e. tourism) elsewhere, leading to rural exodus and the abandonment of the countryside (resulting in geographical imbalances in rural development, as in the example of Greece), or through its extensive exploitation for tourism or other economic activities (i.e. intensive viticulture, alteration of alpine landscapes for winter sports development). Curbing at least some of the adverse impacts of tourism on the landscape and the destination societies may be feasible, through protection and conservation measures—many examples of which best practices abound (Bell 2008; Coccossis 2004). Achieving and harnessing the positive economic potential of tourism may be far more difficult. The latter cautionary conclusion stems from the fact that tourism development depends on successfully meeting global competition, in order to attract tourists at the desired destination, in the first place, while at the same time safeguarding local resources (including the landscape) in ways beneficial for the local communities, as well as assuring their economic growth and development in sustainable ways. We will encounter this challenge most strikingly in the case of terraced landscapes and discuss it extensively further down, in this chapter.

Global flows of mass/organized tourism have predominated in the broader western post-war context (UNWTO 2013–2017). Simultaneously, the past few decades have been witnessing the accelerated growth of scope (special interest and alternative) tourism/leisure, targeting specific niches of the tourism/leisure market (Hall et al. 2014; Rojek and Urry 1997). In this context, we now turn to terraced landscapes, promoted by the supply side on the basis of their competing edge in alternative but affordable forms of leisure, compatible with local sustainable development and consumed by the demand side on the basis of their catering to a variety of broadly accessible tourism/leisure pursuits and activities. For this purpose, we propose an analytical scheme for this twofold relationship, based on a tourism axis connecting tourism variables first with the agricultural system and second with the broader socio-economic system, in order to present, organize and interpret our research questions at stake more clearly and efficiently.

18.3 Tourism Development in Terraced Landscapes: Attractions and Expectations

18.3.1 General Trends and Main Issues

Landscapes characterized by terraces share some common features. Basically, they are rural landscapes of historical origin: some of them have very ancient origins (thousands of years), others are more recent (first half of the twentieth century), but they are all connected to traditional pre-industrial agriculture, before the sector's large-scale mechanization and industrialization. Such circumstances have different consequences, in terms of both local development and tourism trends.

These rural landscapes often suffer from abandonment, caused by marginality (mountain or hilly areas) and by collapsing rural systems, when farmers' incomes become too low. Residual 'heroic' agriculture, then, is often the only way to stem abandonment and re-naturalization and prevent hydrogeological instability and the collapse of terraces. At the same time, terraces create very original vertical landscapes, with spectacular views. They are relevant also from the point of view of heritage, as cultural landscapes, witnesses of ancient practices in the context of humans' interrelationships with their landscapes (see Chap. 15). They may, thus, offer a long series of experiences that cater to various special-interest and alternative forms of tourism, such as agro-tourism, eco-tourism, gastro-tourism, adventure, history, culture. Obviously, tourism trips involve a variety of motives and activities (tourism demand), normally requiring a range of service, commodity and other consumption (tourism supply). As will become obvious in the following sections, tourism in terraced landscapes often combines many or most of such activities, depending, of course, on the profile of the consumer.

We may distinguish between locally versus externally induced tourism, which is tourism that is more instigated and managed by local actors and stakeholders, in contrast to tourism that is propelled by the global tourism industry and system (tour operators, multinational corporations, big industry, etc.). In the former case, we normally include milder forms of terraced landscape tourism, mostly controlled and operated from the bottom-up, whereby profits, costs and benefits often tend to circulate or be recycled in the destination locality or region—either in the local/regional tourism industry itself, or in the local societies, as advocated in Garajonay National Park, Spain, through their Action Plan and Strategy of implementation of the European Charter for Sustainable Tourism in Protected Areas (ECST). In the latter case, profits, costs and benefits normally return to these international actors of the tourism industry and fail to be used towards locally or regionally sustainable development goals and practices. Normally, these terraced landscape tourism activities involve higher demand, larger markets, heavier tourist flows and more intense uses and concentrations of tourist activity at the destination, than the former types of tourism, which tend to cater more to local, regional, internal/domestic, and/or limited tourism or recreation markets.

18.3.2 The Demand Side

We will first turn to the demand side of tourism. Here, we may distinguish between tourism that is less or more involved in local matters and activities, for instance between the all-inclusive (organized or package) type of tourist and milder forms of tourist activities—usually more involved and participating in local community activities and life (i.e. in grape or fruit harvest, drystone wall building). We may distinguish between different types of resource use by the tourists, at the visited terraced landscape destination: natural features, cultural assets, leisure amenities, experiential stimuli, etc. (i.e. rural pilgrimages and cultural routes). We may also postulate that terraced landscapes, by nature and by function, are more amenable to milder, more resource-conscious types of landscape-oriented activities, with the underlying—often unstated, but always implied—the aim of achieving sustainability and multifunctionality (i.e. through the promotion of local products). Accordingly, milder (less intensive, alternative or resource-conscious) types of tourist activity impart fewer adverse impacts on these landscapes, leaving a lighter imprint on them, thus contributing more to their sustainable function and development.

18.3.3 The Supply Side

The local/global levels of tourism development are often interlinked and act concurrently, as they are also interlinked with the supply side of tourism, which we turn to, now. The original features and critical aspects of terraced landscapes introduced above call tourism into question, from two different points of view. From the tourist point of view, these semi-natural rural landscapes, with spectacular views and heritage elements, attract tourists in search of otherness, exotic and original scenery, cultural landscapes and history, and 'authentic' good food and wine, far from the urban chaos, in a 'harmonious coexistence between man and nature' (Zahn and Jin 2015, p. 664). These aspects of the terraced landscapes become poles of tourist attraction. From the point of view of the local community, tourism is seen as an economic activity able to curb terrace abandonment and preserve both the cultural heritage and slope stability, also because it increases local incomes. In fact, terraced landscape tourism can often be practiced as a secondary business, because it produces ancillary revenue, which may sustain people's main source of income (agriculture or other jobs). As will be further elaborated below, tourism and agriculture are not only interrelated, but they are both parts of a broader socio-economic system. In fact, dealing with tourism in difficult contexts, such as terraced landscapes, implies also recognizing that two different rationalities might be employed in developing tourism, both in internally and externally induced 'touristification': one that views the increase of tourism as a benefit in itself (instrumental rationality) and one which aims at preserving and maintaining the ecological and economical system of a terraced landscape (value rationality) (Xiaoyun et al. 2013, p. 273). Combining these two rationalities, local communities—vying for territorial competitiveness in a global market—may take up the challenge of activating latent natural and cultural resources, through the development of small-scale and place-based tourism.

Generally speaking, tourism development in terraced landscape seems an effective win–win strategy, or even 'the goose that brings the golden eggs' (Jansen-Verbeke and McKercher 2013, p. 245). Tourism, however, is not an automatically or unequivocally positive solution for terraced landscapes, as will be shown in Sects. 18.4 and 18.5. Moreover, recognizing the values of the terraces is not a sufficient condition for tourism attraction and stimulation. In order for the supply side of tourism to become competitive and attractive, it also requires investment in various infrastructures, accommodation, facilities, transportation and communication, as well as appropriate management and marketing strategies. The tourism potential may also be recognized either internally or externally. In the first case, tourism tends to be developed and managed more by local or regional actors and attracting mostly local and regional tourism. In the second case, when the terraced landscapes are of outstanding value and high attractiveness, such as in UNESCO heritage sites, tourism develops intrinsically, attracting global flows, in the context of the global tourism industry, and involving also both external and international actors.

18.3.4 Our Approach

Diagrammatically, we may present types of tourism uses of terraced landscapes—including characteristics and repercussions of tourist uses of the terraced landscape—as follows (Fig. 18.1).

Despite the fact that such polarized categorizations may be a bit too simplistic, they may also facilitate our attempt to distinguish between types of tourist (terraced) landscape consumption. Specifically, less involved types of tourist use and activity imply a superficial consumption of the landscape, generally speaking, in the form mainly of scenery or image. This type of consumption is often referred to in terms of 'sightseeing', landscape as a view, a panorama or as unique/striking assemblage of structures or forms. Such partial and uninvolved consumption of the visited landscape, merely as a visual stimulus or set of signs of the visited destination (MacCannell 1992), finds frequent application in the case of terraced landscape destinations, especially when the main form of tourism developed in these destinations is not primarily motivated by or revolving around the terraces themselves. It becomes a stage set, in which tourism takes place—a mere backdrop for tourism activity—by simply providing appropriate, supporting or just beautiful background images for consumption by tourists who may simply be passing through, on their way to other destinations, or targeting their visit and consumption to other resource and amenity uses, rather than acknowledging and enjoying the terraced landscapes

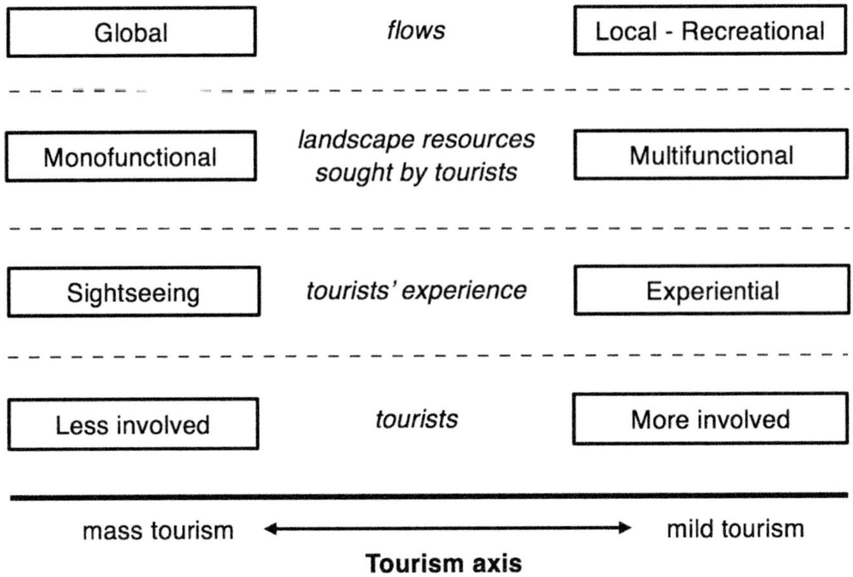

Fig. 18.1 A polarized categorization of types of landscape uses from the demand side of tourism

as complex biophysical and socio-economic systems, worthy of their interest and involvement in and of themselves.

Deeper analysis and a critical approach to the complex relationship of tourism and landscape are, therefore, required, in order to pull out its strengths and weaknesses, as well as its potential and risks. Next, we will focus specifically on the different forms that this relationship may take under various agriculture conditions, in their broader local/regional socio-economic and cultural system.

18.4 Tourism and Local Agricultural System

Various types of tourism landscape consumption, as described previously, relate and impact in different ways on the agricultural system, especially according to the type and intensity of agricultural use, ranging from mechanized terrace cultivation to subsistence uses and abandoned terraces. Several questions arise, in this regard: Can tourism be a solution counter terrace abandonment and decline or is it a direct cause of them? Is it even possible to recuperate abandoned and destroyed terraces through tourism? Generalising, how do different forms of tourism produce different impacts on terraced landscapes, in relation to their level of agricultural productivity? The diagram in Fig. 18.2 represents a critical approach to synthesizing the impacts of tourism on the agricultural use of the terraces.

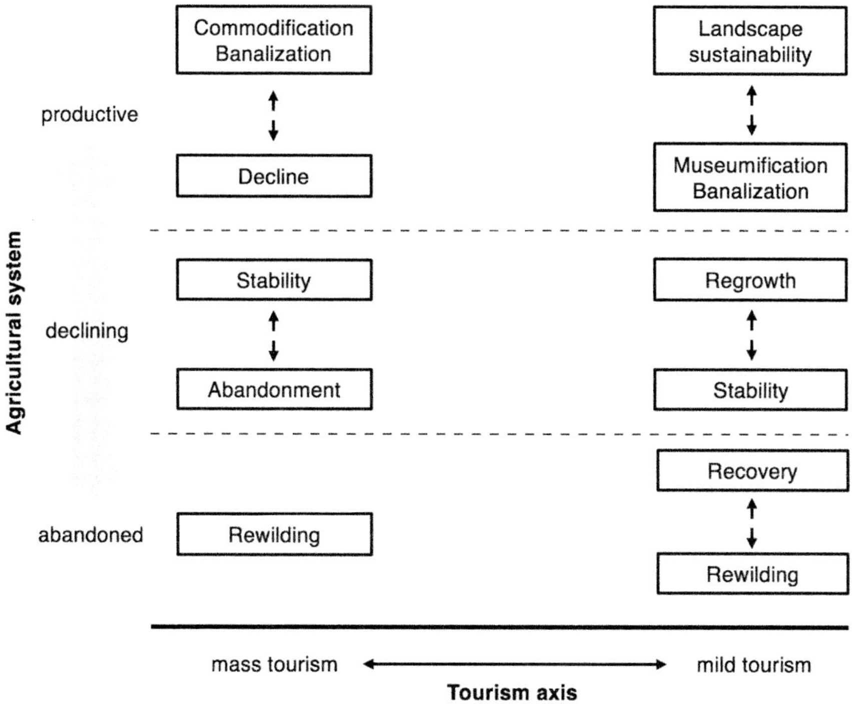

Fig. 18.2 Impacts of different types of tourism on the agricultural system of terraced landscapes, according to agricultural system capacity

Besides terrace demolition, agricultural mechanization might offer some solutions to terrace maintenance and, at the same time, sustenance of high—or at least competitive—levels of production. Thus, the upper part of the scheme corresponds to the productive terraces, encompassing intense cultivations, run and managed mainly for market destinations, affording enough income for farmers/entrepreneurs, so as to be their one and only occupation, as well as terraces cultivated for local consumption, through cooperative distribution or limited to individual/subsistence uses. A more specific and detailed analysis will, undoubtedly, further distinguish these typological categories of productive terraces, taking into consideration also production techniques, goals and targets.

The second (intermediate) level of agricultural use corresponds to terraces in *decline*, where a variety of other variables affect the stability of the production. This occurs mainly where crop yields are not especially valuable in terms of farmers' main source of income, and thus not worthwhile for younger generations to pursue the family's activity or to invest in new farming technologies. As in the case of the productive (previous) category, this category may also be further split or expanded into more levels and types of decline (structural, economic or social decline).

Finally, the lowest level includes abandoned terraced landscapes, which are often found in the same geographical areas of productive or declining terraces but have been exposed to the least agriculturally favourable conditions, in order for them to continue to be profitably or sustainably exploited. These conditions may refer to the physical realm (morphology, solar irradiation, accessibility of the spot), or to the human/social realm (i.e. economic viability, funding, family characteristics, patterns of land ownership, local development strategies), as explored throughout this volume. On the basis of agricultural use levels, the diagram shows the main impacts under mass or mild tourism conditions–also implying all in-between tourism categories.

Turning now to the bottom-left corner of the diagram (abandoned terraces), we found no bibliographical evidence of any case of terrace restoration from the level of abandonment, strictly due to mass tourism. Therefore, when nearby tourism flows in areas surrounding terraced landscapes, do not consider them as a possible destination, the consequence on the terraced landscape, from the point of view of the agricultural system, is a progressive rewilding, especially in the steepest and mountainous areas. However, moving on the bottom-right part of the diagram, special-interest tourism (such as eco-tourism) has shown to have some potential in the recovery of abandoned terraces, where hiking or climbing-related activities offer incentives to clear and rebuild destroyed drystone walls.

Where agriculture on terraces is in decline, tourism may affect it either in negative or positive ways. In some Greek islands, under strained agricultural conditions, 3Ss tourism may be highly stressful to terraced landscape dynamics (Petanidou et al. 2008; Kizos and Koulouri 2006). When mass 3Ss tourism is not interested in the local culture or nature, meanwhile offering more appealing job opportunities for younger generations, vineyards, olive groves and other cultivations are often abandoned (Terkenli 2001; Tsartas 1996).

Alternative or special-interest forms of tourism might, on the contrary, offer some opportunities to farmers, boosting local and regional economies. In these cases, local farmers are often encouraged to rediscover ancient agricultural practices, and, in combination with modern ones, to provide tourists with high-quality products. Mild, adaptive and sustainable forms of tourism are preferable in these areas, where rural populations are small, services and facilities poor, and accommodations often do not satisfy mass tourism standards and demands, preventing large-scale landscape changes which might compromise ecosystem stability and local cultures survival (Du Cros and McKercher 2015, p. 187; Chan et al. 2016; UNESCO 2008).

Finally, where terraced agricultural systems are highly productive and self-sustained (providing place-based quality products), tourism may impact on the agricultural system mainly in terms of commercial dynamics and with the inclusion of other uses of the terraces along the agricultural one, such as wine and gastronomy tours or cycle tourism. In the case of mass tourism flows, there is a risk that, in order to attract greater tourism demand, terrace cultivation declines due to the dominance of agricultural mechanization and other innovations, leading to the neglect of maintenance of terraced slopes and stonewalls; in these cases, what tends to be valued by the tourists are the landscape products, instead of the landscape itself. Other possible impacts might be excessive commodification and banalization of terraced agriculture systems' characteristics, products and services.

In Valtellina (Italy) for example, traditional vineyards used to be oriented perpendicularly to the terraces, while nowadays they tend to be positioned more in parallel to the terraces, to allow small machines to pass through the cultivated rows. This, of course, constitutes a transformation in the landscape, with a definitive loss, in terms of longstanding traditions and product quality (banalization), but it has allowed farmers to implement innovations and stem the decline of terraced agriculture. In addition, the presence of cycling and walking routes through vineyards allows locals to convey cultural knowledge and expose the value of their products to the tourists. In this case, it seems that agriculture and tourism can benefit from one another, operating as a system of assets which concur towards landscape sustainability. Nevertheless, Puleo (2013) warns about a possible 'parasitic' approach, which, on the one hand, through a process of selection and separation, involves the abandonment of terraces not productive or valuable enough, while, on the other hand, through processes of tourist exploitation, relegates the terraces to mere images, for external consumption. In this case, both tourists and tour operators grasp only the scenic dimension of the terraces, their aesthetic value and the impressive effect of the heroic work of generations—accompanied by its appropriate rhetoric—with no further or deeper impact on their knowledge about them or of actual territorial interrelations.

Each of the previous examples is paradigmatic of the main issues that are at stake when agriculture coexists with tourism on the terraces. In general, tourism often seems to conflict with agriculture on terraces, because, here, ecological, social and economic balances are extremely delicate and cannot sustain excessive crowds of visitors. When agriculture survives, it is either in the form of a residual practice, part of the exotic and nostalgic dimension of these landscapes, or it constitutes an innovative activity, as mechanized as possible, and strongly dependent on a market niche, which often uses terraces as a 'brand', but tends to ignore or destroy those less productive terraces, just around the corner.

18.5 Tourism and the Socio-Economic System

If the impacts of tourism on agriculture affect directly the terraced landscape, the effects of tourism on the broad socio-economic system act mostly indirectly on it, altering it significantly (Fig. 18.3).

As already mentioned, tourism primarily brings about an increase in incomes and jobs, instigating their shift from other economic sectors to tourism itself. These processes occur in different ways, at different destinations, depending on the robustness of the other economic sectors, the types and characteristics of tourism promoters (insiders or outsiders; local small entrepreneurs or large tour operators), the intensity of tourist flows and the type of tourism promoted and developed. Of

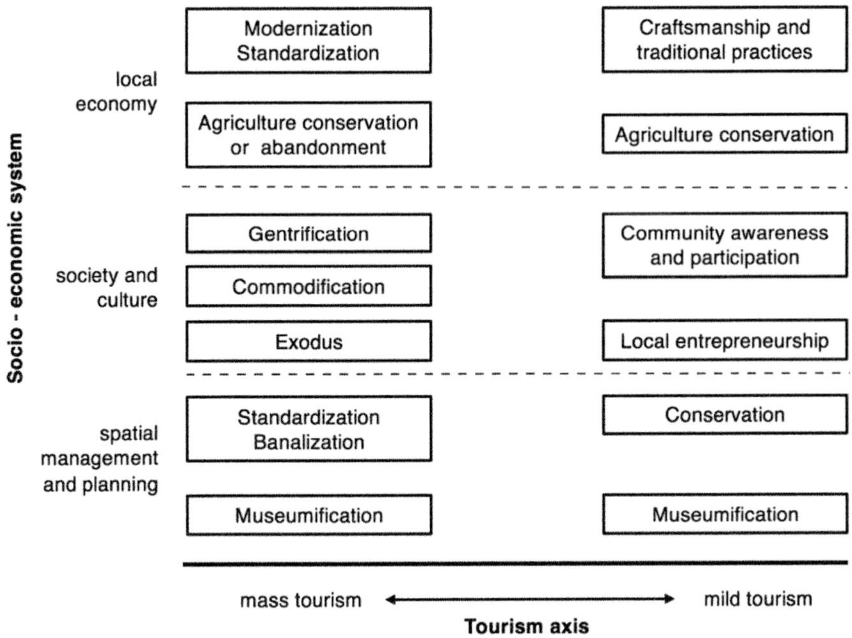

Fig. 18.3 Impacts of different types of tourism on the socio-economic system of terraced landscapes

course, when large tourist flows occur, a large shift towards employment in tourism can carry along a decrease of agricultural jobs, especially among the younger population cohorts, with the risk of consequent abandonment of agricultural practices. This is the case in the Philippine rice terraces of the Ifugao region, a UNESCO World Heritage Site, where young people easily and readily find employment in tourism, and traditional know-how of preservation and upkeep of the terraced landscape is no longer transmitted from elderly farmers to the next generations. Such trends may lead to the destruction of the terraced landscape that represents the very basis of the tourist attraction (Christ 2015). In other situations, where tourism only partially contributes to the family income, farming may maintain its importance and rural exodus may be curbed, with opposite—positive—effects on the conservation of these terraced landscapes, as already shown.

Furthermore, the direct involvement and participation of indigenous people in various tiers and posts of the local tourism industry, but especially as active investors, entrepreneurs and managers, keep tourism in local hands, which tends to recycle profits and benefits (and costs, also, of course) in the local societies, thus enhancing tourism development and/or prompting more general local and regional development. This direct involvement and control of local tourism by the local side contribute to the continuation of the traditional activities (agriculture, but also craftsmanship), which also serve and support tourism, interlinked, or even creating a multiplier effect in the local/regional economy. The same effect may be achieved by keeping tourism inflows under control and basing tourism development more on quality than on quantity, thus preventing the economy from lapsing into tourism monoculture.

Further commonplace indirect impacts of tourism on terraced landscapes and communities are spatial reorganization (settlements, urban layout, gentrification, rural structures and all sorts of tourism or other infrastructures), in order to cater to the needs of the tourism industry. In addition to the direct material modification and landscape reorganization, most commonly encountered impacts relate to the development of new real estate markets, which tend to push away parts of the community, due to rising land values, opening up the ground for high-end external investments. 'Traditional' settlements tend to be reorganized (Terkenli 2001; Tsartas 1996), in order to welcome visitors, losing their 'authenticity'. Marginality, high vulnerability and other particularities of the built elements in terraced landscapes exacerbate these risks. Modern services required by tourists as well as locals, whose standards of living are raised, thanks to tourism development, may contribute to processes of banalization and commodification (UNESCO 2008). Such risk is attenuated when normative systems are in place, protecting 'typical' landscape features, and/or the local communities are aware of and sensitized about their own cultural heritage and keen on preserving it. As a matter of fact, very often tourism development does bring about such expanded acknowledgement of local/regional natural and cultural resources and landscapes by the locals.

In other cases, when 'instrumental rationality' towards the terraced landscape prevails over 'value rationality' (see Sect. 18.3), and tourism is primarily planned and managed in order to meet tourist preferences (Arnberger and Eder 2011), local cultures run the risk of reconstruction according to broadly recognizable stereotypes. Such reconstruction often 'presents tensions between cultural exoticism (tourists' desire for authenticity by freezing the culture in past representations), cultural commodification (selective modification of culture in accordance with tourists' taste) and cultural preservation versus modernity (Indigenous people's desire to achieve modernity)' (Chan et al. 2016, p. 7). Landscapes, thus, risk being hyper-preserved, through processes of commodification and/or museumification.

18.6 Concluding Remarks

Our preceding analysis aimed at highlighting that both risks and opportunities incurred by tourism impacts on visited landscapes are especially pressing in the case of terraced landscapes, running the full range from most negative (i.e. destruction) to most positive (i.e. rejuvenation) possible consequences. Even though the diagrams used in this chapter present mass and mild tourism as two opposite extremes, often with negative and positive connotations, there are actually many degrees and forms of tourism development, in-between, and at various geographically connected scales.

Unregulated heavy use and intense concentration of tourists and tourism infrastructures on the terraces may pose irreversible harm to the visited landscapes, in lieu of boosting local development. On the contrary, the involvement of the local community in relevant decision-making is essential, local tourism factors and entrepreneurs ought to be accountable to their communities, while also professionally equipped, trained and skilled in sustainable tourism planning and management. Moreover, the recycling of tourism profits and benefits should concern not only the tourism sector itself but also the entire community, including farmers.

The existence, value and sustainability of terraced landscape tourism are highly relevant to the scale and intensity of tourism types and activities in time-space, pointing once again to a higher suitability and compatibility of milder tourism uses with terraced landscapes and their resources, amenities and other ecosystem services, as provided both to visitors and to the local/regional societies. Thus, the type of tourism that would be better suited here is reliant on local or regional natural and cultural resources. Moreover, extensive rather than intensive terraced landscape tourism uses—depending also on the type of terrace functions and structures—are more efficient, rewarding and/or profitable for the local/regional societies and economies.

In order to seize tourism as an opportunity and address the above-mentioned risks, a multiscalar approach is required, through strong connections between local community initiatives and sustainable tourism development policies and strategies, at various levels, together with a high awareness of the local communities as regards their natural and cultural heritage, as represented by their terraced landscapes.

References

Aitchison C, MacLeod NE, Shaw SJ (2000) Leisure and tourism landscapes: social and cultural geographies. Routledge, London

Arnberger A, Eder R (2011) Exploring the heterogeneity of rural landscape preferences: an image-based latent class approach. Landsc Res 36(1):19–40. https://doi.org/10.1080/01426397.2010.536204

Bell S (2008) Design for outdoor recreation. Taylor & Francis, London

Carmichael B (1998) Destinations: cultural landscapes of tourism. Routledge, London

Chan JH, Iankova K, Zhang Y, McDonald T, Qi X (2016) The role of self-gentrification in sustainable tourism: indigenous entrepreneurship at Honghe Hani rice terraces world heritage site, China. J Sustain Tour 24(8–9):1262–1279. https://doi.org/10.1080/09669582.2016.1189923

Christ A (2015) Close to heaven—the rice terraces of the Philippines. Christ Media for NDR/ARTE

Coccossis H (2004) Sustainable development, landscape conservation and tourism in the small islands of Greece: the nature conservation-society interface. In: Dieterich M, Van Der Straaten J (eds) Cultural landscapes and land use. Kluwer Academic Publishers, Dordrecht, pp 111–124

Cohen E (1979) A phenomenology of tourist experiences. Sociology 13:179–201

Council of Europe (2000) European landscape convention

Crouch D (ed) (1999) Leisure/tourism geographies: practices and geographical knowledge. Routledge, London

Du Cros H, McKercher B (2015) Cultural tourism, vol 2. Routledge, London

Hall CM, Page SJ (2014) The geography of tourism and recreation: environment, place and space. Routledge, London

Hall CM, Williams AM, Lew AA (2014) Tourism: conceptualizations, disciplinarity, institutions, and issues. In: Lew AA, Hall CM, Williams AM (eds) The Wiley-Blackwell companion to tourism, 1st edn. John Wiley & Sons Ltd., Oxford, pp 3–24

Jansen-Verbeke M, McKercher B (2013) Reflections on the myth of tourism preserving 'traditional' agricultural landscapes. J Resour Ecol 4(3):242–249. https://doi.org/10.5814/j.issn.1674-764x.2013.03.007

Kizos T, Koulouri M (2006) Agricultural landscape dynamics in the Mediterranean: Lesvos (Greece) case study using evidence from the last three centuries. Environ Sci Policy 9(4):330–342. https://doi.org/10.1016/j.envsci.2006.02.002

Löfgren O (1999) On holiday: a history of vacationing. University of California Press, Berkeley

Lorimer H (2005) Cultural geography: the busyness of being more-than-representational. Prog Hum Geogr 29:83–94. https://doi.org/10.1191/0309132505ph531pr

MacCannell D (1992) Empty meeting grounds: the tourist papers. Routledge, London

Pearce D (1995) Tourism today: a geographical analysis, 2nd edn. Longman, Harlow

Petanidou T, Kizos T, Soulakellis N (2008) Socioeconomic dimensions of changes in the agricultural landscape of the Mediterranean basin: a case study of the abandonment of cultivation terraces on Nisyros Island, Greece. Environ Manage 41(2):250–266. https://doi.org/10.1007/s00267-007-9054-6

Puleo TJ (2013) Parasitizing landscape for UNESCO world heritage. Geoforum 45:337–345. https://doi.org/10.1016/j.geoforum.2012.11.024

Rojek C, Urry J (eds) (1997) Touring cultures: transformations of travel and theory. Routledge, London

Rose G (1996) Geography and the science of observation: the landscape, the gaze and masculinity. In: Agnew J, Livingstone DN, Rogers A (eds) Human geography: an essential anthology. Blackwell, Oxford, pp 341–350

Terkenli TS (2001) Local perceptions of tourism impacts on place identity: the case of northern Crete. Tourizam 49(3):229–240

Terkenli TS (2014) Landscapes of tourism. In: Lew AA, Hall CM, Williams AM (eds) The Wiley-Blackwell companion to tourism, 1st edn. John Wiley & Sons Ltd., Oxford, pp 282–293

Tsartas P (1996) Tourists-trips-places: sociological approaches to tourism (in Greek). Exantas Publications, Athens

UNESCO (2008) Sustainable tourism and the preservation of the world heritage site of the Ifugao rice terraces, Philippines. http://unesdoc.unesco.org/images/0018/001826/182647e.pdf. Accessed 2 Apr 2017

UNWTO (2011) Tourism trends 2030: global overview. http://www.wise.co.th/wise/Knowledge_Bank/References/Tourism/UNWTO_Tourism_Toward_2030.pdf. Accessed 2 Apr 2017

UNWTO (2013–2017) Tourism highlights. World Tourism Organization

Urry J (1990) The tourist gaze: leisure and travel in contemporary societies. Sage, London

Urry J (1995) Consuming places. Routledge, London

Xiaoyun T, Xiaohui Y, Dongming Z (2013) A study of agri-cultural heritage tourism impacts based on residents' perception: taking the Longji terrace site in Guilin as an example. J Resour Ecol 4(3):267–274. https://doi.org/10.5814/j.issn.1674-764x.2013.03.010

Zahn G, Jin Z (2015) Hani rice terraces of Honghe—the harmonious landscape of nature and humans. Landsc Res 40(6):655–667. https://doi.org/10.1080/01426397.2015.1060299

Chapter 19
Innovative Practices and Strategic Planning on Terraced Landscapes with a View to Building New Alpine Communities

Federica Corrado and Erwin Durbiano

Abstract Most terraced landscapes are found in so-called fragile areas that today are reconstructing their territorial identity and redefining new forms of territoriality. The transformation that territories are undergoing in general stems, first and foremost, from a new approach to their own resources with a view to building innovative paths to development. Starting from the importance of territorial practices in these fragile areas, as terraced landscapes, the contribution deals with the potentiality of these resources in the building of a specific place awareness. Exercising an awareness of place means supporting processes and policies designed to strengthen the interpretational capacity of a territory and its development. Place awareness, built in this way on short local networks, comes into contact with long networks and hybridises and adjusts in line with modern life and its changes. The case study proposed supports these reflections. It regards the experience of Alto Canavese territory in Piedmont Region (Italy). It is an interesting case study because the terraces, along with other features of the territory, were the basis upon which a heritage framework was reconstructed which could provide a new horizon of meaning for the territory and the sharing of a common heritage that could bring people together.

19.1 Focusing on Practices

Most terraced landscapes are found in so-called fragile areas that today are reconstructing their territorial identity and redefining new forms of territoriality. The transformation that territories are undergoing in general stems, first and fore-

The Sects. 19.1, 19.2 and 19.3.3 are written by Federica Corrado; the Sects. 19.3.1 and 19.3.2 are written by Erwin Durbiano.

F. Corrado (✉) · E. Durbiano
Polytechic of Turin, Turin, Italy
e-mail: federica.corrado@polito.it

M. Varotto et al. (eds.), *World Terraced Landscapes: History, Environment, Quality of Life*, Environmental History 9,
https://doi.org/10.1007/978-3-319-96815-5_19

most, from a new approach to their own resources with a view to building innovative paths to development. Terraces are therefore playing a leading role in this renaissance of rural territories that is being shaped, above all, by current practices (Bevan and Conolly 2013; Varotto and Bonardi 2016).

Once recognised and valued as a resource, we can understand how terraces can support and foster these regenerating processes by focusing on practices. A narrative that explains such practices highlights the visible and invisible relationships that are bound together and characterise the modern landscape, and relationships that give voice to the players that "make territories" (Decandia 2011).

Practices currently represent that action of "making" that tries out new or innovative solutions in order to overcome the crisis that has hit territories and traditional development models. In other words, know-how, local wisdom, historic heritage and, generally speaking, the tangible and intangible potential of places are being reviewed through initiatives, projects and economic, cultural, social and environmental actions. The aim is to encourage the emergence of new ways of appropriating space and new rules for using territories (Decandia 2011), the narrative of a system of practices that "produce identity and meaning thanks to an interactive process of learning" (Pasqui 2008: 59) in order to identify different development models as an alternative to traditional ones. Such models involve taking care of the territory, an activity that "is based on a creative model for growth that focuses on a green economy that features economic, social and territorial collaboration in order to achieve a new short-to-medium-term paradigm of development designed to reduce the consumption of natural resources and control environmental risks, promoting greater energy efficiency, a drastic cut in local pollution, social inclusion and liveability" (Bobbio and Brunetta 2014: 207).

If we look at the recent 2016 "Choosing the Future" Third World Meeting on Terraced Landscapes, it is obvious that the fields of application where local development practices are being implemented in these territories focus on three themes:

- new forms of agriculture as seen in "back to the earth" projects and/or new farm management methods that attempt to try innovative forms of primary sector development;
- new forms of tourism that experiment with alternative development models based on an exchange between visitors and residents that promotes local history and identity through an experiential stay that also involves visitors in the daily work that takes place on terraced land;
- new cultures that combine tradition and innovation, where terraces become places for cultural production. Slow landscapes that, in contrast to the fast pace of urban life, inspire artistic innovation.

Rural territories are providing experimental solutions to these issues, with all the limits that an experiment can have and the difficulties that it can involve. Nevertheless, these practices attribute value to the cultural roots of places, introducing innovation. Such experimentation draws on expert knowledge that sometimes hails from outside and brings with it new ideas, creating new solutions and opportunities thanks to a process of cultural hybridisation with local knowledge. In this particular case, this involves the restoration of many terraced areas that become part of real economic programmes and development projects, even taking their cue from individual life stories (Lodatti 2012).

We therefore find ourselves confronted by a level of experimentation with practices that debunks two paradigms. The first concerns the concept that innovation and experimentation cannot occur in scarcely populated areas characterised by a certain level of social dispersion. As Remotti states (2011), the idea that innovation is something that only belongs to the urban world where there is a greater concentration of networks, opportunities, etc. is called into question, making room for a form of innovation that finds its chance to emerge in just such areas of social dispersion and in the preservation of a natural environment from exploitation. The second paradigm concerns the fact that old and new residents of these fragile rural territories (particularly in mountainous regions) become "problem solvers" (Euromontana 2004), i.e. acting as players who possess the ability to find new solutions when faced with present difficulties.

19.2 Terraces: A Resource for Building Community Awareness in the Alps

In 1972, the United Nations Educational, Scientific and Cultural Organisation produced the Convention concerning the Protection of the World Cultural and Natural Heritage (the World Heritage Convention) where knowledge and tangible and intangible elements of culture all over the world are recognised as universal heritage in order to safeguard their social, cultural, symbolic and economic values. This convention has been reviewed over the years and expanded, first in 2003 and then in 2005, with two other conventions: the Convention for the Safeguarding of the Intangible Cultural Heritage and the Convention on the Protection and Promotion of the Diversity of Cultural Expressions.

The 2003 convention, in particular, includes intangible heritage—such as traditional knowledge regarding things, places, the environment and nature—as elements that should be protected. Indeed, intangible cultural heritage is understood to mean "the practices, representations, expressions, knowledge, skills—as well as the instruments, objects, artefacts and cultural spaces associated therewith—that communities, groups and, in some cases, individuals recognise as part of their cultural heritage". The convention states that intangible heritage is transmitted from generation to generation, is constantly recreated by communities and groups in response to their environment, their interaction with nature and their history, and provides them with a sense of identity and continuity, thus promoting respect for cultural diversity and human creativity.

In 2006, the Alpine Convention drafted a document entitled the Declaration on Population and Culture, signed by all member states within the Alpine region, where it is forcefully stated that the safeguarding and development of the Alpine territory requires the "strengthening [of] community awareness and the identity of the people who live in it".

Now we should not forget that around 16% of terraced areas are in mountainous regions and a high proportion of those are found in the Alps, where the construction of terraces "represents a system of improvements that stem from traditional expertise to do with construction and cultivation techniques and with a perfect understanding of hydrogeological characteristics and climate that can make the most of environmental resources" (Agnoletti 2010). Thus, terraces are entirely involved in that relationship between know-how and local communities that builds an awareness of places: "we need a proactive, informed population that can combine local knowledge with expert knowledge through forms of participatory democracy if we want to care for and reconstruct places in order to attribute value to heritage in a long-term, sustainable manner" (Magnaghi 2006: 4).

As regards this aspect, territorial resources (meaning terraces) should be recognised and reinterpreted by communities using what Bonomi (2009) defines as "place awareness" in contemporary terms. Exercising an awareness of place means supporting processes and policies designed to strengthen the interpretational capacity of a territory and its development. Place awareness, built in this way on short local networks, comes into contact with long networks and hybridises and adjusts in line with modern life and its changes.

If we want to give these territories strategic strength, we must entirely overturn our approach and opt for a debate, a plan, a promotion of active citizenship until we achieve real cooperation at a supra-local level. In this way, we can come to adopt new rationales of knowledge production that are based on the narrative experiences of local communities where practices, innovation and experimentation emerge in an interactive, multidimensional way across the board, allowing us to go on to build a new territorial image, one which remains incomplete. Such an image contributes to building local territorial strategies using agreements that bring together the various different players within strategic local and supra-local frameworks within a circular and trans-scalar process.

19.3 The Alto Canavese Case Study (Piedmont, Italy)

19.3.1 A Territory that Looks to a Future Based on Terraced Landscapes

It is easy to understand the complex nature of the meanings and value of terraced landscapes if we consider the definition found in the Honghe Declaration (2010) where terraces are described as "agricultural, ecological systems" that should be protected due to their biological and cultural diversity. What we are dealing with here are specific features with a strong identity that can have a significant effect on the quality of life of the local community and on sustainable development, agricultural food products and the sense of belonging to a particular place.

To this end, we must keep terraces alive, not just as a piece of history but above all as pieces of territory that make a significant contribution to a territory's stability and production. Terraced landscapes are often places where farm products are grown, sometimes in a heroic fashion; such produce contributes to the creation of economic systems and helps establish specific identities. Many different people work in different ways on terraced landscapes in order to maintain and restore drystone walls, often through daily efforts, and they play an important role in constructing the local community's identity and sense of belonging.

The case study in question concentrates on these aspects, restoring an approach/process towards the reconstruction of identity in a territory located in the north-west Italian Alps. It is an interesting case because the terraces, along with other features of the territory, were the basis upon which a heritage framework was reconstructed which could provide a new horizon of meaning for the territory and the sharing of a common heritage that could bring people together.

To be more precise, the case study concerned the northern end of Piedmont on the border with the Aosta Valley region, which includes the towns of Carema, Settimo Vittone, Nomaglio, Borgofranco d'Ivrea, Lessolo, Andrate, Chiaverano and Montalto Dora. It acts as a link between a number of different elements: firstly, between the Alps and the morainic hills of Serra. This territory extends along an elevation gain of over 2000 m, from the valley floor to the 2371 m of Colma di Mombarone. It is a transit corridor that extends along the Dora Baltea River and features the constant presence of terraces, an element that is immediately noticeable and highly recognisable. The favourable microclimate has encouraged the formation of a particular agricultural landscape featuring stone-walled terraces covered in vineyards. This landscape is characterised by traditional *tupiun* and *pilun*, features of the rural architecture of the past that when viewed as a whole produce a system that is clearly identifiable and boasts a strong identity.

A good number of local subjects already see that these terraces can play a new economic, social and environmental role as one of the territory's most significant resources. Such experience was gained with the use of various different methods and practices due to the specific circumstances of each area. This territory's development began a number of years ago, in 2013, when—faced by town planning choices made by Regione Piemonte, Piedmont's regional government, to locate waste sites in these marginal areas on the edge of Turin's metropolitan area—a group of mayors were inspired to rediscover the strengths of their territory, publicise its beauty and history and attempt to construct a new future from local resources—particularly its terraces—and a new Alpine community, one that, in the past, has been excessively damaged by top-down decisions made in Turin (Fig. 19.1).

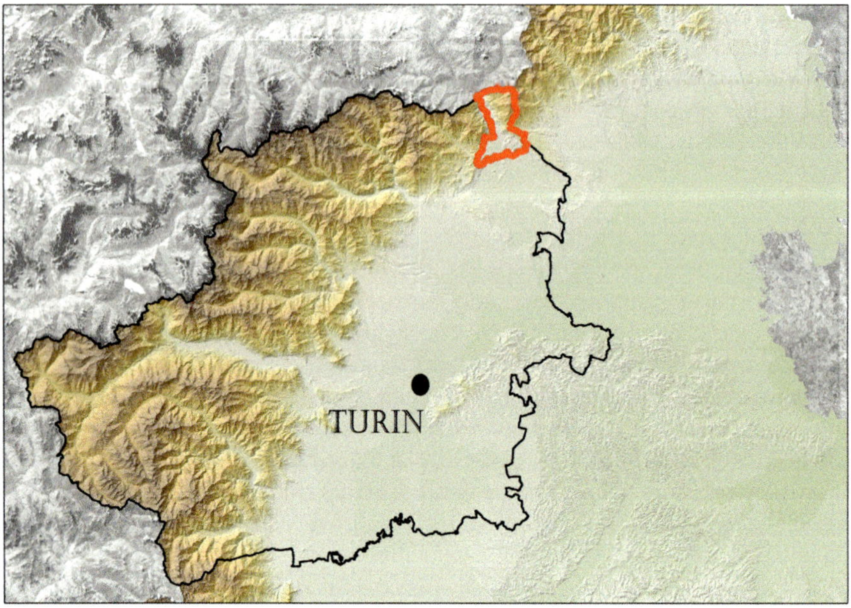

Fig. 19.1 Strategic plan area of Alto Canavese inside Turin metropolitan city

19.3.2 The Community's First Steps: The Practices that Enhance Terraces

Three practices have had a pivotal and decisive role in the reconstruction of this territory's identity and have provided the basis for the local community's initial approach to the terraced landscape.

The first concerns the vineyard trail in the borough of Carema. The Carema project aims to promote the territory using a trail that allows visitors to explore the area's historical and architectural features and its wine-making heritage. Visitors are given the chance to grasp the value of the topiary architecture that has left such an indelible sign on Carema's landscape, and visit the land that produces Carema wine. The route consists of a circuit 4 km long in total with an elevation gain of 100 m that is accessible all year round and takes in all the town's most important features, including—above all—the terraces closest to the town. This practice involves a series of local players who, in different ways, use the terraces (farmers, sports enthusiasts, hikers, grape growers) as well as prominent businesses, particularly the Nebbiolo wine producers of Carema. The local authority has, for some time now, begun to promote this trail through many initiatives that take place throughout the year, and through its support for the labelled recognition of the quality of its wines, particularly D.O.C. and Slow Food certifications.

The second practice concerns the "heroic" farming in Settimo Vittone. The borough council with support from the former provincial government of Turin

(Department of Agriculture, Mountains, Protection of Fauna and Flora, Parks and Protected Areas) has promoted the "Plant an Olive Tree" project designed to increase the number of plants found in the borough and which has resulted in a twofold result: an increase in the amount of cultivated land and the reversal of the trend that saw the abandonment of such land. The project involved assigning olive saplings to local land owners who committed themselves to till abandoned, fallow land that is suitable for growing olive trees and start cultivating olives, a practice that has a long tradition in the area. The requirements that had to be met in order to be awarded the plants included that the land had to be facing south-east to south-west, land that mostly featured terraces. The result was the salvaging of a large portion of territory, the revival of a traditional crop that can generate new economic growth right from the start and a new/renewed interest in the local landscape among residents. This interest has begun to foster a desire to share knowledge and promote common heritage. The third practice concerns Nomaglio's chestnut ecomuseum. In 1996, Nomaglio's borough council launched a project designed to recuperate and promote vestiges of material culture found in the area. To this end, it set up an ecomuseum devoted to the farming of sweet chestnuts, which for centuries was a fundamental crop that provided all kinds of benefits. The town, the surrounding woods, chestnut groves, mule tracks, chestnut drying sheds and the mill are all pieces of a mirror that reflects Nomaglio's community, where people recognise their reflected image and search for the values that underpin it. Nomaglio's chestnut ecomuseum is therefore, unlike others, an open-air ecomuseum whose main theme is the territory; it is an itinerary for the exploration of the sweet chestnut landscape and the tools needed to produce chestnuts as marketable products (Fig. 19.2).

Fig. 19.2 Community activities on terracing of Nomaglio *Photo* Municipality of Settimo Vittone

19.3.3 *The Role of Two Projects*

19.3.3.1 The Strategic Plan

These practices, which are already underway in the territory, have encouraged a reinterpretation of this resource as part of a Strategic Plan. This plan, which was very much desired by the local community, was completed with the help of a scientific team that supported the entire process of analysing and defining the plan itself.[1] The territorial profile produced by the Strategic Plan was based on the definition and fine-tuning of *strategic thematic quadrants* that group the various different resources found during the reconnaissance phase and have them interact in a thorough, integrated way. The definition of these quadrants has permitted the setting up of new strategic perspectives as well as the organisation of resources, perspectives that are the necessary basis for constructing scenarios that can view the various entities involved in the plan as a part of a single identity. The strategic quadrants, which are far-reaching and thematically inclusive, have been divided into the following categories: (a) the landscape and the quality of the environment; (b) the system of linear infrastructures; (c) historic, artistic and architectural heritage; (d) leisure time and outdoor activities; (e) tourist facilities and identity-forming entities. A survey was carried out on each thematic quadrant so as to define the strategic sphere to which each feature of the territory belongs—in order to evaluate the relationship between individual elements and the level to which they belong to existing networks or circuits—and to identify its inclusion in and importance for local policies and mechanisms. The relationships between different resources were examined, taking into account spatial relationships and the territory's morphological and physical conditions; in many cases, territorial resources were placed in more than one quadrant. Once analysed in terms of their strengths and weaknesses, the strategic thematic quadrants permitted the outlining of strategies and contributed to the concise construction of the plan's territorial profile. The process of defining and attributing resources to the various different quadrants involved, first and foremost, the active participation of local authorities as players responsible for governing the many mechanisms at work in the territory.

Following a number of interviews with local groups and perception studies, the importance of terraces emerged, recognised not only for their role in constructing the local landscape but also as an identity-forming feature that qualifies the territory and its economies. Terraces were recognised as a feature that crosses boundaries between spheres and belongs to a number of different strategic thematic quadrants. Their environmental and landscape value was particularly recognised, as was their role as part of historic and artistic heritage and the potential they offer local tourist

---

[1]Scientific support was entrusted to the Italian branch of CIPRA (the International Commission for the Protection of the Alps). The working group's members are Federica Corrado (Politecnico di Torino and CIPRA Italia), Antonio De Rossi (Politecnico di Torino) and Erwin Durbiano (CIPRA Italia).

Fig. 19.3 A typical terraced landscape in Alto Canavese (Piedmont Region) *Photo* Municipality of Settimo Vittone

facilities and identity-forming entities. The actions that were identified on the basis of meetings with local authorities and focus groups held with territorial stakeholders assign terraces two different roles. The first has to do with the promotion of agriculture and high quality crafts as part of a perspective that concentrates on employment and the salvaging of the territory and is applied through actions designed to promote the free leasing of agricultural land, research into the most suitable crops for the area and training courses set up to create new jobs maintaining the territory. The second method for improving terraces involves specific actions and is linked to the feasibility of enhancement work as regards hydrogeological and landscape problems and the application of specific regulations, particularly with a view to encouraging improvements to private land and the protection of the terrain and the landscape through town planning guidelines that are mutually agreed and focus on reuse (Fig. 19.3).

19.3.3.2 Trans-Regionally Initiatives to Define New Actions at a Supra-Local Level

The Strategic Plan and the Third World Meeting on Terraced Landscapes proved to be the chance to look beyond the territory of individual projects, taking in the entire environmental context of terraced landscapes, which go far beyond regional borders and extend as far as the entrance to the Aosta Valley up until the sluices of Bard.

The initiatives launched by a number of local authorities located along the borders of these two regions highlight different approaches to salvaging terraces: the Aosta Valley side has demonstrated a more organised and matured approach to managing local projects and policies, while the Piedmont side is particularly interested in launching initiatives of strategic importance for the future of terraced landscapes.

The Conference has proved to be a particularly beneficial stimulus, inspiring the local community to tackle the challenge by building a route that stretches from Carema to the slopes of Mont Blanc, with the aim of increasing awareness of the terraces and highlighting the activities that are carried out by local people on stone-built landscapes. The involvement of local groups has become essential for the success of this initiative and has become an opportunity to strengthen the network of people that work on stone landscapes. The initiative has involved the following resources and players:

- The terraced landscape and its peculiarities, such as the *tupiun* and *pilun* and relative techniques;
- Museums, particularly ecomuseums, that can explore the theme of terraces and the activities that take place there;
- Places where agricultural produce is processed, particularly wine cellars;
- Aspects to do with popular traditions, particularly as regards farming techniques;
- Historical and architectural landmarks to be considered as both assets worth discovering and places where new activities can take place;
- Traditional local products and supply chains;
- Outdoor activities carried out both spontaneously and in an organised fashion in terraced areas;
- The initiatives that take place during the year in terraced areas and locations.

The success of this initiative, combined with the interactive dialogue-based process of the Strategic Plan, has thus strengthened the idea among members of the local community that terraces can be an enormous resource (active participation process, Virgilio and Imbesi 2007), to the point where there is talk of creating a festival based on this theme and on terraced locations, in the hope of networking the resources and initiatives that today have a beneficial effect on terraced areas, and at the same time strengthening the profile of a landscape that is alive and maintained, capable of manifesting new forms of management that are nevertheless based on tradition. This awareness also becomes the strength needed to make an impact on supra-local visions, starting with the Metropolitan City of Turin's Strategic Plan, which is currently being drafted, right up to experimental trials included in the PPR Regional Landscape Plan.

References

Agnoletti M (ed) (2010) Paesaggi rurali storici. Laterza

Bevan A, Conolly J (2013) Mediterranean Islands, fragile communities and persistent landscape. Cambridge Editor

Bobbio R. Brunetta G (2014) La cura del territorio come forma di sviluppo. Il tema e la discussione. In: Russo M (ed) Urbanistica per una diversa crescita. Progettare il territorio contemporaneo. Donzelli Editore

Bonardi L and Varotto M (2016) Paesaggi terrazzati d'Italia. Eredità storiche e nuove prospettive, Franco Angeli

Bonomi A (2009) La piattaforma alpina nell'ipermodernità. In: La sfida dei territori nella green economy. Borghi E. dir. Il Mulino

Decandia L (2011) L'apprendimento come esperienza estetica. Una comunità di pratiche in azione, Franco Angeli

Euromontana (2004) Dichiarazione di Lillehammer

Lodatti L (2012) Ritorno alle "terre alte": l'adozione di terrazzamenti in abbandono nel canale di Brenta. Agriregioneuropa 33:36

Magnaghi A (2006) Dalla partecipazione all'autogoverno della comunità locale: verso il federalismo municipale solidale. Democrazia e diritto, Franco Angeli

Pasqui G (2008) Città, popolazioni, politiche. Jaca Book

Remotti F (2011) Cultura: dalla complessità all'impoverimento. Editori Laterza

Virgilio D, Imbesi A (2007) Parco nazionale delle Cinque Terre. In: Piani e politiche territoriali in aree di parco, Vinci I. dir. Franco Angeli

Chapter 20
Planning, Policies and Governance for Terraced Landscape: A General View

Enrico Fontanari and Domenico Patassini

Abstract Guidelines for planning, policies and governance are suggested by class of terraced landscapes taking into consideration the main features of geographic domains and environmental contexts. Dealing with irreversibility, reversibility and development processes the guidelines can help the communities to adopt integrated strategies based on an effective institutional design. Input and basic information are provided in the 3rd International Congress on Terraced Landscapes (Italy, 6th–15th October 2016) by the working groups "Rules and policies" held in Trento/Rovereto and "Agronomic and Social Innovation" held in Valstagna, Canale di Brenta (Vicenza).

20.1 Three Landscapes

The terraced landscape (TL) is a heritage of humanity, which needs no awards. It has been around for thousands of years to witness how humans, aware of geographical and climatic conditions, have built basic infrastructure to develop agriculture and the foundations of their settlements.

In considering their life cycle, aside from the constructive, production and management features, recurring censuses or national surveys clearly point to three classes of TL. The first class involving irreversible degradation (TL1), the second with evidence of reversibility (TL2), while the third showing different types of development (TL3).

Irreversible degradation affects a relative small portion of the terraced land where maintenance conditions are very bad, and reconstruction is either unlikely or impossible. Here, the degraded infrastructure is usually accompanied by a

E. Fontanari (✉) · D. Patassini
IUAV University Venice, Venice, Italy
e-mail: enrico.fontanari@iuav.it

D. Patassini
e-mail: domenico.patassini@iuav.it

M. Varotto et al. (eds.), *World Terraced Landscapes: History, Environment, Quality of Life*, Environmental History 9,
https://doi.org/10.1007/978-3-319-96815-5_20

socio-economic or cultural crisis triggered by natural or man-made disasters, or by a final collapse of the local communities (Diamond 2005).

The TL belonging to the second class is often located in a less dramatic contexts: where degradation may be less severe, within contextual conditions that may be either a bit better or worse. The best maintenance conditions can be coupled with a bad socio-economic situation: the TL can in such case be considered as a latent resource, as a source or factor of development. In favourable conditions, the TL can be a source that boosts the local economy: a development factor that can benefit from efficient production combinations.

Attempts to reverse or reuse the existing infrastructural stock can be made for different purposes. In some cases, old mixed cultivation or production practices (with related technologies) can be re-discovered; in others, they may be updated in terms of the mix and the technologies. Reversibility processes should confer a plurality of meanings and values to TLs that in brief can be said to lead to ecological benefits, circular economies (Martins 2016) and the inclusion of TL within spatial frames (ecological, infrastructural or cultural).

The third class is apparently the least problematic, being structurally sound and fully or partially used for productive purposes. In this case, TLs host specialized productions (mostly monocultures) with high yield/surface unit, which continue old traditions or start new ones (replanting, etc.). Yet, due to the monoculture approach, TL3 can generate a negative impact, increasing health and environmental risks as well as seriously compromising the local biodiversity (Weitzman 2000).

From an economic (or better, financial) perspective, TL2 suffer from a relative marginality, being partially or totally "off market". On the contrary, TL3s are economically performative and, based on their production, they supply specific market segments. If marginality usually feeds informality, as well as practices of shared and cooperative economy, the TL3 cultural specialization, to develop, requires advanced forms of management and marketing.

TL2s are influenced by a dual thrust: by those who try to bring them back to the market rules as the only foreseeable chance, while others who prefer to keep them out, recognizing their contribution in terms of environmental performance, circular economy and institutional building: performances that the market finds hard to acknowledge with its price system. The two approaches may trigger conflicts, and the conflicts vary according to the contexts. If within TL3 efficiency is measured on specific crop yields, in TL2 yield is only one of the components. Additionally, the overall efficiency of TL3 should be adjusted (or corrected) by internalizing the above-mentioned impacts at current prices.

Planning, policies and governance issues therefore vary greatly depending on the irreversibility, reversibility or development conditions of the three landscapes.

Physical and strategic planning contributes to designing a spatial structure that inserts TL as a physical component. It also defines the rules, regulations and procedures of land management. From an environmental perspective, and with the help of policies, planning also helps to achieve ecological balance sheets (Mang et al. 2016).

The policies relate to knowledge, training, innovation, financial and fiscal aspects, and to the practices for a circular economy. They may also interact with planning on cross-cutting themes such as land management, accessibility, labour market, welfare and so on.

Governance should help improve the relations between formal and informal activities, community and administrative bodies, but it may help to design an institutional framework capable of connecting planning and policies for each of the three TL classes.

The three types of landscape are rarely recognized as such through ad hoc surveys, local research on agricultural landscapes and by the time series of land cover maps.

20.2 Crisis and Opportunity

We live in a time of crisis (at least in Europe) in which slow growth affects investments in infrastructure and buildings, household consumption, public and private saving. The crisis pushes people to look at the built heritage, its tangible and intangible components, and to acknowledge and extract values from existing assets (see, for evidence, the Annual Reports and Newsletters of the International Centre for the Study of the Preservation and Restauration of Cultural Property, ICCROM).

The different TLs are a great cultural heritage of notable economic value: TL is memory, a "love bond" among generations, often of unique aesthetic value, but also fixed capital with distinct social and physical features. Legislative and operational tools are therefore needed to be able to fully appreciate the values of such heritage, which also encourage revitalization practices that allow existing values to emerge. The reuse and revitalization are particularly urgent in the areas where an economic crisis intersects environmental degradation (correlated to climate changes) and social phenomena such as the new migratory patterns due to geopolitical dynamics.

In order to fully appreciate the role that terraces play as infrastructures for the maintenance of a living territory, TL heritage should be regarded as a real eco-systemic (capital) asset and a component of the natural capital (Costanza et al. 1997).

In designing and operating a possible revitalization scenario, some strategic actions deserve special attention. Firstly, existing regulatory devices need to be simplified to stimulate planning, territorial and economic policies geared at putting the marginal TL at stake. The message here is not "no rules", nor is it "by the rules" (as if every action were an obligation), but "only with rules" that can power a social discourse on TL. A second action that may, to some extent, be linked to the first one, would reclaim (wherever possible) TL1s from a state of neglect and abandonment in which they might be forced forever. The third concerns, in particular, TL2 and TL3: supporting attempts of reversibility in TL2 and qualifying TL3 specializations from a sustainable ecological perspective.

But can legal moves, regulations, and policies help this process?

The work undertaken for the Third World Meeting on Terraced Landscapes, in particular, on innovation, rules and policies in the Brenta Canal (Carpané-

Valstagna) and Rovereto/Trento has produced many ideas. The first problem to solve, at least in Italy (but the problem exists even in other countries), is the fragmentation of administrative powers between the three Ministries[1] in dealing with TL from different perspectives: cultural, environmental and productive. Yet, this issue raises another, perhaps more important point: how to recognize or create renewed interest for the local management, where four major issues (institutional, environmental, economic and aesthetic) intertwine naturally and on a daily basis, but where conflicts between contemporary and customary legal systems also emerge, although in ever-weaker ways. Management of this issue makes up the DNA of the local communities, who live "in" and "off" terraced landscapes, who suffer degradation, but above all who know how to reap the benefits (if any). It is therefore crucial to create an effective "polity" environment and, to do that, it is perhaps necessary to step back. The key issue here is not to recognize the domain of public policies (*policy*), neither the power games (*politics*): but rather, the main problem may be the occupation of a "political space" (*polity*). That may be done by TL communities, with significant cultural side effects. Whenever TL communities make polity, the benefits of advocacy and deliberative planning is likely to become more obvious.

20.3 New Practices, Rules or De-regulation?

In observing the lessons from individual practices of re-appropriation of abandoned places or in catching the signals coming from the market dynamics, planning and policies for terraced landscape (TL) usually come very late (if they are not entirely missing). Local authorities are often inattentive to issues within their competencies, such as the management of public property or civic uses, restoration and maintenance rules, incentives or taxation and the creation of social partnerships.

The delay is due to several factors. Let us consider the two most important causes. The first has to do with the difficulty in interpreting TL values in territorial lifecycles. TL values change economically and geographically, depending on whether their location is central or marginal. If central, TL may often involve intense and specialized processes of exploitation. They can also become part of a very competitive game, whenever value-added agriculture is increasing. If peripheral (marginal), TLs are subject to abandonment and any re-appropriation practice works "off market". In the first case, the protection of fixed capital (its traditional building quality) is often ensured by business strategies, linked to symbolic, aesthetic and brand factors. In the second, only pilot practices can be found; they are rarely self-reliant, requiring aid, incentives to survive, and rules for landscape protection.

---

[1]Ministry of Culture, Ministry of Agriculture and Ministry of Environment.

The presence of different operators therefore creates a TL polarization. The strength of its operators comes from the ways they manage their interpretation of the landscape. As landscape producers, they speak different languages and have diverse relationships with public institutions. If, on the one hand, central TLs impose a market rationale and seek public legitimacy, peripheral or marginal TLs often have difficulty in experimenting new forms of entrepreneurship and cooperation, based on practices of resistance. These, often bold, upstream and critical practices point, in the first place, to the need to define less generic rules that are more attentive to the context. Rules that regulate conservation while, at the same time, facilitating an adaptation of TL to contemporary life and to working conditions.

20.4 Planning, Policy and Governance Guidelines

For each of the TL classes, specific planning, policy and governance guidelines are identifiable, as briefly indicated in the summary table. The guidelines focus on TL as physical infrastructure, set of functions (production, water and land management), source of ecological services, territorial framework and landscape figure.

Even in conditions of irreversibility or quasi-irreversibility, TL1 might be affected by eco-systemic practices of reacquisition based on specific plans and rules. This hypothesis could be coordinated with re-naturalization or recolonization policies (with wooded cultivation, in some cases) identifying the most appropriate governance in terms of gradients of naturalness. In many contexts, it may be useful to rehabilitate the "civic uses" and the "community lands" (see Community Land Advisory Service in UK), especially in mountain areas where they sometimes account for 30–40% of the municipal surface (Alves and Pedro 2009).

Reversibility of TL2 is a great opportunity, which is often overlooked, especially arranging eco-systemic services and the strengthening of spatial frames. In this perspective, planning can significantly contribute to defining the standards and the rules on land cover and land uses but, above all, in appreciating the local energy, food and climate balance sheets. These planning interventions can also affect TL3, whenever negative externalities need to be limited.

Perhaps more than any other class, TL2 requires special attention in terms of policies, since it involves contexts where experimental practices (of production and) are rather common with significant effects on employment and entrepreneurship.

In many cases, the practices, values and principles are incompatible with existing market rules. Indeed, these practices may implicate economic principles as those fostered by alternative processes: for example, forms of circular economy at the local level, forms of food self-sufficiency, barter trade, alternative complementary currency or quasi-money, land and time banking, and so on. Policies should interpret such innovations and support them through incentives, and tax reductions: especially to create jobs, to raise broader awareness of the state of eco-systems or to encourage and strengthen social interactions. Several opportunities have emerged from the dialogue between formal and informal institutions, first, second and third sector, among contemporaries, modern and customary tenure systems.

All these options can pave the way to new governance experiences.

The specialization of TL3 requires policies on product quality, but above all on quality farming practices, in order to contain the negative externalities and protect traditional knowledge and maintenance skills. It is therefore pivotal for market prices of TL3 products to be sensitive to these factors. In such case, forms of public–private partnership for the cultural development of TL can be easily activated (Table 20.1).

Taking into account the topics discussed above on planning, policies and governance, some general guidelines can be summarized as follows.[2]

1. Policies and social innovation should be based on an identification of the variety of meanings, functions and values of TL in terms of agricultural production, history and culture, and environmental components. These functions and values are specific features of TL as eco-systemic infrastructure.
2. Identification and attribution of values should follow evaluation and monitoring, including risk management at the local and large-scale spatial level.
3. Any successful achievement so far, thanks to policies and social innovation, should be properly acknowledged (see, for instance, local practices, experiments, programmes, policies, plans and governance approaches). An international focus should be activated to encourage knowledge transfer and communication on policies and social practices.
4. Governments and public administrations at all levels should be involved. In Europe, since many States share the principles of the Convention of Landscape, they should be actively engaged in promoting actions and practices for the protection and regeneration of TL involving local communities. So far, experience is still very limited and scattered.
5. A TL platform backed by the National Governments should be designed to favour and support integrated and areal projects that place TLs within local eco-systems. This approach will help farmers, local communities and experimental initiatives to initiate a dialogue with institutional powers. Moreover, this will encourage regional and local administrations to include TL into planning strategies into a transcalar perspective. Such integrated approach might help to adopt a multi-functional policy, facilitate access to financial resources and tackle the recurrent ownership and land access issues.
6. To contrast the demographic aging phenomenon and attract younger actors, special attention should be given to technological, institutional and social innovation such as

    a. selection of profitable crop mix with attention to organic and ecological production, not only and necessarily market oriented;

[2]The general guidelines can be drawn from the report "Policies and social innovation—WG 5", III World Meeting on Terraced Landscapes (Padua), by E. Fontanari, D. Patassini and D. Zanotelli.

Table 20.1 General guidelines

	TL1 (irreversibility)	TL2 (reversibility)	TL3 (development)
Planning	– Eco-systemic reacquisition – Risk management – Rules and regulations (i.e. preventive ecological compensation)	– Provision of eco-systemic services – Spatial frames – Hydro-geological efficiency – Energy balance – Rules and regulations	– Reducing negative externalities – Spatial frame – Traditional maintenance skills and technologies – Energy balance – Rules and regulations
Policies	– Recolonization – Re-naturalization	– Market versus "off market" – Circular economy – Parallel currency – Land and time banking – Food balance – Innovation – Taxation	– Mitigate specialization and monoculture – Quality of product and process – Export – Adjustment of market prices (real value, not only parametric rural prices)
Governance	– Management of naturalness gradients, risk and compensation	– Institutional design – Third sector – Integration among land tenure systems	– Design and implementation partnerships

b. multi-sectoral income generating activities (like tourism, slow mobility, local trade, adapted brand procedures and networking, seasonal work);
c. use of a variety of competences, skills and labour force (new-peasants, immigrants, woofing, work tourism and alike) and dialogue between old and new generations to share and pass on practical knowledge and wisdom;
d. provide direct financial support to individual and to groups of farmers (not only anchored to the main job) to support physical infrastructures and supply eco-systemic services, including related small-scale activities.

7. Special attention should also be given to a context-led capacity building that looks at traditional and ancestral knowledge, latent knowledge (for instance, seed conservation and selection, water and land management practices and so forth). Capacity building should also strive to improve statutory educational programmes at all levels involving community and family education. A learning by doing approach should be encouraged through "food for work" and "food for heritage" initiatives.
8. Customary institutions related to land and water rights should be recognized and fully integrated within the policy-making processes. Any effort to dismantle these traditional institutions should be contrasted all over the world. Actually, these institutions should be considered at all effects as a social and economic "leverage" for the local development and the regeneration of TL.
9. Abandonment and degradation of TL should be fined or, at least, strongly discouraged.
10. Often TL (mostly TL2) provide only a limited amount of income which is not enough to cover the maintenance and regeneration of investment costs, except for some monoculture, which has little appeal to young generations. Regeneration projects should therefore integrate legal, institutional, production and eco-systemic components, including easier access to the land (even temporary). A simplification of regulations and procedures is welcome in accordance with the protection of TL heritage and potentialities. A coordinated approach among local authorities and actors is taken as the best way to address local needs and implement the mentioned strategies.

Annex

Summary and adjustment from the document prepared by the working group "Rules and policies", Trento/Rovereto, III World Meeting on Terraced Landscapes.

Foreword

On issues related to the abandonment of TL and aging farmers, there has been renewed interest on the quality and environmental sustainability of agricultural production. Strategies are based on multi-value functions that, beside production, keep such aspects as social interactions, community and cultural development, environment and related activities together, such as compatible tourism and local food markets.

To maintain the TL, the farmer's role is crucial and concerns tangible and intangible values. It requires a constant presence and a strong subsidiarity of the public.

The public actions include empowerment (education and knowledge), rules and regulations, efficiency, facilities and infrastructure, and efforts to design a governance approach.

Here below the actions in detail.

1. *Empowerment*

   Analysis and knowledge
   Geography: general census and sample based surveys on physical-functional assets (atlas, maps, classification and alike)
   Economy/ecology: estimate economic capital values of TL (production, environmental externalities, eco-system services)
   Agronomy: ensure quality of the food production and processing, knowledge and traditional practices

   Education and research
   Agronomy: search for new high-value crops (innovation)
   Development of new technologies that can be adapted to the specific conditions of TL (new machinery appropriate to the nature of the terraces) and/or enhancement of traditional techniques with contemporary (presence of working animals, etc.) and their supply chains
   Promotion, education and training on building and construction techniques, on traditional maintenance (dry stone structures: training courses for farmers, civil engineering specifications to encourage the dissemination of good practices, technical standards for testing drywalls, and so forth).

2. *Rules and regulations*

   Rules, beyond rules, reducing bureaucracy
   Facilitate farmers" work, support their direct intervention for restructuring and recovering abandoned areas (see TL1), income for TL oversight
   Facilitate the use of abandoned terraces through temporary uses, public acquisitions, adverse possession, bank of land, and so forth

Simplification of approval procedures to start works for land improvement and re-cultivate wasteland
Simplification of licensing procedures for small processing activities of local products

3. *Efficiency*

Farming profitability
Encourage interactions among operators (farmers, producers, associations, tour-operators, traders, schools and alike)
Certification of products
Designation of origin (DOC, DOP, IGP, Subzone)
Brand
Introduction of value-added crops
Local processing of primary products

4. *Facilities and infrastructures*

Infrastructural provision (roads, accessibility, irrigation, drainage and alike)

5. *Public-private partnership: funding and taxation*

Project and implementation partnership (co-financing) between public and private to attain sustainable performances
Direct funding (see, for instance, the Japanese model)
Tax benefits
Design and implementation of a project portfolio through programs or strategic plans combining incentives (for environment, landscape and heritage) and direct funding

6. *Governance*

Pro-active role of local authorities
Enhancement of the participation of local communities and associations
Planning-policies interactions
International cooperation for the exchange of good practices (ITLA, Slow Food and alike)

References

Alves G, Pedro J (2009) The artificial simulacrum world. The geopolitical elimination of communitary land use and its effects on our present global condition. Eloquent Books
Costanza R et al (1997) The value of the world's ecosystem services and natural capital. Nature 387:253–260

Diamond J (2005) Collapse: how societies choose to fail or succeed. Viking Press, New York

Mang P, Haggard B, Regenesis Group (2016) Regenerative development and design. A framework for evolving sustainability. Wiley

Martins EO (2016) Ecosystems, strong sustainability and the classical circular economy. Ecol Econ 129:32–39

Weitzman ML (2000) Economic profitability versus ecological entropy. Quart J Econ 115(1): 237–263

Chapter 21
Integrated Policies for Terraces: The Role of Landscape Observatories

Anna Marson

Abstract So far, we do not have a systematic evidence about the effectiveness of single policies, nor of their different combinations applied in diverse contexts for safeguarding and bringing into new life terraced landscapes. We just know that policies that are more diffuse include regulations about how terraced landscapes should be preserved and restored, and some financial compensative measures in the frame of agricultural policies. From an academic point of view, if we consider, according to Lowi (1972), the four types of policies usually employed (distributive, redistributive, regulatory and constituent), it becomes quite clear that successful policies need a sensible combination of these different "resources". Just as an example: a restriction (like a rule about how to restore terraces) is more easily accepted if combined with some kind of benefit (financial, but also of other nature) and will have an easier implementation if the diverse institutions share it. However, such an approach is seldom practised, since terraced landscapes do not represent a sectoral interest, and no financial or economic lobby pushes for redirecting public policies towards effectiveness in preserving this collective heritage. In such a general context, local landscape observatories represent a new and great resource, since they are reframing the issue from below, adopting de facto an integrated point of view and a collective interest. In Italy, many of them have done an important job in raising consciousness about terraced landscapes heritage value, but also acting directly for repairing them, for preventing abandonment, for promoting new sustainable agriculture, for spreading good practices. After an introduction dealing with policies for terraced landscapes, the paper will therefore explore, with reference mainly to Italy, what landscape observatories are, how they work both on a voluntary or institutional base, how they are helping to reframe the terraced landscapes preservation issue, looking into new ways to reconcile economy and landscape.

A. Marson (✉)
IUAV University Venice, Venice, Italy
e-mail: anna.marson@iuav.it

M. Varotto et al. (eds.), *World Terraced Landscapes: History, Environment, Quality of Life*, Environmental History 9,
https://doi.org/10.1007/978-3-319-96815-5_21

21.1 About Policies: Why Current Ones Do not Work?

In Italy, as well as for a large extension of the Mediterranean region, the heritage of terraced landscapes is extremely diffuse and altogether little known. Generally, terraces are located in steep and therefore less "developed" areas, i.e. areas less transformed and often almost spared by twentieth-century modernization processes. They represent therefore a valuable cultural heritage, incorporating a lot of contextual knowledge about how to deal with nature in a sustainable (*durable*, i.e. lasting) way. At the same time, they are the result of human creation perhaps more than any other rural landscape, requiring continuous human work to survive, without which nature comes back but terraces will definitely die. As Magnaghi (2005, p. 62) writes

> a terraced hill system is a living system, a highly complex 'neo-ecosystem' constantly evolving over time, produced through the constant work of construction, transformation and maintenance of the stone terraces, determining new and more complex balances between human action and nature: exposure to the sun, and the yield and fertility of the soil increase; water is controlled and channelled, microclimates and hydro-geological safeguards are created. This leads to special farming techniques and crops and therefore is an information, technical and cultural heritage. In short, an anthropic landscape is created.
>
> If abandoned, however (in other words if it is not continually used, 'nourished', and cared for), the terracing is degraded and ultimately dies as a territory with its value as a heritage [...]. The landscape thus reverts back to 'nature' through processes of washing away, erosion, subsidence, land caving in, larger landslides, and the growth of vegetation and wild woods with new fauna and so on until it reaches a new hydro-geological natural 'climax'.

Regulatory policies alone, in the case of terraced landscapes do not work at all. As it happens usually for heritage in general, and more specifically for this peculiar landscape, the additional charges due to compliance with rules meant to protect it, without any compensation measure, bring most peasants or economic users to abandon cultivation and maintenance, or to avoid rules and transform, sometimes destroy, old terraces. Only in a few cases, agricultural enterprises succeed in creating benefits for their production thanks to the image of terraces, and therefore producing a self-incentive for maintaining and restoring them. Rules alone are not enough to protect terraced landscapes, if new productive uses, coherent with their characters, do not become the object of a specific focus, thanks also to public policies, to keep them alive.[1]

Since local and regional governments, nor to speak about national ones, often have a vague idea of how many terraces there are on their territories, or even about the fact that there are terraces on their territories[2], public proactive policies for

[1]On this issue of giving back new productive uses to terraced landscapes, see, for instance, Baldeschi (2001).

[2]For the Regione Piemonte terraces, for instance, the only panel presented in the ITLA exhibition on terraced landscapes (Venice-Padua, October 2016) has been submitted by a local NGO, *La prima langa. Osservatorio per il paesaggio delle valli Alta Bormida e Uzzone*.

maintaining or repairing terraces are seldom present. Also when existing, they do not happen to be the result of a proper policy design process.

This lack of institutional action is only partly counterbalanced by a number of pioneer initiatives, taken both by people moving to the countryside from metropolitan areas and by local young people, investing time, work and sometimes money to bring terraces back to specific cultivations, for which they are far better than normal terrains: vineyards, saffron, aromatic plants, etc. Moreover, by a number of efforts by civic associations, ecomuseums handcraft schools and analogous dedicated institutions to foster the knowledge and the maintenance of such a heritage.

At the same time, these many bottom-up initiatives to keep terraced landscapes alive would be much more effective if institutional policies were sympathetic and collaborative in fostering the maintenance of terraced landscapes with new and coherent forms of multifunctional life.

Looking outside Italy, some policies seem to be more robust,[3] but in international literature it is not easy to find proper case studies, dealing in depth with the diverse policies developed in the different contexts.

A major issue, which appears quite evident from Italian experience, is anyhow the fact that policies tend to be sectorial, rather than integrated, with regulatory policies disjointed from other types of action.

In this sense, it may be useful to go back to Lowi (1972) classification, which identifies four types of policies: *distributive*, *redistributive*, *regulatory and constituent*.

Still used in policy analysis, although the latter has today often changed in «governance», this classification points out that rules are just one specific kind of policy, usually not sufficient to promote the minimum maintenance of terraced landscapes, even less to guarantee the multifunctional role traditionally performed by these carefully built landscapes.

In fact, the role of terraces in centuries (*la longue durée*) has been at least a triple one:

- slowing down hydrogeological dynamics;
- creating new fertile soil;
- bettering existing microclimate.

Not by chance, the Register of Historical Rural Landscapes, promoted by the Italian Ministry of "Politiche Agricole, Alimentari e Forestali" (Agnoletti 2010), refers to many terraced areas as rural landscapes with an important heritage value. In fact, terraced landscapes represent the result of good and useful practices selection, able to last in time. Today, we look at landscape no longer as just an aesthetical experience, but first of all as a structure produced by a number of

[3]Among many others, I would quote the action taken for the French Parc des Cevennes, the Suisse Domleschg project and even the Trento Landscape Observatory/STEP, an Italian case that so far represents a positive exception.

positive relations: in this case, a successful and lasting synthesis of Vitruvious' *firmitas* and *utilitas* (in multifunctional terms, of course), and therefore *venustas*.

Indeed, these terraced areas represent not only a rich depository of applied knowledge, but also great places for a diverse, wiser, development of human well-being, with specific reference to the so-called interior areas and the issue of re-inhabiting them (Barca 2015).

How do different policy's types interact with this potential perspective?

Distributive policies are usually the most easy and less conflictual policies, although financial constraints growingly experienced by public institutions have reduced a lot the possibility to activate them. Moreover, contribution alone, to restore a terrace or something alike, usually does not produce a lasting or a replication effect, although it can help the maintenance of some terraces otherwise destroyed by natural processes or men action.

Redistributive policies are always more conflictual than distributive policies, but also easier to last in the medium-long term. There are not many examples applied to terraced landscapes, although they might define some interesting perspective. For instance, since terraces are hydrogeological devices, other landowner benefitting from them might be asked to pay a small tax, like those usual collected for drainage services.

Terraces also might be "redistributed", since the municipalities might entrust the abandoned ones to people caring for them, taking them away from the absent or even unknown owners.[4]

Regulatory policies are usually highly conflictual, when not integrated with distributive or redistributive policies, compensating the burden of complying with more or less heavy rules. At the same time, since they do not require a direct financing capacity, when there are no other possibilities at hand for dealing with a critical issue, both civil society and institutions ask for or propose new rules. Because of the lacking integration with other measures, usually rules suffer an implementation deficit.

Constituent (governance) policies consist of setting up special institutions, like, for instance, institutional landscape observatories, but also building new collective actors, creating arenas able to bring forward plus than zero-sum games together with and for the community.

Practice is of course more messy, but successful policies are usually made by a sensible combination of different types of policies (distributive or redistributive, regulatory, constituent or governance). A restriction (like a rule about how to restore terraces) is more easily accepted if combined with some kind of benefit (financial, but also of other nature) and will have an easier implementation if a collective institution (in the wider term) shares it and takes proactive action besides it.

Unluckily, in most contexts, the scheme "protection" versus "no care at all" still prevails in public policies for terraced landscapes. In some regions, terraced

---

[4]In marginal areas, many rural plots of land are property of emigrated people, whose descendant have never visited or reclaimed the property, and are often unknown to local public institutions.

landscapes are within protected area; this helps to avoid great alteration, but not dereliction (Cinque Terre[5], Amalfi coast, etc.), despite the tourism interest for these coastal areas. On the other hand, in a larger number of places outside these areas, terraced landscapes are still undergoing rough destruction processes, in order to plant new hazel groves or vineyards or become the base for building new houses. No collective action to adapt tractors to places or to preserve the *genius loci*: owners are just looking for the most rentable function in the short term.

Of course financing matters too: the Suisse landscape Fond has 5,000,000 Sfr. a year for non-repayable incentives, while in Italy, we have only a few cases of some public (Fondo paesaggio Trentino), private (Fondo Ambiente Italiano) or private–public (Distretto Culturale Valtellinese) financing (but no national direct financing measures) for landscape. At the same time, no European Union policy in this field exists.[6]

What is even worst that no financing at all is having on the same territory both rules and incentives, but unrelated one to the other (like landscape rules and agriculture payments available for terraces, but without any «*if, then*» ratio; even worst landscape prescriptions and agricultural payments for modernization processes destroying terraces).

Anyhow, in many cases, there are no financial resources at all, and therefore discovering or even creating new resources is quite important. Useful action may include: new labelling and traceability for terrace products; promoting walking groups, artistic performers, voluntary groups, etc., for guaranteeing at least a minimal maintenance of terraces able to slow down their ruin; reframing the vision diverse local actors (from mayors to people working in forestry services; from local experts to school students) have about potentialities of terraced landscapes.

About reframing the vision, dry stone terraces can represent a limitation to agricultural modernization (the standard one) or as a resource for context-specific agricultural practices, less subject to concurrence than other agriculture productions. Similarly, spontaneous vegetation growing on terraces can be seen as a positive trend (by forestry service and by some ecologist) or as a very negative one, because it threatens terraces.[7]

To be effective in bringing derelict terraces a new life, policies should:

- approach preservation with a pragmatic approach open to sensitive innovation in techniques, materials and dimensions/shapes, bridging codified and local knowledge(s);
- ensure a specific labelling for terraces rural production (restricting it to existing terraces);

[5]A multidisciplinary work on potential new perspectives for Cinque Terre terraces in Besio (2002).

[6]This means that the implementation of instruments like the European Landscape Convention, promoted by the Council of Europe, is a national stuff, while EU policies pay no attention to these objectives.

[7]For instance, in some regions of Italy (like Liguria), the permit to cut vegetation on terraces is easy to obtain, in others (like Piedmont) much more difficult.

- tune public payments to multi-functionality (agriculture but also hydrogeological equilibrium: avoiding the current state of affairs, with agriculture payments that often help to destroy important rural landscapes);
- redistribute costs and benefits (activities which benefit from the landscape quality, like tourism, but also other forms of local production, should contribute to its maintenance);
- enlarge policy communities (keeping it wide enough to avoid interest group lobbying).

All the issues mentioned so far recall the need of an appropriate mix of policies, able to bring new life into terraced landscape without destroying them by dereliction or by mono-functional transformation. However, such an approach is uncommon, since terraced landscapes do not represent a sectoral interest, and no financial or economic lobby pushes for redirecting public policies towards effectiveness in preserving this collective heritage.

In this general context, landscape observatories, and local ones more specifically, represent a new and great resource, since they are reframing the issue from below, adopting de facto an integrated point of view and a collective interest.

21.2 What Landscape Observatories Are?

Landscape observatories are among recommended instruments for the implementation of the European Landscape Convention (ELC). Their mission, although operationally varying with their scale and their institutional or civic nature, is landscape awareness raising, recognizing in landscape an essential component of people's Identity and monitoring both the integration of landscape into policies and changes occurring to landscapes.

In Italy, landscape action has grown in later years mainly thanks to the development of regional landscape plans and to the birth of many local landscape observatories.

A 2004 National Law (*Codice dei beni culturali e del paesaggio*) has introduced landscape plans for the regional territories as mandatory planning instruments. So far, only those for Apulia and Tuscany have been approved, both in 2015,[8] but at least ten more regions have already done a good deal of work on them. The two landscape plans already approved offer a deeper knowledge of the different landscapes of the region and rules to govern transformation without destroying landscapes, but have been also an important stimulus for citizens and their associations to participate into the debate about landscape, landscape policies and the governance of landscape transformation.

[8]Both plans can be downloaded by the two regions' institutional Web pages. For the plan of Tuscany, see also Marson (2016).

At the same time, both the dissemination of ELC principles and the process of landscape plans redaction have given birth, in the last decade, to quite a number of local landscape observatories, as a way to promote bottom-up action by people and their local associations. Their activity, made by voluntary work, mostly directed to raise place-consciousness, shows that better practices are feasible, stop "ignorant" destruction, etc, using mobilization of inhabitants for rediscovering local know-how, bridging codified and contextual knowledge (like in dry stone-building workshops organized by some of them).

Concerning the need of more integrated action, the role of landscape observatories really gives to this perspective some roots, at least at bottom level.

At national level in 2015 has finally become operational the national observatory for the quality of landscape, an institutional body headed by Ministry of Culture but formally representing both other institutional competences (Ministry of Agriculture), professional organizations and environmental associations.

The national observatory is currently the only institution in Italy addressing the issue of integrated landscape policies at the national level. At the same time, having no financial or operational resources, it can just suggest action or denounce the lack of consideration for landscape issues, with a larger influence on policies promoted by the Ministry of Culture than by other ministries. Among the action so far accomplished or on track:

- giving advice on a number of controversial disputes regarding landscape;
- fostering the making of regional landscape plans;
- posing the issue of developing specific landscape policies as a result of inter-sectorial action (for instance, making landscape a key point of the new strategic plan for tourism);
- editing the first *Report on landscape policies* in Italy.[9]

"Observing" in this case means to not only look at, but also bringing forward reflexive action, although with the limits already remarked. The risk for the national observatory, as for many institutional bodies, is to become a bureaucratic instrument of top-down action, instead than a reference point for the web of actors interested into and capable of good practices and policies. This kind of risk can be minimized, only if the national observatory will play its specific role keeping alive the network of regional and local observatories.

Regional observatories are just a few, so far. Even less are operational. Regional governments, according to the specifications of regional laws, should establish them: in both Apulia and Tuscany, for instance, they shall promote a regional network of observatories. Veneto gave an important institutional impulse to local observatories a few years ago and then stopped. Sardinia had produced a quite interesting work in the past. The autonomous Province of Trento has the interesting experience of a landscape observatory together with a school of territorial governance. Among these many differentiated experiences, many of which have in

[9]https://box.beniculturali.it/index.php/s/6HDyFyQyZL9n8ic#pdfviewer.

common a recurrent "stop and go" regime, quite peculiar is the case of Piedmont, with no regional observatory at all but the largest network of local landscape observatories in Italy.

A closer look to this network can help to understand what local landscape observatories really are and what they do. First, it is made of quite heterogeneous entities: ten observatories, relating to larger or smaller territories, sometimes grouping together many associations, in other cases just a group of citizens, of course with many intermediate cases. Some among them have Web pages, scientific committees, and bring forward both studies and direct action with local people. Many of them have a watching role upon landscape offences; a few are born from the opportunity to bring forward a project financed by EU, or by local administrations; others would like to promote a different, more sensible to heritage, development.

Altogether, and acting within a coordinated network, these observatories represent a bottom-up huge opportunity for experimenting integrated action for landscape.

21.3 Why Landscape Observatories Can Play an Important Role in Fostering Effective Policies for Terraced Landscape

Which relation exist between terraced landscapes and local landscape observatories? In Italy, a number of local observatories—Canale di Brenta (Castiglioni and Varotto 2013; www.osservatorio-canaledibrenta.it) probably has been the first one —have grown out of a specific interest for bringing new life into derelict terraced landscapes, fostering collective action aimed to merge landscape restoration and new uses. At the same time, if we look at landscape as a structure produced by a number of positive relations lasting a long time, as a synthesis of Vitruvius's *firmitas*, *utilitas* and *venustas*, terraced landscapes represent an excellent "litmus test" for public policies to properly deal with landscape issues.

The issue most of the time facing terraced landscape is namely how to preserve them without banning context's sensible innovation. It is not an easy task, since when terraces are subject to protection rules these are usually difficult to comply with, and therefore most people prefer to abandon terraces to wilderness. On the other hand, inappropriate transformation can easily destroy these landscapes. The knowledge and experience of local landscape observatories could therefore be precious for bridging the gap between the often too rigid rules of "protection" and the "no care at all" situations, helping to understand what could and should be done with reference to each specific landscape, in relation to the main socio-economic opportunities and threats.

At the same time, top-down resources for action suffer not only knowledge deficits, but also growing financing and legitimacy limits. In general, public financial resources are strongly diminishing. When available, their expenditure too often goes to the interest of few, rather than to those of the community in proper sense. Even institutional actors as municipalities have to mediate between many issues and short-term economic interests, which too often result in a view of landscape as an interest of visitors opposed to those of local people.

Local landscape observatories, as local collective actors, may help to reverse this view, deepening the local knowledge about each place-related heritage, revaluing its long-term economic value, building a shared consciousness about the perspectives offered by a proper management of their landscapes. Consequently, they may monitor and influence local policies and private investments, merging volunteer bottom-up action with other kinds of resources, but also starting and bringing forward direct action on and about terraced landscapes when public bodies are absent or silent. In doing this, they adopt de facto an integrated point of view, since bottom-up action is not affected by sectoral bureaucratic organization.

A couple of examples from Canale di Brenta and La prima Langa (www.laprimalanga.it) local observatories, the first—a pioneer—established more than a decade ago, the second in 2015, help understanding the multiple roles that such collective actors can play.

The well-known Canale di Brenta experience, currently a local observatory nested in the Comunità Montana[10] offices and officially recognized by Veneto Regional Government, was born first as a research activity, engaging university students to measure and represent in cartographies the real consistence of terraces.[11] In the same years (2005–2008), an Interreg Alpine space research project (ALPTER) gave an impulse to a wider and deeper consideration of terraced landscapes in the Alpine region, networking a number of case studies, among those the Canale di Brenta one. Since then, activities brought forward include educational activities with local schools; expert and participatory mapping; specific training for designers and local planners; and many others. In 2016, the project "Adopt a terrace" received a special mention by Ministry of Culture Commission in charge of selecting the Italian candidate to the Landscape European Prize, for "social inclusion, new generation coming back to place-care".

The high Bormida and Uzzone valleys local observatory is the last one born within the Piedmont network of local landscape observatories and is entirely bottom-up promoted, by single citizens and civic or environmental associations. Located at the regional border with Liguria, in the so-called Alta Langa, these valleys' landscape heritage comes from a complex overlapping of cultures: Romans, Goths, Longbards, Saracens and many feudal properties until the eighteenth century. Oriented in south–north direction, these valleys are characterized by a complex system of ridges and slopes, those facing south usually steeper. Made by

[10]The local Municipalities Union, with special competences since it is a territory in "mountain" area.

[11]The technical cartography (CTR), produced by the regional administration, represented just a limited number of terraces, out of the many present on the terrain. Second cycle diploma work by Tres and Zatta, tutor prof. Domenico Patassini, 2006. See also Fontanari and Patassini (2008).

Miocene Marl, in layers of sands, sandstones and clays, are altogether inconsistent formations, as shown by the many badlands, detrital bands and landslides, these latter prevented by terraced works. The terraced system connects and integrates

- a network of compact urban centres both on the ridge and at the bottom of the valleys;
- a denser network of rural hamlets, usually built on slopes;
- a thick grid of farmsteads, as well as of smaller *cascinotti* and *ciabot*.

all connected by roads often terraced, and functioning all together as a polycultural rural economy (Figs. 21.1, 21.2 and 21.3).

The historical landscape structure is still in place, except for some new industrial settlements. Since the 1950s, the exodus of many peasants towards the new industrial centres, sped up by the diffuse pollution generated by Cengio's ACNA industry, has produced a large expansion of woods, still going on in areas more difficult to reach, like terraced slopes. Since a few years, a resettlement process is going on, regenerating parts of historical heritage with new rural, cultural and tourism initiatives of quality.

Fig. 21.1 Bergolo, extreme terraces highlighted by snow (*Photo* Anna Marson). For many abandoned terraces, the snow helps to understand what a huge artificial system they represented

Fig. 21.2 Valle Uzzone, Contrada Valentini, vineyard on terraces with its small *ciabòt* (*Photo* Anna Marson). Some small-scale traditional production is still in use. Recently Slow Food has designed as "presidium" the Dolcetto wine produced on terraces, although for a small area around Cortemilia only

Fig. 21.3 Valle Bormida, stone terraces between Cortemilia and Torre Bormida, once used for wheat cultivation, now with peach trees (*Photo* Anna Marson). In a number of cases, a limited number of terraces are cared for as gardens near residential units

Since its birth, the landscape observatory has started a documentation work,[12] thanks also to a solid scientific committee,[13] and already organized a couple of educational days with diverse order schools.[14] It is currently promoting a number of events to make both inhabitants and tourists more aware of the relations between rural cultivated landscapes and dry stone traditional handcrafted buildings, besides presenting other experiences to show that new economies are possible often thanks to specific traditional landscapes.

Within the near perspective of the Piedmont regional landscape plan approval, the role not just of this single landscape observatory but of the whole regional network might be quite important in feeding back information about implementation issues in general, but also with specific regard to terraced landscapes.

These two shortly summarized examples of local landscape observatories give just an idea of the kind of action that such actors can bring forward and of their diversity according to the local institutional context as well as to the diverse social actors and networks in place. At the same time, in a country like Italy, where civic grassroots have such a long history and vivacity (Putnam 1993), it would be quite stupid to base in official public institutions alone the hope for a better care and consideration of our terraced landscapes.

Nevertheless, it is urgent that official public institutions care also more for terraced landscapes. It is about 60–70 years now that most peasants abandoned terraces,[15] and without sufficient maintenance and renewal in a close future these terraced landscapes will massively ruin, forever lost.

Among official institutions, the national observatory for landscape quality could be quite important for promoting a specific focus on terraced landscapes at the institutional level, and a consequent integration among different public policies. Having among its actors a representative of the Ministry for Agricultural Policies, the national observatory first might raise the issue of an appropriate relation between agriculture payments and landscape protection, promoting a fairer deal between the diverse interests behind them.

Integration to consider and foster is therefore more than twofold: integration among different public policies, integration between top-down and bottom-up action, integration among public, private and collective interests.

Recognizing terraced landscapes as a heritage for rethinking our future, as a mean of new life and well-being, means at least, and first, to make the effort to

[12]The only presence of Piedmont at the International exhibition of terraced landscapes (ITLA 2016 in Venice and Padua) has been one of the High Bormida and Uzzone valleys, thanks to the homonymous local landscape observatory.

[13]The scientific committee hosts researchers and professors by all the nearby universities, including Genua, Turin, Alessandria and Pollenzo (www.laprimalanga.it).

[14]Primary schools but also the local rural school for adults.

[15]First, in the 1950s, for emigration towards larger towns industries offering a better salaries perspective, and a couple of decades later for the formal dismissal of agricultural pacts (mezzadria) between landowner and agricultural workers. With this last change, most agricultural areas less apt to industrialized agriculture suffered dismissal.

better relate the different public policies acting upon them. Even to obtain this, the action of landscape observatories is essential, since policies do not reform by themselves.

References

Agnoletti M (ed) (2010) Paesaggi rurali storici. Per un catalogo nazionale, Laterza

Baldeschi P (2001) Il Chianti fiorentino. Un progetto per la tutela del paesaggio. Laterza

Barca F (2015) Un progetto per le "aree interne" dell'Italia. In: Meloni B (ed) Aree interne e progetti d'area. Rosenberg & Sellier

Besio M (ed) (2002) Il vino del mare. Marsilio

Castiglioni B, Varotto M (2013) Paesaggio e osservatori locali. L'esperienza del Canale di Brenta. Franco Angeli

Fontanari E, Patassini D (eds) (2008) Paesaggi terrazzati dell'arco alpino. Esperienze di progetto, Marsilio

Lowi T (1972) Four systems of policy, politics, and choice. Public Adm Rev 32(4):298–310

Magnaghi A (2005) The urban village, a charter for democracy and local self-sustainable development. Zed Books

Marson A (2016). La struttura del paesaggio. Una sperimentazione multidisciplinare per il piano della Toscana. Laterza

Putnam R (1993) Making democracy work. Civic traditions in modern Italy. Princeton University Press

Manifesto
Choosing the Future for Terraced Landscapes

M. Varotto et al. (eds.), *World Terraced Landscapes: History, Environment, Quality of Life*, Environmental History 9,
https://doi.org/10.1007/978-3-319-96815-5

ITALIAN MANIFESTO FOR TERRACED LANDSCAPES
6th - 15th OCTOBER 2016
ITALY
SHARING EXPERIENCES
RURAL HOSPITALITY
HERITAGE
THE LANDSCAPE IS THE DOOR!
NO THANKS!
OPEN SOURCE TECHNOLOGY
MOSAIC
DIVERSITY
EVERY STONE HAS ITS PLACE!
ALERT!
BARTER
FARMER & ARTISAN PRIDE
COMMUNITY AWARENESS
CALA CALA!
LESS BUREAUCRACY

TERRACED
LANDSCAPES
CHOOSING
THE FUTURE

International Terraced Landscapes Alliance
3rd World Meeting on Terraced Landscapes
Italy, October 2016

We, 250 people from over 20 countries, meeting in Padova in the final session of the 3rd World Meeting on Terraced Landscapes, express warm thanks to our Italian hosts for the stimulating experience of plenary discussions in Venice and Padova. We also draw encouragement and many practical ideas from the exchanges during the field visits in Costiera triestina, Topolò-Dordolla, Valstagna, Valpolicella-Valdobbiadene, Trentino, Ossola valley, Canavese-Aosta valley, Chiavari-Lavagna-Vernazza, Ischia-Costiera amalfitana, Pantelleria.

We have focused our attention on the remarkable history and regional variety of terraces around the world, with a particular focus on those in Italy and Europe and have reflected with high concern and high hope upon the future of terraced landscapes. Looking across the broad heritage of terraced landscapes, we believe that the present time—after decades of abandonment and degradation processes—should be the nadir of a great U-shaped curve in the state of that heritage. We commit ourselves to start the upward trend of revival in the use and sustained management of terrace systems.

This bold commitment is justified in our view by a new valuation of the inherited stock of terraces as key part of the environmental and cultural richness of landscapes, and now capable of being used in a new and multi-functional perspective. The solutions can be based upon, and justified by, the long-term public benefits that terraced hillsides can yield: the prevention of soil erosion, the mitigation of climate change, the control of flooding and effective management of water systems; the protection and enrichment of agrobiodiversity and ecosystems; the diversification of rural economies by adding value to different food, seeds and crops; the educational value of heritage and cultural landscapes, and of the human and natural stories which they embody; the beauty and attraction of these landscapes for leisure and tourism. This rich range of benefits or 'public goods' justifies a new governance and the injection of communal resources—at local, regional, national, continental, or global scales—into the maintenance, and where necessary the redemption, of terraced lands.

We note with high interest the approach to analysis and appreciation of landscapes embodied in the European Landscape Convention, principles we can apply elsewhere in the world. By adopting the Convention, many European governments have committed themselves to identify, evaluate, and prepare objectives with full public consultation related to all the landscapes in their territory and to secure protection and management of all landscapes and promote redemptive action where that is needed. It points to the need for research and policy-making to draw on a wide range of disciplines and knowledge systems. Mapping and researches by universities and regional governments are laying the groundwork for such actions.

We recognize the central role of old and new forms of sustainable agriculture and craftsmanship (farmers, dry-stone wall builders, hobby and part-time farmers, cooperatives, and associations) in the maintenance of terraces. The adaptive knowledges of these stakeholders (o figures) should be fully recognized in debates and decision-making about the future of terraced lands.

We welcome the increasing interest in revival of abandoned terraced system. In a growing number of places, communities, local authorities and farmers are leading this revival. It is refreshing to see the commitment among young people in farming the land and call for action by local authorities and civil society to support them in these ambitions through vocational education, on-the-job training, financial support for newcomers, and other techniques. The challenge, in such regions, is to find new ways of life and viable activities, which will make good use of these resources. The viability of farms in terraced areas can be greatly enhanced by the strengthening of new economies, by adding value to local food products and the introduction of more circular systems in local economies. The quality of food production, the heritage qualities, and the environmental quality of terraces can also offer a unique experience to visitors and bring complementary income to rural economies.

We call upon international organizations, governments, local authorities, landowners, commercial interests, rural communities, peasants, artisans and other 'hands-on' producers, educators, researchers and all relevant stakeholders to commit themselves to the protection and long-term maintenance of terraces. They should incorporate respect for terraced landscapes in all policies and programs, notably those supporting agriculture, rural development, environmental protection, water and river catchment management, spatial and territorial planning, and regional development.

We entrust the International Terraced Landscapes Alliance to take the lead in promoting multinational exchanges of ideas and experiences and to assist the strengthening of initiatives and networks in this field.

Padua, 15th October 2016

Index

M. Varotto et al. (eds.), *World Terraced Landscapes: History, Environment, Quality of Life*, Environmental History 9,
https://doi.org/10.1007/978-3-319-96815-5

Printed by Printforce, the Netherlands